ed
ARBORISTS' CERTIFICATION STUDY GUIDE

4TH EDITION

ARBORISTS' CERTIFICATION STUDY GUIDE

CORE DEVELOPMENT TEAM

SHARON J. LILLY, LEAD

CORINNE G. BASSETT

JAMES KOMEN

LINDSEY PURCELL

ISA
International Society of **Arboriculture**

The *Arborists' Certification Study Guide* is intended to serve as a recommended program of study. It is not intended to be the only program of study to obtain the ISA Certified Arborist® credential. This *Guide* and the ISA Certified Arborist® certification program are developed through separate processes and entities. The practices and recommendations contained in this *Guide* should be used in practice only by those properly trained, educated, and experienced in the field of arboriculture. ISA is responsible only for the educational program contained in this *Guide* and not for the use or misuse of these ideas in specific field situations or by inexperienced or improperly trained individuals.

ISA
International Society of
Arboriculture

International Society of Arboriculture
270 Peachtree St. NW, Suite 1900
Atlanta, GA 30303
United States

+1 (678) 367-0981
www.isa-arbor.com

Copyright ©2022 by International Society of Arboriculture.
All rights reserved. Except as permitted under the United States Copyright Act of 1976, no part of this publication may be reproduced or distributed in any form or by any means or stored in a database or retrieval system without the prior written permission of the International Society of Arboriculture (ISA).

Cover Design: Shawna Armstrong
Page Design and Composition: Bookbright Media
Printed by Premier Print Group, Champaign, IL

Print ISBN: 978-1-943378-21-0
Digital ISBN: 978-1-943378-26-5

10 9 8 7 6 5

0425-NY-6000

CONTENTS

Acknowledgments . . . . *vii*
   Contributing Authors . . . . *vii*
   Technical Reviewers. . . . . *vii*
Image Credits . . . . *ix*
Notice to Exam Candidates . . . . *xi*
Introduction . . . . *xiii*

   Chapter 1: Tree Biology. 1
   Chapter 2: Tree Identification. 25
   Chapter 3: Soil Science 41
   Chapter 4: Water Management 63
   Chapter 5: Tree Nutrition and Fertilization. 81
   Chapter 6: Tree Selection 97
   Chapter 7: Installation and Establishment 113
   Chapter 8: Pruning 139
   Chapter 9: Tree Support and Lightning Protection 165
   Chapter 10: Diagnosis and Plant Disorders. 189
   Chapter 11: Plant Health Care 219
   Chapter 12: Tree Risk Assessment and Management . . 245
   Chapter 13: Trees and Construction 273
   Chapter 14: Urban Forestry 293
   Chapter 15: Tree Worker Safety 317
   Chapter 16: Climbing and Working in Trees 343

Answers to Workbook Questions. *375*
Glossary of Terms *419*
Recommended Resources *445*
Bibliography *449*
Index *453*

ACKNOWLEDGMENTS

The International Society of Arboriculture gratefully acknowledges the generous contributions of the many individuals who assisted in the development of this publication. As a professional society committed to the advancement of professional knowledge and education in the field of arboriculture, ISA understands the importance and value of academic excellence and professional consensus in its publications and educational offerings. ISA can achieve this level of excellence only with the dedication, knowledge, and experience contributed by its many volunteers who represent the best of the profession.

The International Society of Arboriculture and the development team would like to thank the contributing authors and reviewers listed below for contributing their time and expertise to making this fourth edition of the *Guide* a better book. We also acknowledge the experts—too numerous to list here—who contributed many hours to reviewing and improving the previous editions of this *Guide*. A special thanks is extended to Dr. Beth Brantley for her careful attention to reviewing the manuscript of this edition in its entirety and offering valuable feedback.

Contributing Authors

Dr. John Ball
Corinne G. Bassett
Dr. Beth Brantley
Dr. Susan D. Day
Dr. Kelby Fite
Dr. Ed Gilman
Dr. Richard Hauer

Alex Julius
Dr. Brian Kane
Dr. Andrew Koeser
James Komen
Sharon J. Lilly
Nelda Matheny
Lindsey Purcell

Dr. Michael Raupp
Dr. Bryant Scharenbroch
Dr. E. Thomas Smiley
Dr. Glen Stanosz
Dr. Gary Watson
Dr. Les Werner
Dr. P. Eric Wiseman

Technical Reviewers

Mark Adams
Dr. Stephanie Adams
Dr. A.D. Ali
Dr. Mike Arnold
Dr. David Ball
Dr. John Ball
Dr. Laurence Ball
Dr. Nina Bassuk
Joe Boggs
Dr. Beth Brantley
Dr. Daniel Burcham
Dr. Len Burkhart
Rob Calley
Dr. Tenley Conway
Dr. Jeff Dawson

Dr. Susan D. Day
Dr. Dirk Dujesiefken
Dr. Julian Dunster
Mark Duntemann
Dr. Francesco Ferrini
Dr. Alessio Fini
Dr. Bill Fountain
Dr. Richard Hauer
Dr. Dan Herms
Dr. Andrew Hirons
Jeff Jepson
Dr. Brian Kane
Dr. Cecil Konijnendijk van den Bosch
Dr. Igor Lacan

Dr. Leena Lindén
Dr. Chris Luley
Dr. Sreetheran Maruthaveeran
Nelda Matheny
Dr. Darby McGrath
Dr. Jason Miesbauer
Dr. Gregory Moore
Dr. Justin Morgenroth
Dr. Eric North
Dr. Glynn Percival
Brian Phelan
Scott Prophett
Dr. Thomas Randrup
Dr. Chad Rigsby
Dr. Chris Riley

Dr. Matt Ritter
Dr. Bryant Scharenbroch
Dr. Henrik Sjöman
Dr. Duncan Slater

Dr. E. Thomas Smiley
Dr. Glen Stanosz
Jim Urban
Keith Warren

Dr. Gary Watson
Mike Wendt
Dr. Les Werner
Dr. Kathleen Wolf

IMAGE CREDITS

The following are gratefully acknowledged for their image contributions to previous editions of this *Guide*. The artwork and photographs they provided remain relevant where they appear in the pages of this edition.

Mark Adams
Steve Adams
Dr. Bonnie Appleton
ArborCare, Inc.
Arbor Day Foundation
Asplundh
Autumn Tree Care
Dr. John Ball
Bartlett Tree Research Laboratories
Dr. Rex Bastian
Mike Berg
Stacey Borden
Ian Bruce
Bugwood.org
Casey Trees
Dr. William Chaney
Steve Chisholm
Steve Cieslewicz
Dr. Jim Clark
Dr. Larry Costello
Davey Tree Expert Company
Dr. Michael Dirr
Kevin Eckert
Dr. Bill Fountain
Dr. Ed Gilman
Bruce Hagen
David Hanson
Neil Hendrickson
Dr. Don Hodel
Dr. Harvey Holt
Horticopia
Dr. Gary Johnson
Dr. Warren Johnson
Dr. Bobby Joyner
Steven Katovich
Kingstowne Landscaping
Dr. Gary Kling
Dr. Andrew Koeser
Bryan Kotwica
Bill Kruidenier
Lauren Lanphear
Sharon J. Lilly
Dr. John Lloyd
Dr. Chris Luley
Nelda Matheny
Alan McBeth
Dr. Fredric Miller
Dr. Larry Morris
Joe Murray
Natural Resource Conservation Service
Joseph O'Brien
Ohio Landscape Association
Ohio State University
PA Dept. of Conservation and Natural Resources
Scott Prophett
Karolina Przybysz
Dr. Michael Raupp
Tobe Sherrill
Dr. Dave Shetlar
Dr. Alex Shigo
Jim Skiera
Dr. E. Thomas Smiley
Dr. T. Davis Sydnor
United States Department of Agriculture
United States Forest Service
University of Georgia
Jim Urban
Luana Vargas
Vermeer Manufacturing
Dr. Gary Watson
Dr. Todd Watson
West Coast Arborists
Dr. P. Eric Wiseman

Thank you to the contributors listed below for providing new art for this edition. The artwork they provided is indicated by figure numbers in parentheses.

Nick Araya (15.17b)
Dr. John Ball (15.4, 15.17, 15.18, 15.19, 15.20, 15.21, 15.22, 15.23)
Bartlett Tree Research Laboratories (13.8)
Corinne G. Bassett (10.5, 10.21, 12.3, 14.18)
Simon Berger/Flickr (11.6b)
Joe Boggs (10.3, 10.4, 10.6, 10.11, 10.19a, 10.22, 10.26, 11.14, 11.18)
Dr. Beau Brodbeck (13.11)
Elliott Brown/Flickr (4.11)

Tim Bushnell (15.16)
Dr. Jim Clark (13.8)
Clark/Matheny/Hortscience/Bartlett (13.8)
Dr. L.R. Costello (4.3, 4.5, 4.6)
Davey Tree Expert Company (15.16, 16.1)
Dr. Julian Dunster (12.22)
Mark Duntemann (12.2, 14.14)
John Wayne Farber (16.3)
Dr. Francesco Ferrini (2.10, 2.11)
Freepik.com (3.8)
John Gauthier (15.17a)
Ryan Gilpin (4.6)
Laura Grant (10.14)
GW, Eric Hirsch (1.18, 3.5, 14.19)
Jeff Harris (8.1, 8.23, 8.24)
Dr. Dan Herms (11.1, 11.17)
Dr. Don Hodel (1.24, 1.25, 4.7, 6.11)
Hortscience (13.8)
Bob Houser (5.10)
Mike James (3.18)
Joe Kellerhals (14.17)
James Komen (4.4)
Will Koomjian/Emergent Tree Works (Front Cover Climber Photo, Chapter 16 Opener)
Bryan Kotwica (1.5, 3.3, 3.19, 4.15, 5.2, 5.5, 5.7, 7.7, 7.9, 7.16, 7.24, 7.25, 7.27, 7.28, 8.2, 9.7, 9.8, 9.12, 9.13, 12.6, 13.5, 13.9, 14.9, 15.3, 15.6, 15.7, 15.12, 15.14, 16.2, 16.6, 16.8, 16.9, 16.22, Table 9.1 Figures, Chapter 16 Knots)
Dave Leonard (13.16)
Avery Lewis (Front Cover Inspection Photo)
Dr. Chris Luley (10.7)
Dr. Roger Magarey (6.10)
Gale Martel (5.10)
H.G. Marx (1.23)
Nelda Matheny (4.3, 4.5, 4.6, 13.8)
Matheny, Costello, Randisi (4.3, 4.5)
Matheny, Costello, Randisi, Gilpin (4.6)
Sophie Nitoslawski (14.7)
Jonathan Picker (11.8, 12.4)
Lindsey Purcell (1.10, 1.20, 2.2, 2.16, 3.1, 3.16, 3.17, 4.12, 5.4, 6.3, 6.4, 6.12, 6.13, 7.1, 7.5, 7.11, 8.3, 8.5, 8.7, 8.8, 8.11, 8.13, 8.21, 9.2, 10.1, 12.1, 12.11, 13.4, 13.7, 13.14, 13.15, 14.1, 14.6, 14.12)
Carol Randisi (4.3, 4.5, 4.6)
Dr. Bryant Scharenbroch (4.13, 5.6)
A.L. Shigo (1.23)
Bodo Siegert (12.23)
Dr. Duncan Slater (8.14)
Dr. E. Thomas Smiley (4.9, 5.1, 5.8, 5.12, 9.1, 9.22, 12.9, 12.10, 12.13, 12.21, 12.24, Chapter 14 Opener)
United States Forest Service (1.23)
Jim Urban (6.1)
Luana Vargas (12.7)
Travis Vickerson (15.1)
Dr. Gary Watson (7.12)
Wonderlane/Flickr (11.6a)

NOTICE TO EXAM CANDIDATES

The purpose of the ISA Certified Arborist® examination is to assess a candidate's knowledge and skills in the field of arboriculture. The examination is designed to assess the fundamental knowledge and skills that all tree care professionals should have, regardless of their area of practice. Some knowledge is gained cumulatively through time and experience in the field and through the accumulation of scientific information related to the care of trees. Other things can be learned through the study of texts or in the classroom. The breadth of knowledge required to pass the exam is great; thus, candidates should begin preparing for the exam well in advance of the scheduled test administration date.

The ISA *Arborists' Certification Study Guide* is designed to help ISA Certified Arborist® candidates review the topics covered on the examination. The *Study Guide* provides useful information on examination content, practice items, and preparation for testing. All candidates should also be aware, however, that the *Guide* does not necessarily represent the full range of examination content, nor does it necessarily present the full range of individual question difficulty included on any actual examination.

The *Study Guide* also includes a recommended resource list that is intended to further assist candidates in developing a knowledge base useful to a professional arborist. However, all certification candidates should be aware that the reference list does not attempt to include all acceptable reference material, nor is it suggested that any of ISA's certification examinations are based on these references.

Any educational or preparatory material, whether published by ISA or not, should not be considered the sole source of information for an ISA certification examination. This publication and the ISA certification program exams are developed through separate processes and entities.

INTRODUCTION

Each chapter of the *Study Guide* consists of five sections: narrative, workbook questions, challenge questions, sample test questions, and other sources of information. In addition, each chapter includes a list of objectives and a list of the key terms introduced.

Before beginning to read the narrative section, review the objectives and look over the list of key terms. Doing so provides a focus and orients you toward the concepts that you should learn in that chapter. The next step is to read through the narrative section. The purpose of this section is to serve as a primer and provide the basic concepts of the topic. Many illustrations, photographs, and charts are included to help you understand the material.

The workbook section follows the narrative. The questions in this section are designed to reinforce the basic concepts presented, and to review the terminology presented. These short-answer questions help you determine which parts of the chapter require more study.

The goal of the challenge questions is to strengthen your understanding of the subject areas and help you apply the concepts learned. These questions go beyond the scope of the narrative. To answer them, you must demonstrate a more detailed understanding of the subject and be able to project the concepts into practical applications.

The next section is the sample test questions. These questions are typical of those that might appear on the certification exam. Their purpose is to help familiarize you with the style and scope of the exam. The questions on the certification exam may correspond to, but will not be limited to, the material presented in the *Study Guide*. No single guide can represent the full range of examination content. Exam questions may reflect other sources of information, such as the references in the recommended resources section of this book, or knowledge an arborist would be expected to gain from experience. By dedicating time to working through the sample test questions, you will get used to the types of questions that you will encounter on the exam.

Once you have successfully completed the workbook section, you should proceed to the recommended resources section. These suggested materials provide greater depth and detail on each subject and help you apply the basic concepts to field situations.

CHAPTER 1

TREE BIOLOGY

▼ OBJECTIVES

- Explain the structures and functions of the buds, leaves, wood, and roots of a tree.
- Describe the relationship of plant structures and their functions.
- Describe the basic composition of a tree's vascular system, and explain how water, minerals, carbohydrates, and plant growth regulators are transported within this system.
- Describe the relationship roots have with mycorrhizae. Explain how the soil environment affects root growth and distribution.
- Discuss the processes of photosynthesis and respiration and the factors affecting these processes.
- Discuss how the growth and development of a tree are the results of the interaction between its genetic potential and the environmental surroundings.
- Explain the concept of Compartmentalization of Decay (Damage) in Trees (CODIT).
- Describe the basic anatomy and physiology of palms, including how they grow.

KEY TERMS

abscission zone
absorbing roots
adventitious bud
aerial roots
angiosperm
anthocyanin
apical bud
apical dominance
apical meristem
auxins
axial transport
axillary bud
barrier zone
branch bark ridge
branch collar
branch union
bud
buttress roots
cambium
carbohydrate
carotenoid
cell turgor
cellulose
chlorophyll
chloroplast
CODIT
compartmentalization
cork cambium
cuticle
cytokinins
deciduous
decurrent

differentiation
diffuse porous
dormant
earlywood
epicormic shoot
eudicotyledon (eudicot)
evergreen
excurrent
fiber
frond
geotropism
growth rings
guard cells
gymnosperm
heartwood
included bark
inflorescence
internode
lateral root
latewood
leaf axil
lenticel
lignin
meristem
monocotyledon (monocot)
mycorrhizae
node
osmosis
parenchyma cells
petiole
phloem
photosynthate

photosynthesis
phototropism
plant growth regulator
plant hormone
primary growth
propagation
radial transport
ray
reaction zone
respiration
ring porous
root crown
root initiation zone
root mat
sapwood
secondary growth
shake
sink
sinker roots
source
stomata
symbiosis
taproot
temperate
terminal bud
tracheid
transpiration
tropism
vessel
xylem

Introduction

A tree, by definition, is a woody perennial plant with a single or multiple trunks. To paraphrase Alex Shigo, trees are woody, long-lived, compartmentalizing perennials. These unique characteristics have contributed to the fact that trees can be successful, long-lived organisms, despite environmental challenges.

Arboriculture is both an art and a science. It combines skill and craft with knowledge and fact. Thus, a foundation for the practice of arboriculture is a thorough understanding of how trees grow. The skilled arborist learns how a tree grows and how to care for and manage it in a way that supports its growth and development. Like physicians, arborists use knowledge of growth and development to diagnose health issues, assess genetic potential, and prescribe treatments that are ethical and appropriate. The arborist must understand tree biology to prescribe a program of care.

The study of tree biology is the study of structure and function and the relationship between them. It also includes study of the tree within its environment. Anatomy and morphology are the studies of the structure and form of a tree. Physiology is the study of the biological, physical, and chemical processes within these structures, providing the basis for function. This text will focus primarily on the biology of eudicotyledonous **angiosperm** trees, but some attention will be given to conifers (which are **gymnosperms**) and palms (which are monocotyledonous angiosperms). More information about these classifications is found in Chapter 2, Tree Identification.

Tree Anatomy

Basic Structure: Cells and Tissues

All living organisms share a basic organizational theme based on cells, tissues, and organs. Cells are the basic building blocks of structure. New cells arise from the division of existing cells. In trees, this process occurs in specialized growth initiation areas called **meristems**. Following division, cells undergo

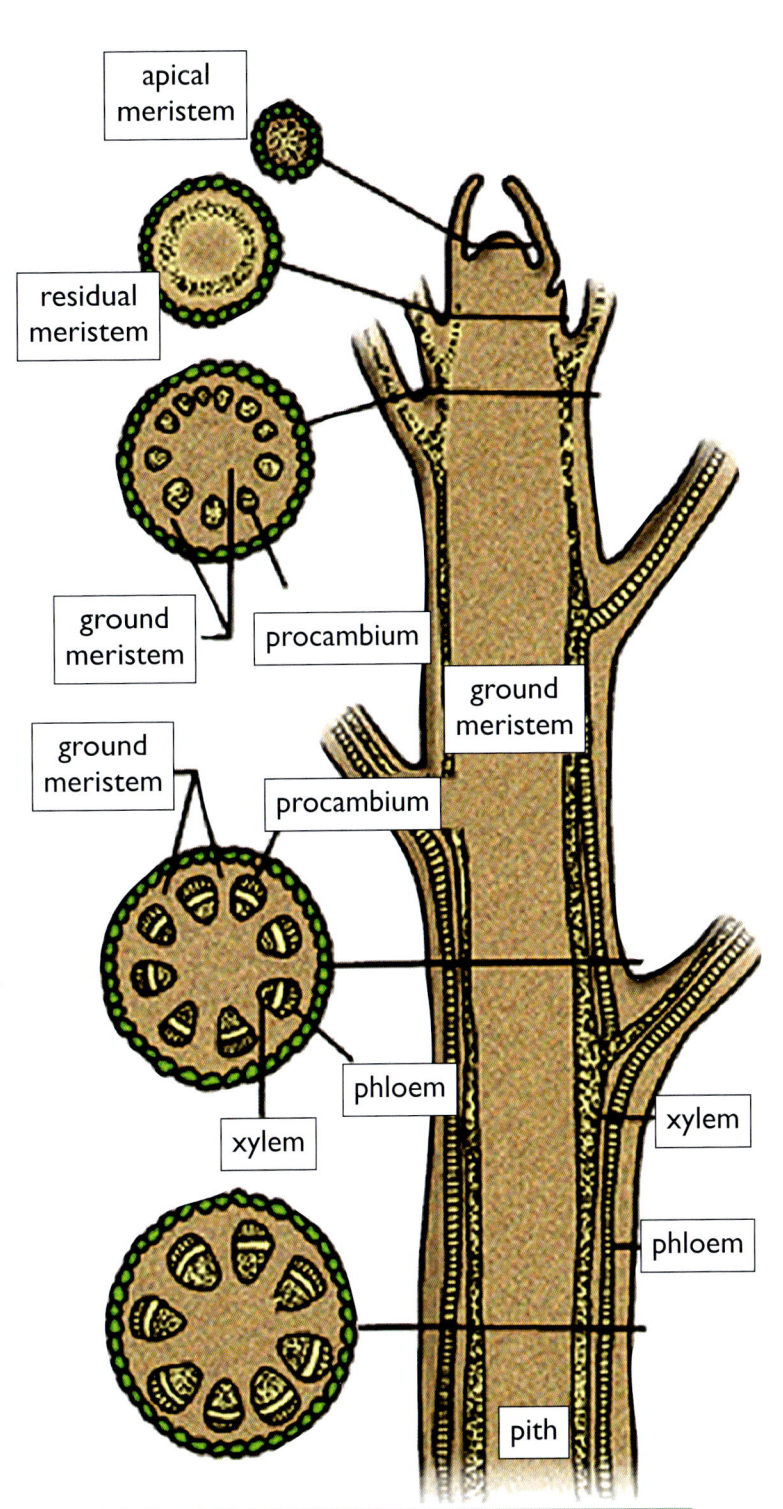

Figure 1.1 Longitudinal and cross sections through a shoot tip. New cells arise from the division of existing cells in specialized growth initiation areas called meristems. The cambium layer is not yet established. Following division, cells undergo differentiation, which changes their structure and permits them to assume a variety of specific functions.

differentiation, which changes their structure and permits them to assume a variety of specific functions. Groups of cells—some with quite specialized functions—are organized into tissues that have specific roles within the tree. Tissues are then organized into organs, of which plants have five: leaves, stems, roots, flowers, and cones or fruit. Finally, organs are organized into intact, fully functional organisms—trees.

Tree growth occurs in two ways. Growth from the root and shoot tips, resulting in increases in height and length, is called **primary growth**. Growth that increases the thickness of stems, branches, and roots is called **secondary growth**. Primary growth occurs in small areas at the tips of roots and shoots called **apical meristems**. Leaf expansion and increases in the length of stems and roots are the result of growth at the apical meristems.

In shoots, the apical meristems are found inside **buds**. In most temperate trees, the overlapping scales or modified leaves of buds protect both the meristematic region and the developing shoot. However, in a few temperate species and most tropical trees, the buds do not have protective scales and are referred to as naked buds. In root tips, the apical meristem is protected by a root cap.

Trees have two lateral meristems, which produce the secondary growth. The first lateral meristem is the vascular **cambium**. The cambium layer is a thin sheath of dividing cells located just under the bark. It produces the cells that will become the vascular system of the tree. The vascular cambium produces two kinds of tissue: **xylem** to the inside and **phloem** to the outside. The second lateral meristem is the **cork cambium**, which produces the outer tissues (periderm) and, ultimately, the bark. The exceptions to this growth pattern are palms, which lack secondary growth.

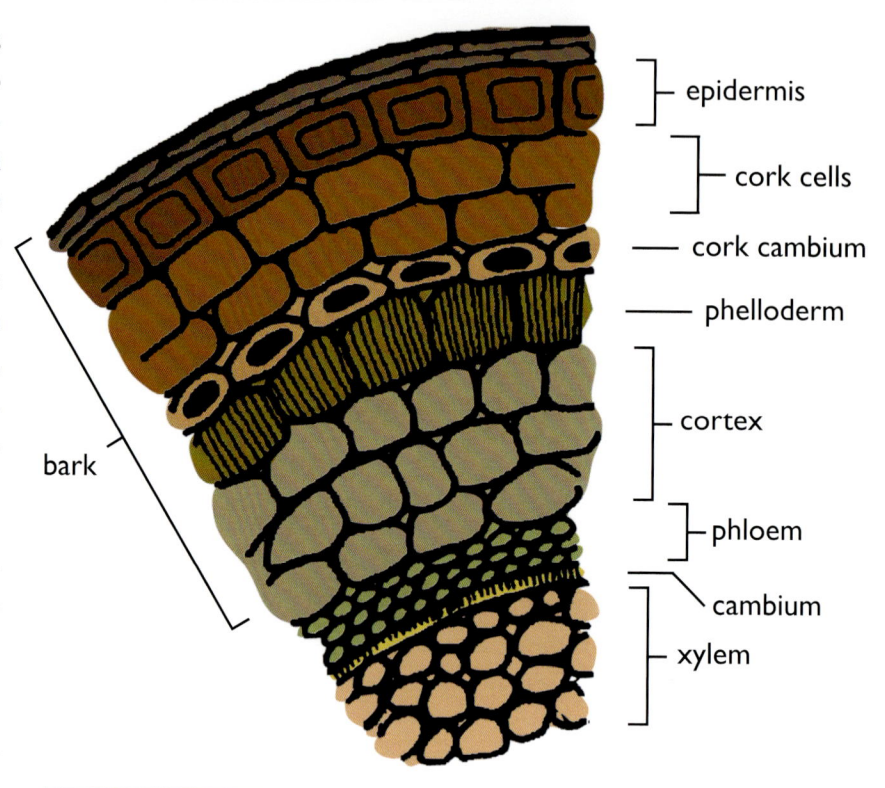

Figure 1.2 Partial cross section through the bark, phloem, and cambium, and slightly into the wood of a broad-leaved tree.

The structural component of the primary cell wall is **cellulose**, which plays a role in providing the architecture of different types of cells. **Lignin** is another, more rigid component of plant cells formed in the cell walls of the wood. The lignin forms a matrix in which microfibrils (long chains of cellulose) are embedded; this forms a reinforcing structure similar to that achieved by rebar in concrete. This reinforcement provides the strength and rigidity that allows trees to grow tall.

Xylem and Phloem

The xylem produced by the vascular cambium during secondary growth is called secondary xylem and is more commonly referred to as wood. It has four primary functions: (1) conduction of water and dissolved minerals (elements)—collectively known as sap, (2) mechanical support for the tree, (3) storage of **carbohydrate** (starch) reserves, and (4) defense against the spread of dysfunction, disease, and decay. Xylem is a complex tissue, composed of both dead and living cells. The cell walls of the dead cells are comprised of cellulose, lignin, and a cellulose-like compound called hemicellulose. Together, these components provide the strength that characterizes wood.

The wood of gymnosperms (for example, pines and spruces) is composed mostly of **tracheids** and few **parenchyma cells**. Tracheids, which conduct water and provide mechanical support, are elongated, closed-ended, dead cells with pointed ends and thickened walls. Parenchyma cells are living cells interspersed among the other cells. They are responsible for storing carbohydrates and defending against decay.

The wood of **eudicotyledon (eudicot)** trees is made up of vessel elements, **fibers**, and parenchyma cells; in some species, tracheids may also be present. **Vessels** are the primary conducting elements in angiosperms such as maple and oak trees. The xylem vessels can be thought of as stacks of dead, open-ended, hollow cells that form long tubes for conducting water, like straws. Vessels are much more efficient in water conduction than tracheids are. Another differentiating factor between gymnosperms and angiosperms is that parenchyma cells are more abundant in angiosperm trees.

Many physical and biological properties of different types of trees are related to the size and distribution of the cell types within the xylem. Some trees form wide vessels early in the growing season and narrower vessels later in the season. These trees are said to be **ring porous**, and they include elm (*Ulmus* spp.), oak (*Quercus* spp.), and ash (*Fraxinus* spp.), among others. Other species produce vessels of uniform size throughout the growing season and are called **diffuse porous**. Examples include maple (*Acer* spp.), planetree (*Platanus* spp.), linden (*Tilia* spp.), and others.

When a tree has been cut and can be viewed in cross section, **growth rings** of xylem are visible in the wood. These rings are the result of the cambium's seasonal xylem production. They appear as rings because the relative size and density of the vascular tissues change throughout the growing season. As the season progresses, cells become smaller in diameter. Thus, the contrast between cells produced early in the season (**earlywood**) and those produced later (**latewood**) shows the diameter increase within an individual year. As with most things in nature, there are exceptions. Conifers do not have vessels and are said to have nonporous wood. The growth rings of tropical trees, if

Figure 1.3 Cross section through a California black oak (*Quercus kelloggii*), a ring-porous tree. Vessels formed in the spring are large; vessels formed later are smaller. Note the prominent rays and star-shaped pith, characteristic of oak species.

Figure 1.4 Cross section through American beech (*Fagus grandifolia*), a diffuse-porous species. The vessels are scattered evenly throughout each yearly increment of growth.

The heartwood no longer plays a physiological role for the tree.

The phloem moves photosynthates such as carbohydrates (sugars) produced in the leaves throughout the plant for storage or consumption. Phloem transport is active and requires energy, whereas xylem vessels are passive in their transport. The old phloem tissue becomes "crushed"; living cell contents are reabsorbed into the tree, and cell walls are incorporated into the bark layers.

In addition to the axial (longitudinal) transport system of the phloem and xylem, which transports materials along the stems and branches, trees also have radial transport cells arranged in **rays**. Rays are made up of parenchyma cells that grow radially, like spokes on a bicycle wheel, and extend across the growth increments of xylem and into the phloem. Ray cells transport carbohydrates and other compounds into and out of sapwood, store carbohydrates as starch, and assist in restricting decay in wood tissue.

The protective layer of the tree is the bark. It is the outer covering of branches, stems, and, in some cases, roots. It is composed of protective tissues that moderate the temperature inside the stem, offer defense against injury, and reduce water loss. Outer bark is composed of corky tissue. The cell walls are impregnated with wax and oil that minimize water loss. **Lenticels**, small openings in the bark, permit gas exchange of oxygen and carbon dioxide. Many types of bark develop in trees. Beech (*Fagus*) trees have very smooth bark with little corky material, whereas cork oak (*Quercus suber*) produces thick layers of cork.

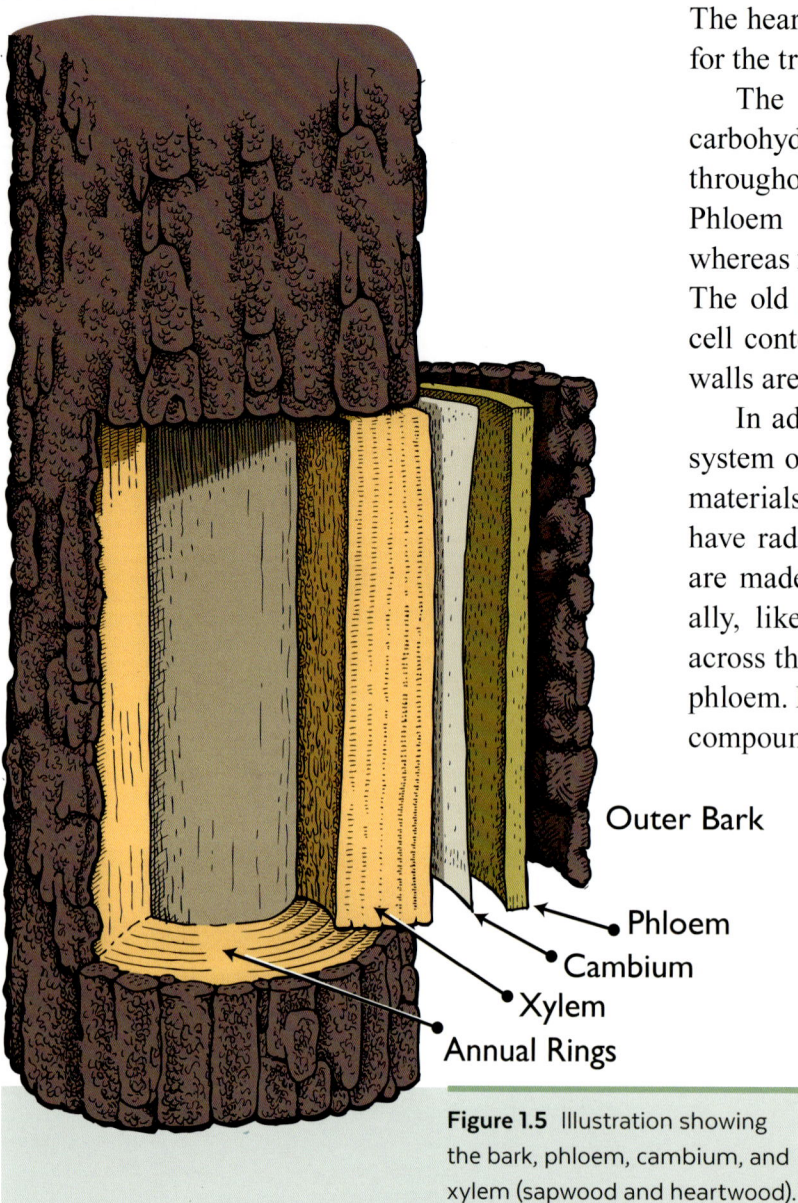

Figure 1.5 Illustration showing the bark, phloem, cambium, and xylem (sapwood and heartwood).

present, may not be annual due to their constant growing conditions, and palms do not form rings of xylem at all.

Xylem that conducts water is called **sapwood**. There are many living parenchyma cells in the sapwood. Deeper within the tree, old layers of sapwood cease to conduct water and transition into **heartwood**—nonconducting xylem that contains no living cells and is sometimes darker in color than the sapwood. The heartwood contributes to mechanical support of the tree, can resist invasion by microorganisms, and is an important site of stored carbon.

Stems

Twigs are small stems that provide support structure for leaves, flowers, and fruit. Branches support twigs, and the trunk supports the entire crown. Both leaf and flower buds are typically found at the end of existing shoots or at leaf bases.

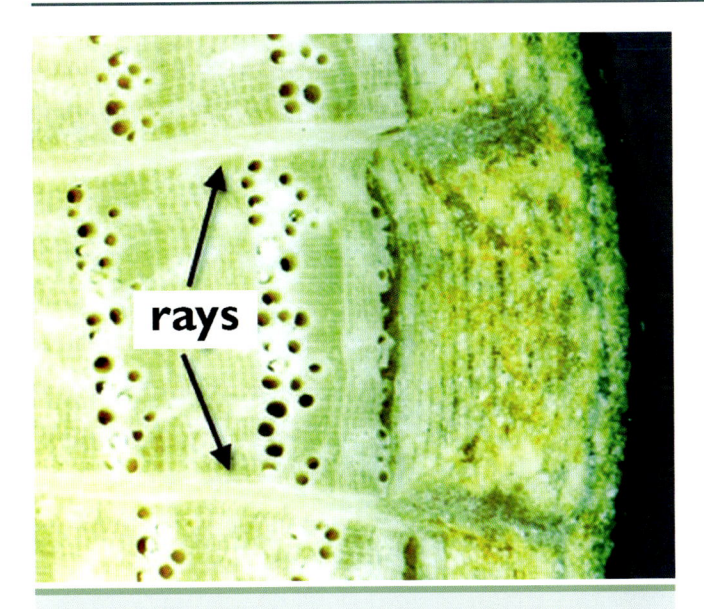

Figure 1.6 Cross section of an angiosperm stem showing the rays. Rays are made up of parenchyma cells that grow radially, like spokes on the wheel of a bicycle, that extend across the growth increments of xylem and into the phloem.

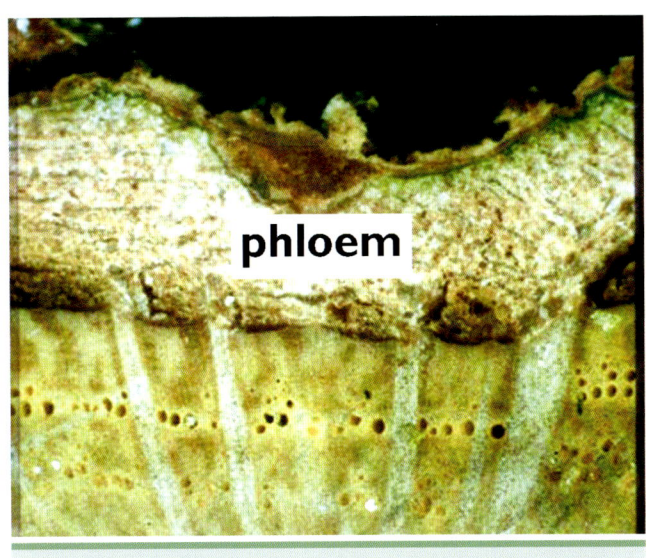

Figure 1.7 Note the connection of the rays from the phloem through the xylem. Ray cells transport carbohydrates and other compounds into and out of sapwood, store carbohydrates as starch, and assist in restricting decay in wood tissue.

Buds located at the end of a shoot are called the **apical buds**, or **terminal buds**. Buds that occur along the stem are typically **axillary buds**. Normally, the terminal bud is the most active on each branch or twig. Axillary, or lateral, buds are often **dormant**. Their growth may be inhibited by the **apical dominance** of the terminal bud, whereby the terminal bud chemically inhibits the growth and development of laterals on the same shoot. As shoots lengthen and age, or when pruning removes terminal buds, lateral and dormant buds may become more active, leading to new shoot development.

Adventitious buds are produced along stems or roots where primary meristems are not normally found. Their development may be stimulated by the loss of apical buds and the plant hormones they produce.

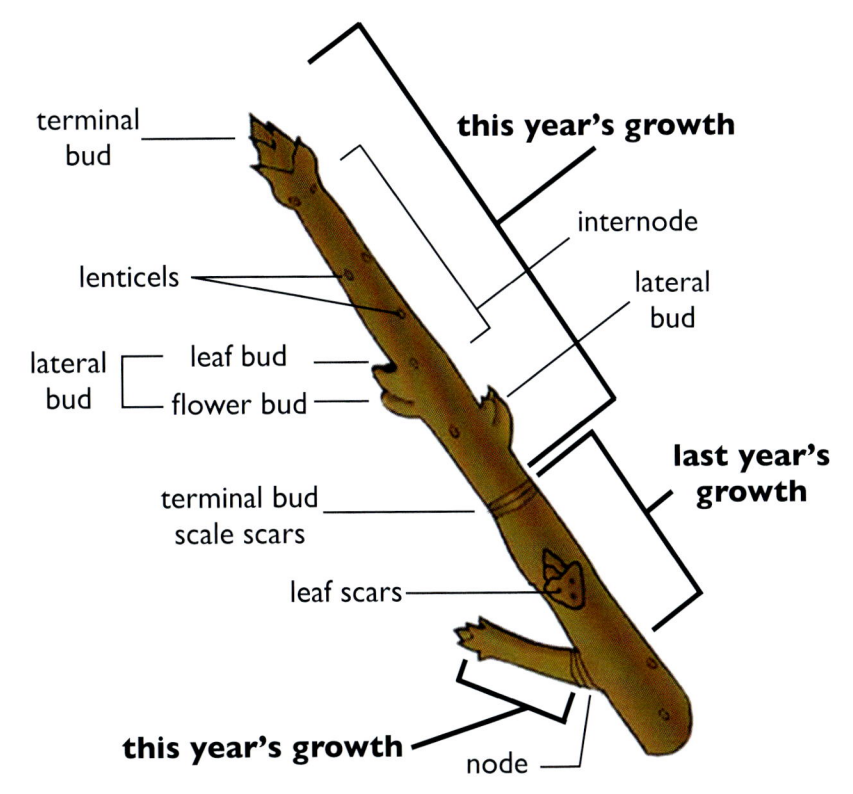

Figure 1.8 Twig anatomy showing annual growth. Buds located at the end of a shoot are called the apical, or terminal buds. Buds that occur along the stem are lateral (axillary) or flower buds.

Figure 1.9 A dormant growth point below the bark. Growth could be initiated in response to a hormonal signal triggered by excessive pruning, storm damage, or other changes.

Some tree species such as quaking aspen trees (*Populus tremuloides*) grow in groups in which individual trees develop from adventitious buds on the roots, resulting in many trees with a common root system. Latent buds are suppressed within growing tissues, originating with the first year's shoot and remaining suppressed beneath the bark until growth is triggered by increased light or tree injury. When latent buds elongate and produce shoots, these are termed **epicormic shoots** or sprouts.

A **node** is a slightly enlarged portion of the twig where leaves and buds develop. The **internode** is the area between the nodes and is important as a diagnostic tool. Leaf scars and terminal bud scale scars are visible on new twigs and are useful in measuring annual twig elongation in many species. On the outer surface of the twig, lenticels permit the exchange of gases.

Each branch of the tree is considered somewhat autonomous; branches are capable of producing and storing enough carbohydrates to sustain themselves. Generally, the carbohydrates used for growth and other plant processes are obtained from storage cells close by. However, long-distance transport of simple carbohydrates also occurs, such as transport to the trunk and root system.

Each branch is similar in structure and function to the entire tree crown, yet branches are not simply outgrowths of the trunk. Instead, branches and trunks have a unique attachment form that is critical to the application of arboricultural practices

Figure 1.10 The annual production of layers of tissue at the junction of the branch to the stem is seen as a shoulder or bulge around the branch base called the branch collar. In the branch union, specialized wood is formed that is typically much denser and exhibits twisted and whirled wood grain. An external sign of the development of this specialized wood is the junction's branch bark ridge.

such as pruning. The annual production of layers of tissue at the junction of the branch to the stem is seen as a shoulder or bulge around the branch base called the **branch collar**. In the junction, called the **branch union**, specialized wood is formed that is typically much denser and exhibits twisted and whirled wood grain, central to the connection between the trunk and the branch. An external sign of the development of this specialized wood is the junction's **branch bark ridge**. If bark has become embedded into the junction during its development, it is referred to as **included bark**. Included bark typically weakens the branch attachment because the normally strong branch-to-trunk attachment is compromised.

Leaves

Leaves are the food producers of the tree. Leaves have cells with **chloroplasts** that contain a green pigment called **chlorophyll**. Chlorophyll is the primary leaf pigment that absorbs sunlight. The energy of the sunlight is collected in the chloroplasts, where, in a reaction called **photosynthesis**, it is converted to chemical energy in the form of carbohydrates.

A second role of leaves is **transpiration**. Transpiration is the loss of water through the foliage in the form of water vapor, which helps cool the leaf. The process of transpiration is the primary factor in drawing water up through the xylem from the roots, which is discussed later in this chapter.

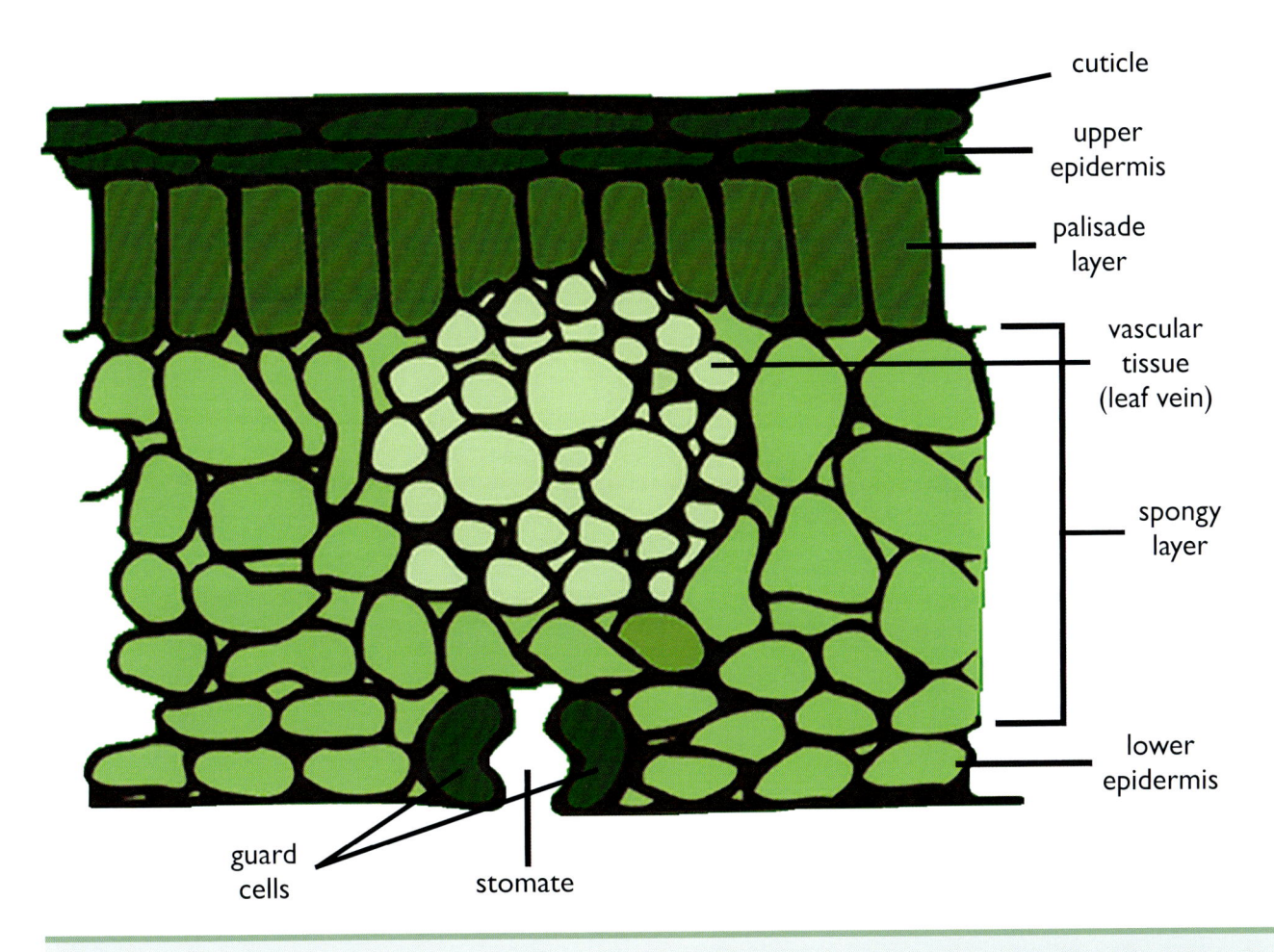

Figure 1.11 Cross section of a leaf blade. Stomata, small openings mostly found on the underside of the leaf surface, control the loss of water vapor and the exchange of gases. Carbon dioxide is absorbed into the leaf, while oxygen and water vapor are released.

The structure of leaves is uniquely adapted to carry out the roles of photosynthesis and transpiration. Leaf blades provide a large surface area for the absorption of sunlight and carbon dioxide needed for photosynthesis. Because leaves are thin, no cells are far from the surface. This structure facilitates the exchange of gases and absorption of light.

The outer surface of a leaf is covered by a waxy layer called the **cuticle**. The cuticle functions to minimize desiccation (drying out) of the leaf. **Stomata**, small openings mostly on the underside of the leaf surface, control the loss of water vapor and the exchange of gases. Carbon dioxide is absorbed into the leaf, while oxygen and water vapor are released. Trees regulate carbon dioxide input and water vapor output through the **guard cells**, which regulate the opening and closing of the stomata in response to environmental stimuli such as light, temperature, and humidity.

Leaves have a network of conducting tissues comprising the veins, or vascular bundles. These veins contain phloem and xylem tissues. They transport water and essential elements and carry carbohydrates produced in the leaf cells to other parts of the tree.

Trees that shed their leaves periodically, typically every year, are called **deciduous**. Trees that hold their leaves for more than one year are called **evergreen**. Deciduous trees generally lose their leaves in response to periodic environmental changes, such as day length, temperature, and/or rainfall, as a result of cell changes and growth regulators that combine to form an **abscission zone** at the base of the leaf stalk, or **petiole** in the autumn. The abscission zone has two functions: (1) to enable leaf drop and (2) to protect the region of the stem from which the leaf has fallen against desiccation and pathogen entry.

Fall foliage color in deciduous trees results from the breakdown of chlorophyll, which gives other pigments more prominence in the leaf. Short, sunny days combined with cold nights enhance the accumulation of sugars and trigger a decrease in chlorophyll production. This process allows other pigments, including **anthocyanins** (reds and purples) and **carotenoids** (yellows, oranges, and reds), to be produced or unmasked, resulting in fall color. These pigments protect leaf cells from ultraviolet radiation while sugars and amino acids are reabsorbed for storage to be used in the spring.

Roots

The roots of trees serve four primary functions: anchorage, storage, absorption, and conduction. Larger roots are similar to the trunk and branches in structure. **Absorbing roots** are the small, fibrous, primary tissues growing at the ends of and along the

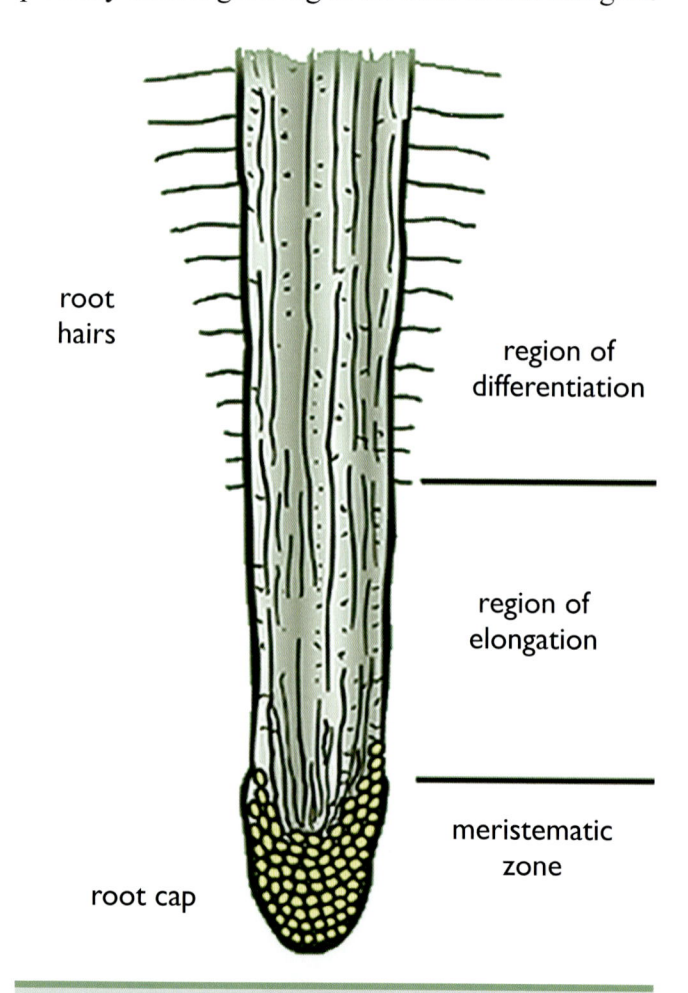

Figure 1.12 Root tip anatomy. The absorbing roots have epidermal cells that may be modified into root hairs, which aid in the uptake of water and minerals. As with shoot tips, root tips contain a meristematic zone where the cells divide and elongate, allowing roots to grow in length.

Figure 1.13 Roots grow where water, oxygen, and space are available, often well beyond the drip line of the crown.

Figure 1.14 Absorbing roots branch into extensive networks to maximize the ability to absorb water and minerals.

main, woody roots. The absorbing roots have epidermal cells that may be modified into root hairs, which aid in the uptake of water and minerals. As with shoot tips, root tips contain a meristematic zone where the cells divide and elongate, allowing roots to grow in length.

Roots grow best where adequate moisture and oxygen are available. Most absorbing roots are found in the upper 30 cm (12 in) of soil. Horizontal **lateral roots** are also usually near the soil surface. Some trees form **sinker roots**, which grow vertically downward off the lateral roots, providing improved anchorage and access to available water deeper in the soil profile. The downward-growing **taproot** of young trees is usually replaced by the expansion of roots around it or is diverted from its downward growth by unfavorable growing conditions such as compacted soil layers. Few mature trees have taproots.

The area where the roots join the main stem is known as the **root crown** (trunk flare). From there the roots spread out and decrease rather quickly in diameter in the "zone of rapid taper" to long, spreading, branching roots, 2.5 to 5 cm (1 to 2 in) in diameter. Roots may extend laterally for considerable distances, depending on the tree and soil conditions. Roots of trees grown in the open

Figure 1.15 Many roots live in a symbiotic relationship with certain fungi. The result of the association is termed mycorrhizae (fungus roots). The fungi derive nourishment from the roots of the tree. In turn, the fungi aid the roots in the absorption of water and essential mineral elements.

without any impediments often extend two to three times the radius of the tree's crown. The extent and direction of root growth is more a function of environment such as available soil volume than genetics of the tree.

Many roots coexist in a symbiotic relationship with certain fungi. The result of the association is termed **mycorrhizae** (fungus roots). Mycorrhizae are present in nearly all soils and are beneficial to plants. In this form of **symbiosis**, both organisms (the tree and the fungus, in this case) benefit from the living arrangement. The fungi derive nourishment from the roots of the tree. In turn, the fungi aid the roots in the absorption of water and essential mineral elements, thus increasing the effective root network and providing benefits to both the fungus and the tree.

Tree Physiology

Photosynthesis

Photosynthesis is the process by which plants use light energy to build carbon molecules that make carbohydrates. Literally, photosynthesis means "putting together with light." Photosynthesis takes place within cells that contain chloroplasts. Chloroplasts contain molecules of chlorophyll, the light-absorbing pigment that gives plants their green color. The two essential components necessary for photosynthesis are carbon dioxide and water. The tree absorbs carbon dioxide from the atmosphere through the stomata in the leaves. Light energy is collected in the chloroplasts. The light energy is converted to chemical energy (carbohydrates) and used for growth and development or stored as starch for later use. Oxygen, a byproduct of photosynthesis, is released through the stomata.

The sugar products of photosynthesis are sometimes referred to as **photosynthates**. Photosynthates are the building blocks for many other compounds required by the plant. Proteins, starch, fat, cellulose, lignin, growth regulators, amino acids, and other important compounds are produced from photosynthates when combined with other essential elements such as nitrogen, potassium, sulfur, and

Figure 1.16 The light-absorbing, green pigment, chlorophyll, is located in chloroplasts within some leaf cells. The light energy is converted to chemical energy (carbohydrates) and used for growth and development or stored as starch for later uses.

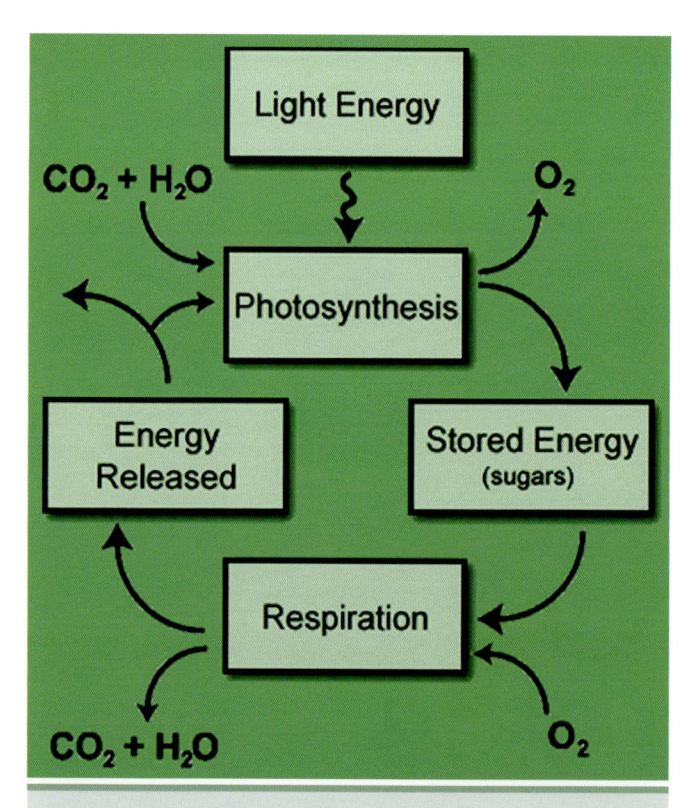

Figure 1.17 An overview of the chemical processes of photosynthesis and respiration.

iron. Photosynthates are primarily stored by the tree in the form of starch for later energy requirements when not put to immediate use.

Respiration

Respiration is the process by which the carbohydrates are converted in a controlled manner into energy; it is independent of light. Plants produce their own energy resources, and it is important that overall photosynthesis (energy production) exceeds respiration (energy use). At night, respiration continues in the absence of photosynthesis, but if respiration exceeds photosynthesis over a long period of time, the tree must rely on stored carbohydrates. A practical example is a tree that is repeatedly defoliated by pests or severe storm events. Without foliage, photosynthesis stops or is reduced dramatically, and the tree cannot produce carbohydrates and store them as starch.

Transpiration

Transpiration is the loss of water from leaf surfaces in the form of water vapor. The evaporation of water from the leaf surface not only cools the leaves but also creates a transpirational pull that moves water up through the xylem. The waxy cuticle layer helps prevent uncontrolled water loss from the leaf surface. Gas exchange, where water vapor and oxygen are released and carbon dioxide is absorbed, takes place through stomata. Each stomatal pore is lined by two guard cells that regulate the aperture of the stomata. Stomatal opening is influenced by environmental conditions such as light, temperature, **cell turgor** (the pressure of water inside the guard cells), and humidity. In trees, stomata usually open in the light and close in the dark, but they also respond to environmental conditions such as high temperatures and low humidity.

Temperature, humidity, and available water affect the rate of transpiration. Transpirational water loss is also affected by anatomical features such as cuticle thickness, the presence of hairs on the leaf surface, and the number and

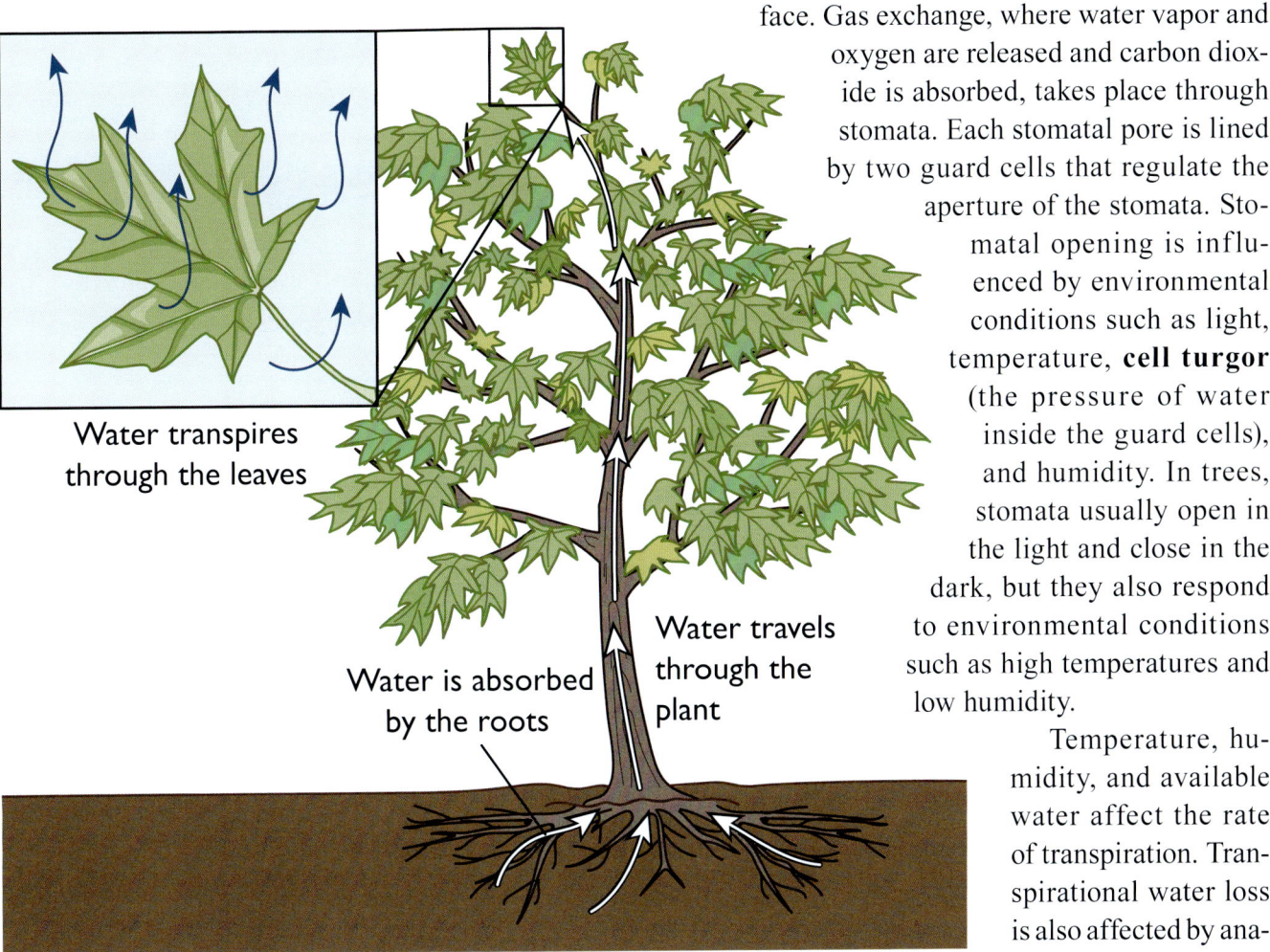

Water transpires through the leaves

Water travels through the plant

Water is absorbed by the roots

Figure 1.18 The xylem can be thought of as a series of continuous and tiny conduits for water that extend from the outermost roots to the tips of the shoots, where the transpiration of water molecules from the leaves into the air pulls water up through the tree from the roots.

location of stomata. Some plants with a thick cuticle, small or hairy leaves, and sunken stomata are adapted to hot and dry conditions.

Absorption, Translocation, and the Vascular System

Water is essential for all living cells. Most biological reactions, including photosynthesis, require water. Water maintains cell turgidity (fullness and firmness) and is necessary to transport essential elements within the xylem. Water and mineral elements are absorbed from the soil by the roots. The tree uses some of this water for growth and metabolism, but most water is lost through transpiration. The xylem can be thought of as a series of continuous and tiny conduits for water that extend from the outermost roots to the tips of the shoots, where the evaporation of water molecules from the leaves pulls water up through the tree from the roots.

Water enters young roots or mycorrhizal roots, in part, by a process called **osmosis**. Osmosis is the movement of water through a membrane from a region of high water potential (water concentration) to a region of low water potential. In the soil and root areas, pure water has the highest potential; adding anything such as minerals or sugar lowers the potential. Water normally moves into roots where the water potential is lower than in the surrounding soil. If the water potential is lower in the soil than the root cells, water will actually move out of the roots into the soil. An example is when salt concentrations are high in the soil, such as from deicing or excessive fertilizer application.

Trees move their food products through the living transport system of the phloem. Carbohydrates are actively pumped through the phloem, a process that requires energy. In discussing phloem transport, the terms **source** and **sink** are often used. Leaves are the source of photosynthates. The photosynthates move through the phloem in a direction from source to sink—that is, from areas high in carbohydrate concentration to areas where more is required. Sinks are plant parts that use more energy than they produce. Almost all plant parts, including young leaves, are sinks at some time. It is sometimes thought that carbohydrates are produced in the leaves and transported, through the phloem, exclusively to the roots for storage. Most photosynthate is either utilized or stored in proximity to where it is manufactured, although it can move in either direction in the phloem.

The movement of water in the xylem and photosynthate in the phloem are examples of longitudinal transport, or **axial transport**. **Radial transport** is

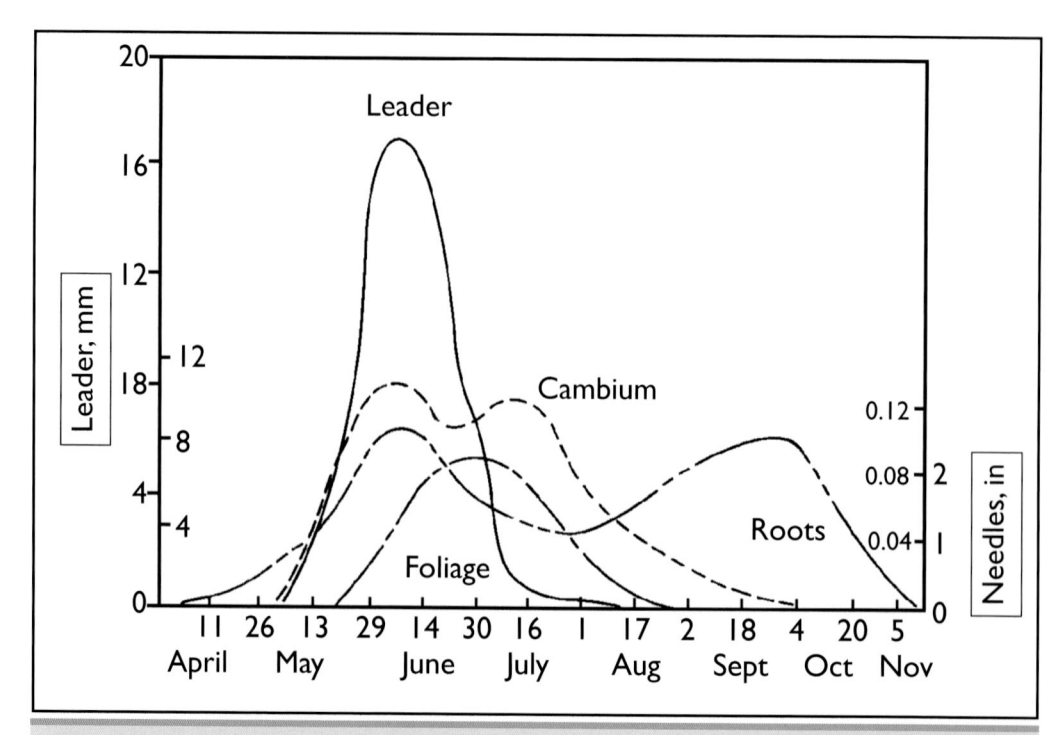

Figure 1.19 Typical seasonal growth of a pine tree (*Pinus* spp.) in a temperate climate (Northern hemisphere). Note the late-season growth of roots.

the movement of water or nutrients within the tree between cells of different ages, primarily through ray cells. Rays are living channels of cells through which water, elements, and carbohydrates move laterally.

Control of Growth and Development

The growth and development of a tree is the result of the interaction between its genetic potential and the surrounding environmental conditions. There are many examples of this interaction. In nature, sweetgum (*Liquidambar styraciflua*) may attain a height of 46 to 53 m (150 to 175 ft). Yet rarely does the species grow that large in urban areas. Although sweetgum may have the genetic potential to grow 46 m tall, the urban environment limits the expression of that potential. Other genetic components work similarly. For example, the range of size, fall color, and form seen in many cultivars of red maple (*Acer rubrum*) represents the breadth of genetic potential within the species. Yet the red maple cultivar 'Somerset' will not look like 'Brandywine', even when grown under the same environmental conditions, because the genetic makeup of these two cultivars is different.

Plant systems, like all living organisms, respond to environmental stimuli. Developmental responses to light, gravity, and temperature can be essential to the survival of a tree. For example, a long period of exposure to cold may be necessary to induce budbreak, flowering, or seed germination. The coordination of processes in trees is controlled in part by plant growth substances (sometimes called **plant growth regulators** or **plant hormones**). Plant growth regulators are chemical messengers that act in small quantities to regulate plant growth and development in many different ways. Plant processes are regulated by the major hormone groups including auxins, gibberellins, cytokinins, ethylene, and abscisic acid. These plant growth substances work together to control such functions as cell division, cell elongation, flowering, fruit ripening, leaf drop, dormancy, and root development.

Auxins are plant growth regulators linked to several developmental processes. Although auxins are produced primarily in shoot tips, they are known to be important in root development. Synthetic forms of auxin are sold commercially to enhance the rooting of cuttings; some synthetic auxins can even be used as herbicides. The fact that auxin is produced in the shoot tips may partially explain why heavy crown pruning to compensate for root loss during transplant has not proved to be effective. Removing the shoot tips decreases auxin flow, which has the effect of inhibiting root growth and also triggers lateral shoot development. Conversely, **cytokinins**, produced in the roots, are instrumental in shoot initiation and growth. Auxins and cytokinins exist in a delicate balance in plants, regulating shoot and root growth.

Auxins are also involved in plant responses called tropisms. **Tropism** is the directional growth of a plant in response to an external stimulus such as light or gravity. An example of a tropism is the upward orientation of stem growth or the downward direction of root growth. These phenomena are responses to the Earth's gravitational pull and

Figure 1.20 Phototropism. Plants grow toward light due to differential distribution of auxins.

are examples of **geotropism**. Light also has a strong influence on the direction of plant growth. A tree that grows at an angle toward the sunlight exhibits **phototropism**.

Apical dominance is a result of internal plant growth regulators as well. Growth regulators present in terminal buds (auxins) inhibit the growth and development of lateral buds on the same shoot. Strong dominance is confined primarily to the current season's shoot growth. During the following season, lateral buds start growing. If the new lateral shoots outgrow the original terminal shoot year after year, a round-headed, or **decurrent**, tree will result. **Excurrent** trees tend to have strong apical control, resulting in upright trees with strong central leaders. Although some species tend toward one of these forms throughout their lives, all trees start out with excurrent traits as juveniles, and most become more decurrent as they mature. Sweetgum (*Liquidambar styraciflua*), some species of *Araucaria*, and most coniferous trees have excurrent forms. In decurrent trees, the ability to maintain the strong central leader is lost, and a rounded form develops.

Maturity and environmental conditions can also influence the growth form of trees. Many pines maintain a strong excurrent form throughout most of their lives but develop a rounded top as they age. Limited access to sunlight, such as in forest understory conditions, can cause a normally rounded tree to grow upright.

It is important to take a holistic view of tree growth and development. A delicate and changing balance of chemical signals controls the metabolic processes of photosynthesis, respiration, and all biological functions within a tree. Tree growth regulators impact each other and remain in a dynamic equilibrium. External stimuli, including those caused

Figure 1.21 Excurrent growth form.

Figure 1.22 Decurrent growth form.

by humans, trigger changes in regulator concentrations that, in turn, stimulate changes in resource allocation, which results in a response in the tree's growth or development.

A System of Defense

Trees cannot actively fight or move away from harm, but they aren't completely defenseless. They can have a number of features that serve as protection: thick bark, thorns, leaf hairs, thick cuticles, and many others. In addition, certain cellular materials may resist decay or may be indigestible by insects. Another defense mechanism is the production of chemicals that resist insect feeding, pathogen infection, or decay.

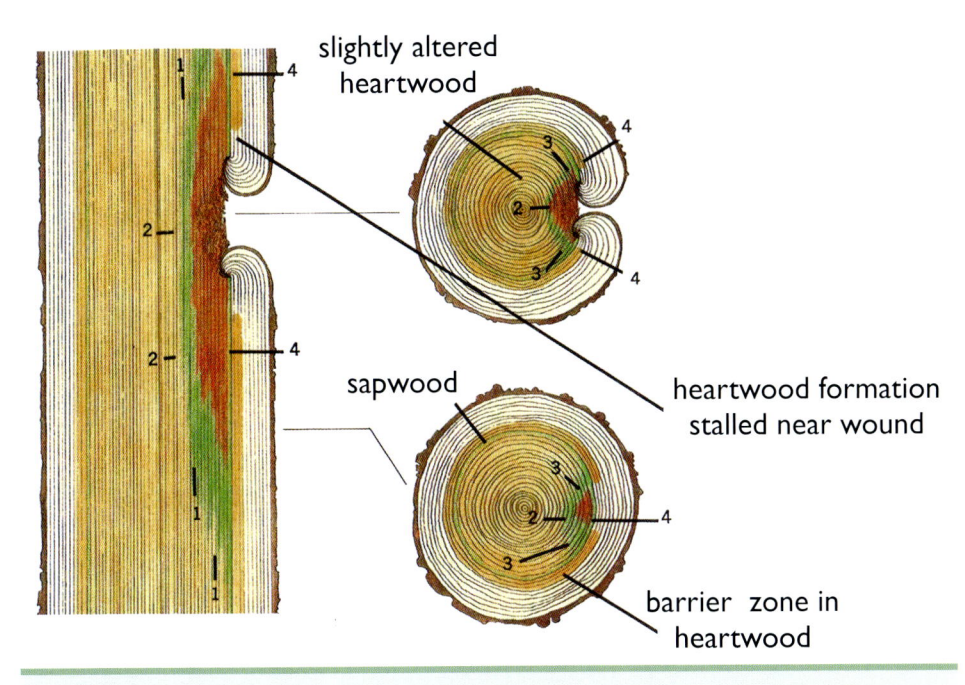

Figure 1.23 CODIT model. Numbers indicate the walls associated with CODIT. Brown represents decay processes. Green represents decay microorganisms' invasion and spread. Red areas indicate the physical and chemical response of the tree to decay.

A process unique to trees is the ability to compartmentalize, or wall off, decay and damage. **Compartmentalization** is the process by which trees can limit the spread of dysfunction and decay. After a tree has been wounded, reactions are triggered internally that cause the tree to form physical and chemical boundaries around the affected area.

Alex Shigo proposed a model of this compartmentalization process called **CODIT** (Compartmentalization of Decay in Trees). More recently, the *D* of CODIT has been more broadly interpreted as "damage" or "dysfunction" by other authors. In Shigo's model, the tree creates four "walls" in the vicinity of a wound. Wall 1 resists longitudinal spread of decay organisms by plugging xylem vessels or by blocking pits in tracheids. Wall 2 attempts to resist inward spread by developing dense latewood cells and by depositing specialized chemicals in these cells. Wall 3 inhibits lateral spread around the stem by activating the ray cells to resist decay. These three walls form the **reaction zone**. Wall 4 is the next layer of wood to form after injury (new wood tissue), and it protects against the outward spread of decay. This is the **barrier zone**. Wall 1 is the weakest of the boundary walls; Wall 4 is the strongest barrier against decay.

At times, the tree cannot resist the spread of aggressive decay fungi due to poor health or genetic capability. It is fairly common for Walls 1, 2, and 3 to fail, allowing decay to spread inside the tree, forming a hollow cavity. Wall 4 fails less commonly, but it can be breached by certain fungi. Also, in some cases, Wall 4 never develops. The barrier zone is strong chemically but weak structurally because Wall 4 interrupts the structural role of the parenchyma rays. As a result, the process that resists the spread of decay into new growth can also lead to other issues such as **shakes** and cracks. Shakes are lengthwise separation of the wood along the grain, usually occurring between or through the annual growth rings.

Tropical Trees and Palms

Tropical Trees

Tens of thousands of species of tropical trees have been identified, many in the rain forests of the world. The tremendous diversity among the species makes it very difficult to make general statements about their biology. Their anatomy varies greatly—much more than that of **temperate** species—and their physiology sometimes breaks the "rules" that temperate-climate arborists have come to accept. The great variety of architecture, reproductive and defensive strategies, and leaf morphology can challenge even scientists and educators.

Tropical trees may lack annual growth rings because growth can be more or less continuous, or it may be highly regulated by wet and dry seasons. Many species do not experience dormancy except, sometimes, at the seed stage. Some species have extremely large foliage, flowers, or fruit. Some may also have a wide-spreading array of **buttress roots**—roots at the trunk base that help support the tree and distribute mechanical stress.

Some species produce roots from stems or branches above the ground, called **aerial roots**. Aerial roots are thought to have one or more of several functions, including support (serving as props), gas exchange, and **propagation** (creation of new plants). New plants produced in this way are clones of the parent tree.

Because of the favorable growing conditions in tropical climates, plants tend to have a rapid growth rate compared to those in temperate regions. In some ways, these favorable conditions have a "forgiving" effect that aids in ease of transplanting, wound closure, and regrowth after injury. But just as with temperate species, tree managers must understand that when tropical trees are removed from their native ecosystems, challenges with growth and health are likely to occur.

Palms

Palms are **monocotyledons (monocots)** and have more in common biologically with grasses than with eudicot trees. Palms do not have a cambium layer or growth rings of xylem. Instead, they have vascular bundles of phloem and xylem sheathed in exceptionally strong, fibrous tissues, embedded in a matrix of parenchyma cells and interspersed within the stem. The stem develops all the vascular bundles it requires for life behind the primary bud as it grows, and these vascular bundles typically retain their conductivity throughout the life of the plant. The stem is considered "woody," although it differs anatomically from the wood of eudicot and conifer trees. Cells in the stem thicken and strengthen as they age, and the stem is capable of storing starch in the parenchyma cells.

Figure 1.24 Cross section of a palm stem (*Licuala spinosa*). The very thin outermost layer is the pseudobark, and the next layer with very few vascular bundles is the cortex. The largest part is the central cylinder with the vascular bundles (dark or black dots) concentrated near the periphery.

As the vascular tissues are dispersed throughout the stem, movement of water and minerals is not restricted to a few rings of xylem toward the outside of the trunk. With no cambium layer toward the periphery of the stem, there is no secondary growth resulting in increased girth. These traits are thought to reduce the vulnerability from injury to the stem; however, palms do not have the ability to repair (grow over) wounds. They are not believed to have any process analogous to CODIT, so wounds to the stem are permanent. At the same time, many palms are considered to be resistant to decay, probably as a result of the strong, hard sheathing fibers of the vascular bundles.

Palm stems are characterized by having a single apical meristem or growing point, which is also referred to as the bud or heart. All new leaves (**fronds**) and flowers develop from the apical meristem. If this bud is lost due to injury or pest damage, the palm (or at least that stem) will die. During a palm's establishment period, the number and size of vascular bundles increases. Palms attain their maximum girth during this stage, mostly prior to beginning vertical elongation of the stem.

Photosynthesis takes place in the fronds, which dominate the palm crown. Fronds are the leaves of palms, consisting of a blade, a petiole, and a leaf base. Leaves are produced sequentially: as the innermost, newest leaves emerge, the outer ones are "pushed" outward and down. Palms have the largest leaves of any trees, growing up to 18 m (60 ft) long in some species, with great variation in color, form, features, and thickness.

The reproductive structures (flowers and fruit) emerge interspersed among the fronds or below the crown from the **leaf axils**, leaf scars, or nodes. Palm flowers are typically clustered into large aggregates of many small flowers. On some palms, these **inflorescences** are inconspicuous, but on others they can be quite showy. A few species of palms flower once in their life and then die. Because the inflorescences consist of many individual flowers, palms usually produce fruit in large numbers, although there is great variation in the fruit size among species.

The root system of palms is also very different from that of eudicot trees. Like the stems, palm roots lack secondary growth (lateral meristems), and, other than the primary seedling root, they are adventitious—developing from an area, typically at the base of the stem, called the **root initiation zone**. Most palm roots are in the upper 30 to 45 cm (12 to 18 in) of soil, and most are close to the stem in a densely packed network called the **root mat**. It is not unusual for a portion of the root mat to be above ground. The roots of palms generally have a good capacity to resprout when cut, which, along with their adventitious nature and root initiation zone, contributes to their typical ease of transplanting. Palms, like most woody plants, form mycorrhizal associations.

Figure 1.25 Palm stems are characterized by having a single apical meristem or growing point, which is also referred to as the bud or heart.

CHAPTER 1 WORKBOOK

1. Sites of rapid cell division in the shoot tips, root tips, and cambium are called _____ .

2. Meristems located at the end of the roots and shoots are called primary, or _____ , meristems.

3. The tendency for terminal buds to inhibit the growth of lateral buds is called _____ _____ .

4. The "food factories" of trees are the _____ .

5. The process of _____ combines carbon dioxide and water in a reaction driven by light to produce sugars. _____ is also a product of this reaction.

6. The green color of leaves is created by the presence of the pigment _____ , which is necessary for photosynthesis to take place.

7. _____ is the loss of water vapor from the leaves.

8. The opening and closing of _____ on the undersides of leaves allow for gas exchange.

9. Water and dissolved essential minerals are transported within the tree in the _____ . The _____ conducts carbohydrates.

10. The _____ is a layer of meristematic cells located between the phloem and the xylem.

11. The _____ _____ is formed when trunk tissue grows around branch tissues. As the branch and trunk tissues expand against each other in the branch union, the _____ _____ _____ is formed.

12. _____ protects the branches and trunk of a tree from mechanical injury and desiccation.

13. Name four functions of the root system.

    a.

    b.

    c.

    d.

Chapter 1: Tree Biology

14. The sugar products of photosynthesis are sometimes referred to as _____ .

15. The orientation of growth in response to an external stimulus is called _____ . Two examples are _____ and _____ .

16. CODIT stands for C_____ o____ D_____ i____ T_____ .

17. Trees with upright growth and a strong, central leader are said to exhibit _____ growth. More rounded trees, which are often broader than they are tall, have _____ growth habits.

18. Roots and fungi form _____ , which are a symbiotic relationship, aiding in the uptake of water and minerals.

19. The process by which chemical energy, stored as sugar and starch, is released is called _____ .

20. Trees that lose their leaves in the autumn are called _____ . Trees that maintain their leaves for more than one year are called _____ .

Label the Following Diagrams

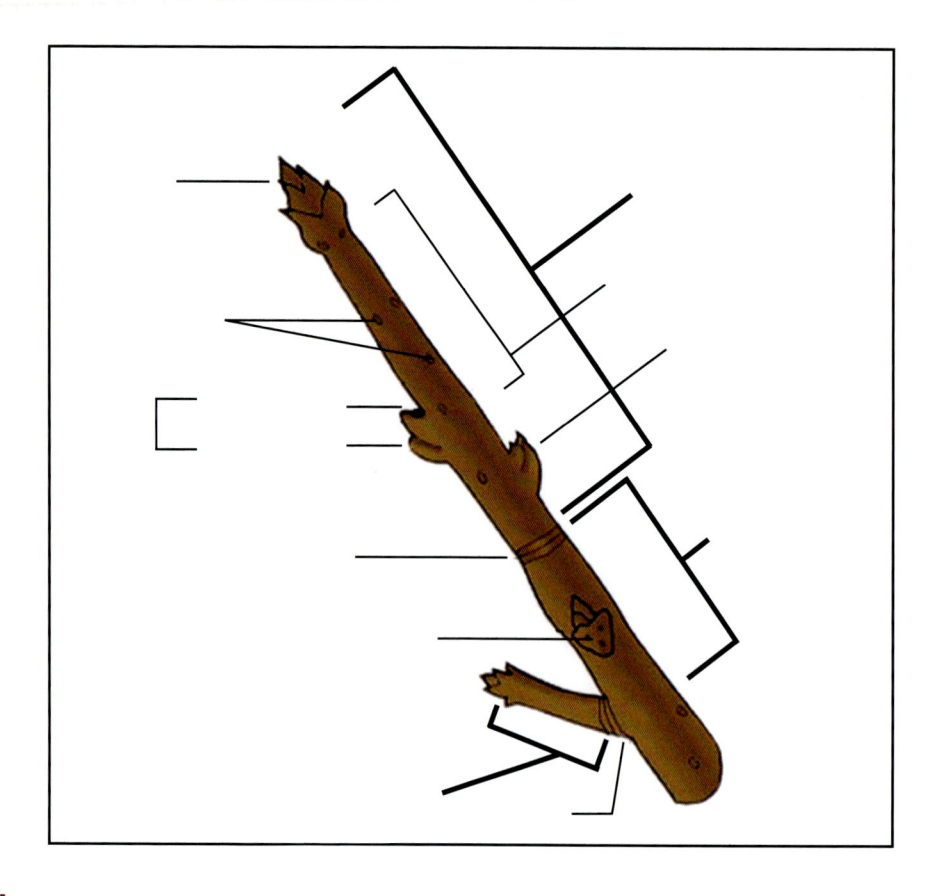

Matching

____ auxin   A. uses more energy than it produces

____ chlorophyll   B. mostly located in the upper 30 cm (12 in) of soil

____ cuticle   C. "stalk" of a leaf

____ petiole   D. cells that cross the phloem and xylem for radial transport

____ internode   E. waxy covering of a leaf

____ lenticel   F. small openings in stems for gas exchange

____ ray   G. plant growth regulator

____ absorbing roots H. between the nodes of a stem

____ source   I. mature, green leaves—sugar producers

____ sink   J. green pigment

Challenge Questions

1. Describe how growth rings are formed. Are they always annual? What information can be obtained by examining growth increments, and how might the rings be useful in diagnosis?

2. What is the relationship between photosynthesis and respiration? What is net photosynthesis? What might be the effect on a tree if respiration exceeds photosynthesis for an extended period of time? How could this occur?

3. Explain the process of Compartmentalization of Decay in Trees. What are the four walls, and how do they function? How can trees be hollow and still remain vigorous?

4. Describe the differences and similarities between trees (eudicots) and palms, and consider how the different growth strategies of palms may be advantageous or limiting.

Sample Test Questions

1. When cutting through a tree with a chain saw or drilling into a tree, you would pass through which structures (in order)?

    a. bark, cambium, phloem, xylem
    b. bark, phloem, cambium, xylem
    c. bark, cambium, xylem, phloem
    d. bark, xylem, phloem, cambium

2. If the terminal bud is removed in pruning,

    a. growth may be stimulated in lateral buds
    b. flowering is stimulated to enhance fruit production
    c. the branch will die back
    d. all of the above

3. The growth rings of many trees

    a. are visible because of anatomical differences between earlywood and latewood
    b. are distinguished by the rays that separate them
    c. contain water-conducting xylem, regardless of age
    d. are visible in angiosperms but not in gymnosperms

4. Which layer of cells is responsible for outward trunk growth and increased girth of a tree?

    a. cambium
    b. pith
    c. epidermis
    d. cortex

5. Mycorrhizae are

    a. collar-rot fungi
    b. elongated underground stems producing sucker sprouts
    c. a symbiotic relationship between fungi and roots
    d. cells in which photosynthesis takes place

Recommended Resources

(See Recommended Resources in back of book for detailed information.)

Applied Tree Biology (Hirons and Thomas 2018)

The CODIT Principle (Dujesiefken and Liese 2015)

Introduction to Arboriculture: Tree Anatomy online course (ISA)

Introduction to Arboriculture: Tree Physiology online course (ISA)

A New Tree Biology (Shigo 1986)

CHAPTER 2

TREE IDENTIFICATION

▼ OBJECTIVES

- Describe how plants are classified, including how botanical names are based on the classification system.

- Explain what a botanical name is, why botanical names are used, and how they are written.

- Explain how characteristics such as reproductive structures, growth habit, texture, and color can be used in tree identification.

- Describe how leaf arrangement and morphology are used to help identify trees.

- Compare various leaf shapes and types of leaf margins, bases, and apices.

KEY TERMS

alternate
angiosperm
bipinnate
broad-leaved
class
compound leaf
conifer
cultivar
cultivation
deciduous
entire
eudicotyledon (eudicot)
family
foliage
form (forma)

genus
gymnosperm
hybrid
identification key
internode
kingdom
leaf apex
leaf base
leaf margin
monocotyledon (monocot)
morphology
node
nomenclature
opposite
order

palmate
phylum
pinnate
pith
propagation
serrate
simple leaf
species
specific epithet
subspecies
taxonomy
variety
vascular plant
whorled

Introduction

Identification of tree species is the first step before attempting to diagnose problems and prescribe tree care. This is because diseases, pests, and cultural requirements can vary substantially depending on the species; also, plant species identification is a requirement in order to legally apply pesticides in many countries. Accurate identification requires a combination of knowledge and experience. Identification skills can be taught and learned, but proficiency requires practice and repeated exposure to woody landscape plants at different times during the year.

Figure 2.1 A good arborist learns to identify trees using many characteristics, including flowers, fruit, form, bark, buds, twigs, leaves, and even scent.

Plant Classification

Taxonomy is the science of identifying, naming, and classifying organisms. It has two purposes: (1) to help us communicate accurately about plants and (2) to represent our understanding of how they are related to each other. Plants are classified and named to represent how they are related to each other. Over time, taxonomic name changes occur when advances in botanical knowledge lead to a reclassification of plants. With DNA analysis, changes have been accelerated and, in some cases, have led to controversies in classification.

The highest classification level for living organisms is the domain, followed by the **kingdom**; trees are in the plant kingdom. Within the plant kingdom is the classification level **phylum**, which further separates most **vascular plants** (plants with xylem and phloem) from plants lacking vascular tissue.

The two largest groups of plants with vascular tissue are those with seeds covered by an ovary (fruit), or **angiosperms**, and those with "naked seeds," or **gymnosperms**. Gymnosperms include **conifers** (cone-bearing plants) and ginkgo. Most **deciduous** trees (trees that shed their leaves each year) and **broad-leaved** evergreens are angiosperms.

Angiosperms include several major groups of plants, the largest being **eudicotyledons (eudicots)**, **monocotyledons (monocots)**, and basal angiosperms (for example, *Magnolia* spp. and *Cinnamomum* spp.). Aside from gymnosperms and palms, most common tree species are eudicots. Monocots include grasses, lilies, orchids, and palms. The vascular tissues of monocots are in bundles, scattered throughout the stem. Because these bundles do not increase in girth, palm stems have little ability to increase in diameter.

Phyla (plural of phylum) are separated into **classes**, followed by **orders** and then **families**. Plants in the same family typically have common characteristics, most notably their types of flowers and fruits. For example, alders (*Alnus* spp.) and birches (*Betula* spp.) are both in the birch family

Classification of *Acer saccharum*, Sugar Maple	
Kingdom:	Plantae
Phylum (division):	Magnoliophyta
Class:	Magnoliopsida
Order:	Sapindales
Family:	Sapindaceae
Genus:	*Acer*
Specific epithet:	*saccharum*
Species = Genus + specific epithet:	
	Acer saccharum

(Betulaceae), and their flowers are morphologically similar.

Plants that are very closely related will show similar characteristics, particularly in their reproductive structures, and may be classified in the same **genus** (plural is genera). For example, all trees in the genus *Quercus* produce acorns. In general, plants within a species will freely interbreed with each other but not as easily, or at all, with plants outside of the species.

The **species** is the level that identifies the particular plant. The word "species" is both singular and plural. The species name is a combination of the genus name and the **specific epithet** (name). Some people use a common saying to help them remember the taxonomic hierarchy from kingdom down to species: **K**ings **p**lay **c**hess **o**n **f**at, **g**ray **s**tumps.

A general understanding of plant classification can help an arborist learn to identify plants. Knowledge of plants and their characteristics can be helpful in diagnosis because trees in the same family or genus are often susceptible to the same diseases, pests, and other disorders.

Plant Nomenclature

Plant **nomenclature** is the naming of plants. Arborists are often familiar with common names of trees because they have learned to identify them through years of field experience. However, the exclusive use of common names can lead to confusion and misunderstanding.

One tree may have several common names. *Carpinus caroliniana* is known as American hornbeam, blue beech, ironwood, and musclewood. Several tree species may have the same common name. For instance, *Magnolia × soulangiana*, *Spathodea campanulata*, and *Liriodendron tulipifera* are all called tuliptree in different areas of the world. Common names can also be misleading. For example, Douglas-fir is not a fir, baldcypress is not a cypress, screw-pine is not a pine, and mountainash is not a species of ash.

Each plant has a unique botanical name, sometimes called a scientific name or a Latin name, that is the same throughout the world. Botanical names of plants are based on a species classification system, and each botanical name has at least two parts. The first part of a botanical name is the genus, which is written with a capitalized first letter. The second part of the name identifies the specific epithet and is not capitalized. The species is composed of the genus and specific epithet. If the specific epithet is not known, "sp." will follow the genus if it is only one species. The letters "spp." will follow the genus name when referring to two or more species of the same genus. Both the genus and specific epithet are written in italics or underlined; sp. and spp. are not.

Common names should not be capitalized unless they include a proper name such as "European," and they should not be italicized. Botanical names should be written with either underlined or italicized text; italics is preferred. Examples of properly written botanical and common names are

Lagerstroemia indica:	crapemyrtle
Washingtonia robusta:	Mexican fan palm
Acer saccharum:	sugar maple
Magnolia grandiflora:	southern magnolia
Eucalyptus globulus:	blue gum eucalyptus

Hybrids are the result of crossbreeding between two different species or, less commonly, genera.

Names of hybrid species are written with an "×" between the genus and specific epithet. Hybrid genera are written with an "×" before the genus. The "×" should not be in italics or underlined.

Some species are further divided into subspecies, varieties, forma, and/or cultivars. A **subspecies** is a naturally occurring, closely related group within a species that has some distinctly different characteristics. The sugar maple (*Acer saccharum*) and black maple (*Acer saccharum* ssp. *nigrum*) are one example. The abbreviation for subspecies ("ssp.") is not italicized or underlined.

A **variety** is a subdivision of a species that has a trait distinctly different from the other plants within the species and naturally breeds true to that trait. Variety names are not capitalized, but they are italicized or underlined. The abbreviation for variety or varieties is "v."

A **form** (plural is forma) is similar to a subspecies, but the differences are less obvious and more sporadic. These plants may have slightly different levels of cold hardiness or unique flower colors that randomly appear. *Benthamidia florida* f. *rubra* (formerly *Cornus florida* f. *rubra*), the pink flowering dogwood, is an example. The abbreviation for form and forma ("f.") is not italicized or underlined.

Cultivars are cultivated varieties that require human intervention (**propagation** or **cultivation**) to maintain a trait. Cultivars are genetically uniform. Cultivar names are written within single quotation marks or with the abbreviation "cv." between the specific epithet and the cultivar name, but not both. The first letter of each word in a cultivar name is capitalized.

Corymbia citriodora—lemon-scented gum
Gleditsia triacanthos f. *inermis*—thornless common honeylocust
Acer platanoides 'Crimson King'—Crimson King Norway maple
Amelanchier × *grandiflora*—apple serviceberry

Today, plant nomenclature of cultivated plants is further complicated by the use of trademark names and patented cultivar names. An example of this is *Betula nigra* 'Cully', known by the trademark name of Heritage® river birch (*Betula nigra* Heritage®). Cultivar names cannot be trademarked. As a result, the rules of botanical nomenclature now state that the cultivar name must be different from the trademark name. Trademark names are never written using single quotation marks (for example, *Betula nigra* 'Heritage' is incorrect).

Basic ID Principles

Woody plant identification is based on **morphology**, which is the size, shape, and external appearance of plant parts. A fundamental knowledge of woody plant morphology is therefore essential. Although

Figure 2.2 At times, it is possible to use multiple features (leaves, flowers, fruit, bark, and buds) present on a tree at the same time to identify it.

arborists usually concentrate on the leaves and overall form when identifying trees and shrubs, botanical classification is often based more on the reproductive characteristics: the flowers and the fruit. A good arborist learns to identify trees using many characteristics, including form, bark, buds, twigs, leaves, flowers, fruit, and even scent. This will enable identification in any season. An arborist who learns tree identification by leaves alone will be able to identify deciduous trees only during a portion of the year.

Many trees can be identified at a distance based on their form and branching characteristics (growth habit). The American elm (*Ulmus americana*) may be identified by its vase-shaped growth habit and overarching branches. Sugar maples (*Acer saccharum*) in the autumn can be identified from a great distance because of their brilliant fall colors. Live oak (*Quercus virginiana*) is commonly identified by its spreading habit and Spanish moss. In other instances, however, a group of arborists may find themselves gathered around a plant, closely examining the details of a twig in order to identify it.

Recognizing the difference between a **simple leaf** and a **compound leaf** is an important first step in identifying many trees. A simple leaf is a single, one-part (one blade, one needle, etc.) leaf that is not subdivided into leaflets. A compound leaf has two or more leaflets but only a single bud or cluster of buds at the base of the petiole. A leaf is determined by the presence of a bud where it attaches to the stem. A **pinnately** compound leaf has small leaflets arranged along its central leaf vein and has a similar appearance to a feather. A **palmately** compound leaf has small leaflets joined at a common center point on the leaf, much like fingers on a hand. A

Figure 2.3 A simple leaf is a single, one-part (one blade, needle, etc.) leaf that is not subdivided into leaflets.

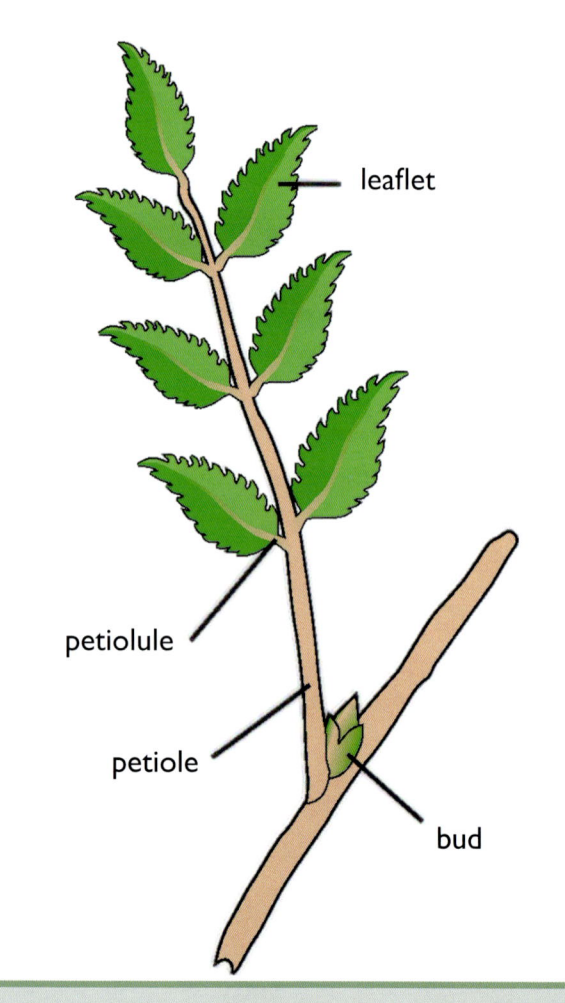

Figure 2.4 A compound leaf has two or more leaflets but only a single bud or cluster of buds at the base of the petiole.

pinnate • palmate • bipinnate

Figure 2.5 A pinnately compound leaf has leaflets arranged along the midrib, the central leaf vein. A palmately compound leaf has small leaflets joined at a common center point on the leaf, much like fingers on a hand. A bipinnately compound leaf has a second order of smaller leaflets comprising each larger leaflet.

bipinnately compound leaf has a second order of smaller leaflets comprising each larger leaflet.

Identifying some tree species requires recognizing relatively minor leaf, bud, or twig characteristics. For example, the type of **leaf margin**, shape of the **leaf base** or **leaf apex**, presence of hairs on the upper or lower leaf surfaces, or the color of young twigs may be required to distinguish between two closely related species. For winter identification of certain deciduous trees, arborists must be familiar with the characteristics of the bark, branching habit, twigs, buds, fruit, and **pith**.

Trees may also be identified by the arrangement of their leaves on the stem. A **node** is the point on a stem where one or more leaves arise. An **internode** is the region of the stem between nodes. **Opposite** leaf arrangements have two leaves emerging at the same node on the stem across from each

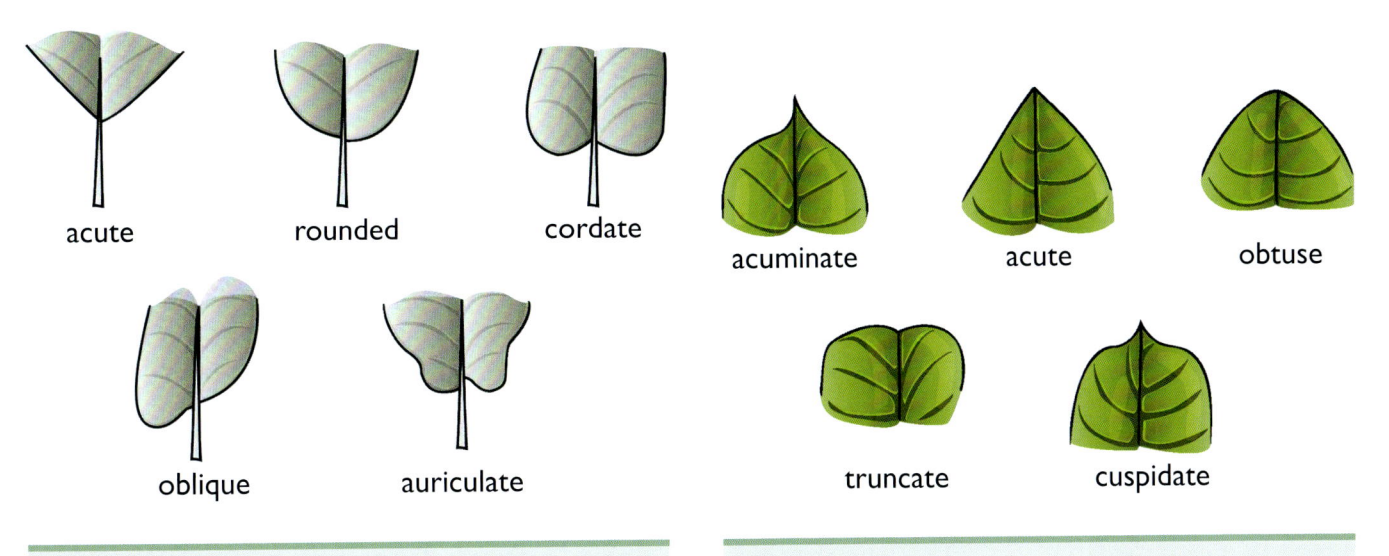

acute • rounded • cordate
oblique • auriculate

acuminate • acute • obtuse
truncate • cuspidate

Figure 2.6 Leaf bases.

Figure 2.7 Leaf apices.

other. **Alternate** leaf arrangements, which are the most common, have a single leaf at each node. Trees with **whorled** leaf arrangements have three or more leaves arising at the same node.

Arborists sometimes use little tricks to help them identify certain species. One such trick helps narrow down the genus based on whether a tree has an opposite or alternate leaf and bud arrangement. For compound leaves, the arrangement of whole leaves should be considered, not the arrangement of leaflets. In temperate North America, Europe, and Asia, most of the trees with opposite leaf arrangement fall into four genera represented by the memory device "MAD Horse," for **m**aple, **a**sh, **d**ogwood, and **horse**chestnut.

There are also a few simple ways to distinguish among the major groups of conifers, or cone-bearing trees. Pines (*Pinus* spp.) have needles usually bundled in clusters of two, three, or five. Counting the needles can help identify the species. Spruces (*Picea* spp.) and firs (*Abies* spp.) produce single needles. Spruce starts with an "s," which helps us to remember that these plants have needles that are **s**hort, **s**harp, **s**ingle, and **s**quare. The needles of firs detach from the stem, leaving a circular "pad,"

Figure 2.8 Leaf margins.

Figure 2.9 Opposite leaf arrangements have two leaves emerging at the same node on the stem across from each other. Alternate leaf arrangements have a single bud or leaf at each node. Whorled leaf arrangements have three or more leaves arising at the same node.

whereas spruce needles leave a tiny stalk. Also, a fir tree—like a fur coat—has needles that are softer than spruce needles. Other conifers may have awl-like or scalelike **foliage**.

Figure 2.10 Five-needle pine. Some species of pine have needles borne in clusters of five.

Figure 2.11 Two-needle pine. Many species of pine have needles borne in clusters of two. A few can have needles in clusters of three.

Figure 2.12 Single needles. A: Fir (*Abies* spp.); B: Spruce (*Picea* spp.); C: Hemlock (*Tsuga* spp.).

Figure 2.13 Multiple needles, larch (*Larix* spp.).

Figure 2.14 Scalelike foliage, arborvitae (*Thuja* spp.).

Figure 2.15 Awl-like foliage, juniper (*Juniperus* spp.).

Palms

Palms are classified as monocots and, as such, are related to grasses. The variety within palms is remarkable. There are thousands of species representing more than 200 genera, and they are native to all of the continents except Antarctica. Most of the common species are single-stemmed, while others are multi-stemmed; some are shrubby, and others are vinelike. There are species that can grow more than 65 m (200 ft). Coconut palm produces the largest seeds of any plant, and they float, allowing them to travel from island to island. Most palm species are native to tropical or subtropical rain forests, but some species are grown as crops (such as coconut, oil palm, ornamentals). Several tropical plants are palm-like, and are even named palms, but are technically not palms (for example, sago palm, ponytail palm, dracaena, and yucca). Most palm leaves (fronds) are compound, either palmate or pinnate.

Using a Key

Many tree and shrub reference books or databases contain **identification keys**. A key is a step-by-step method for unlocking the identity of a plant. Identification keys use terminology that describe the shape, texture, and arrangement of the leaves; bud characteristics; twig characteristics; and the morphology of flowers and fruits. An identification key may be used to systematically determine the identity of a plant. Most keys consist of a series of yes or no questions. The user narrows down the possibilities by reviewing characteristics of the plant in question. For example, leaf characteristics, such as determining whether the leaf (bud) arrangement is opposite or alternate, whether the plant has simple or compound leaves, whether the leaf margins are **serrate** or **entire**, and so on, may be used.

A few notes of caution about relying on keys:

- A certain level of expertise and understanding of the terms used is necessary to identify a plant correctly.

- The plant being identified may not match its written characteristics exactly due to seasonal differences in plant color and morphology.
- Not all keys are complete with all trees that might be found in a given region.
- Genetic variability, plant condition and location, and the environment can affect the size of plant parts or result in some irregularities.
- The characteristic described in the key may be seasonal and not present at the time.

Nevertheless, a key can be a helpful tool in determining a tree's identity.

One drawback to using identification keys is that the user can get stuck if unable to answer a question. If this happens, the user can proceed down one path and then return to the question if no matching plants are found in the key the first time through. Either way, an understanding of identification terminology is vital.

Advances in technology have aided practical plant identification. Some widely available applications have made identification keys available for electronic use. Others have algorithms that can identify trees based on uploaded photos. But while electronic applications can be another useful tool for plant identification, they have many of the same limitations as printed identification keys.

Figure 2.16 Most palm leaves (fronds) are compound, either palmate or pinnate.

CHAPTER 2 WORKBOOK

1. The classification of living organisms, including plants, is called _____ .

2. List the levels of classification. The first letter of each term is given.

    K

    P

    C

    O

    F

    G

    S

3. _____ are vascular plants whose seeds are covered (by an ovary).

    _____ are vascular plants with "naked seeds."

4. Eudicotyledon (eudicot) refers to plants that have two seed leaves at germination. Grasses, banana, and palms belong to another group called _____ and have only one seed leaf.

5. The naming of plants is called _____ .

6. Name five plant characteristics used to identify trees.

    a.

    b.

    c.

    d.

    e.

Chapter 2: Tree Identification

7. Draw a twig with the following leaf arrangements.

   opposite   alternate   whorled

8. Name a tree with palmately compound leaves: _____ .

   Name a tree with pinnately compound leaves: _____ .

9. Draw a simple leaf with a lobed leaf margin.

10. Draw a compound leaf with serrate margins on the leaflets.

11. A compound leaf with multiple leaflets will have _____ bud(s).

12. Give an example of a tree species that has more than one common name: _____ .

13. In the botanical name *Acer saccharum*, *Acer* identifies the _____ , and *saccharum* identifies the _____ _____ .

14. Species may be subdivided into _____ , _____ , _____ , and/or _____ that have distinct differences from the general species.

15. A _____ is a cultivated variety.

Challenge Questions

1. Name three nonvisual tree characteristics that can be used in identification.

2. Draw a bipinnately compound leaf attached to a twig, and label the bud, petiole, and a leaflet.

3. Write a tree's botanical name and label its genus, specific epithet, variety, and cultivar.

Sample Test Questions

1. Douglas-fir (*Pseudotsuga menziesii*) differs from balsam fir (*Abies balsamea*) in that
   a. they are not in the same genus
   b. they are not in the same family
   c. Douglas-fir is actually a type of hemlock
   d. balsam fir is not a conifer

2. When two leaves and/or buds are located at the same node on a twig, the arrangement is called

   a. opposite
   b. alternate
   c. whorled
   d. bicompound

3. Select the botanical name that is written correctly.

   a. *Quercus Rubra*
   b. *Quercus rubra*
   c. *quercus Rubra*
   d. *quercus rubra*

4. Which genus of trees usually does *not* have an opposite leaf arrangement?

   a. *Acer* (maples)
   b. *Fraxinus* (ashes)
   c. *Quercus* (oaks)
   d. *Aesculus* (horsechestnuts/buckeyes)

5. Which conifers have needles in bundles?

   a. hemlocks
   b. firs
   c. pines
   d. spruces

Recommended Resources

(See Recommended Resources in back of book for detailed information.)

Manual of Woody Landscape Plants (Dirr 2009)

Introduction to Arboriculture: Identification Principles online course (ISA)

Missouri Botanical Garden Plant Finder website

SelecTree: A Tree Selection Guide website

CHAPTER 3

SOIL SCIENCE

▼ OBJECTIVES

- Explain how soil texture and structure affect soil physical properties such as porosity and water movement.

- Define pH, cation exchange capacity (CEC), and buffering capacity. Describe their effects on the availability of essential elements to trees.

- Define rhizosphere, and describe the importance of soil organisms and organic matter to tree roots.

- Explain the relationship between soil moisture, nutrient uptake, and root growth.

- Define and describe the concepts of gravitational water, field capacity, permanent wilting point, soil structure, and infiltration rate.

- Discuss approaches and limitations to soil improvement.

KEY TERMS

- aggregate
- air-excavation device
- anion
- buffering capacity
- bulk density
- capillary water
- cation
- cation exchange capacity (CEC)
- clay
- compost
- field capacity
- gravitational water
- ion
- leach
- loam

- macropore
- micropore
- mineralization
- mycorrhizae
- nutrient cycling
- organic layer
- organic matter
- parent material
- permanent wilting point
- pH
- pore space
- rhizosphere
- root exudates
- saline soil
- sand
- silt

- sodic soils
- soil biological properties
- soil chemical properties
- soil compaction
- soil food web
- soil horizon
- soil physical properties
- soil profile
- soil structure
- soil texture
- structural cells
- structural soil
- suspended pavement
- symbiotic
- water-holding capacity

Introduction

Trees and soils are so ecologically interdependent that it is hard to imagine separating them from one another. Yet the processes involved with urban development disrupt this ecological balance, creating growing conditions that range from suitable to unfavorable to antagonistic. It has been said that the vast majority of tree decline situations can be attributed to problems originating below the ground.

Understanding soil science is vital to arboriculture because the relationship between trees and the soils in which they grow has a major influence on their health and performance. By knowing more about **soil physical**, **chemical**, and **biological properties** as well as common urban soil conditions, the arborist will be better equipped to manage urban trees.

Soil is much more than a substrate of rocks, sand, silt, clay, and organic matter that anchors roots and provides water and minerals to trees. Rather, soil is an ecosystem inhabited by insects, earthworms, nematodes, fungi, bacteria, and other microbes all living together in a delicate balance. A handful of soil can contain tens of millions of living microbes, each with a function in the soil's ecology.

The ecology of soils varies with geographic regions, climates, the underlying geology, and the types of plants growing there. The mix of plant and animal microorganisms below a grass lawn may be very different from that found beneath a forest floor.

Soils are generally described in terms of their physical, chemical, and biological properties. Physical properties include the characteristics of the minerals that make up the soil, how they are arranged, and other physical features, such as how water moves through the soil. Chemical properties include characteristics such as soil acidity, salinity, nutrient status, and other elements, while biological properties include fungi, bacteria, and many other life forms that make their homes in soils. Although it is convenient to separate these characteristics when describing or evaluating soils, in reality they are all interconnected. Understanding these relationships can help explain many phenomena observed by arborists.

Physical Properties

Native soils are the result of hundreds to millions of years of biological, chemical, and physical weathering of **parent material**, or underlying bedrock. Soil types are usually determined by the geology of the soil parent material. Some soils develop from weathered rock; some develop from sediment deposited along waterways or the beds of ancient seas and lakes; and some develop from a combination of

Figure 3.1 Horizons are visible layers of soil oriented parallel to the soil surface. The arrangement, depth, characteristics, and number of horizons vary with different soils.

both. The nature of the parent materials is an important factor that affects soil characteristics.

Over time, soils develop horizontal layers below the surface called **soil horizons**, due to rainfall, leaching, heating and cooling, chemical reactions, biological activities, and accumulation of different elements and materials. Horizons are visible layers of soil oriented parallel to the soil surface. The nature, arrangement, depth, and number of horizons vary with different soils. Together they form the **soil profile**. Soil profiles can vary significantly by region from the "typical" profile.

The typical soil profile normally consists of five major horizons (O, A, E, B, and C), although soil scientists classify a number of sublayers and transitional layers. These layers can be distinguished by differences in color, texture, and smell, which can indicate variations in drainage, organic and mineral content, and other characteristic changes.

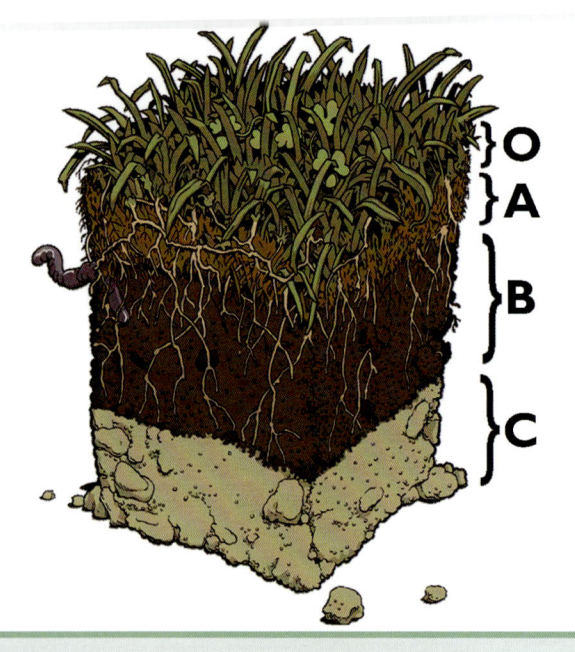

Figure 3.2 The soil profile shows the different horizons. Soil profiles can vary significantly by region. The E layer is not present in this sample.

Figure 3.3 A Dutch or bucket auger is commonly used to assess soil profiles in urban areas.

The O horizon, known as the **organic layer**, is a layer at the top of the soil profile that comprises organic material in various stages of decomposition. It provides a source of nutrients for plants, and it buffers the soil from climatic extremes. The next layer down is the A horizon. Its organic matter content is significantly less than that of the O horizon (less than 20 percent organic matter). The A horizon contains most of the fine roots of trees and is typically biologically active. Although it is primarily composed of inorganic material (sand, silt, and/or clay), the A horizon is normally rich in organic matter, which gives this horizon its characteristic dark color. The E horizon, when it exists, is found below the O and A horizons. It lacks the organic matter found in these surface horizons and is lighter in color.

The B horizon occurs below the A horizon (and the E horizon, if present). It is a zone of accumulation where materials that have leached from the surface mix with soil particles from the lower parent material. The uppermost layers (O, A, E, and sometimes the upper part of the B horizon) are generally referred to as the topsoil, although there is no technical definition of this term. In urban situations, these layers are often disturbed, mixed, or buried and may not be easily identifiable. Additionally, a large portion of the topsoil may have been stripped away or damaged during construction and land development. The C horizon is the deepest layer—just above the bedrock—and is composed of partially weathered parent material. Soil in this horizon is continually forming and changing through the physical, chemical, and biological weathering of the parent material.

Soil **organic matter** consists of dead plant and animal material in various stages of decomposition. A significant portion of soil organic material comes from fine root turnover—the decaying of fine roots that cease to function. Organic matter shrinks and swells with changes in moisture content, which helps to form **pore space**, the voids between soil particles, within the soil. Some organic matter helps to bind soil particles together to form larger groups of soil particles, which improve soil structure. Organic matter also serves as a food (carbon) source for soil organisms. The growth and movement of these organisms through the soil are important for improving aeration, fertility, and soil structure.

Tree roots grow where soil conditions are favorable. Roots require space among soil particles, organic materials and essential mineral elements, and

Figure 3.4 Forest soils tend to be rich in organic matter from the decomposition of plants and animals contributing to nutrient cycling.

adequate oxygen and water. Most of the fine, absorbing roots are found in the upper 15 to 25 cm (6 to 10 in) of soil. Relatively few tree roots grow deeper than 1 m (3 ft), but depending on the tree species and environmental conditions, trees may develop deeper root systems.

Soil Texture

Soil texture refers to the relative fineness or coarseness of the inorganic, mineral soil particles—specifically the relative proportions of **sand**, **silt**, and **clay**. Sand particles are relatively large, resulting in coarser-textured soils. Soils that have a high percentage of clay, the smallest soil particles, are considered fine textured. Silt particles are intermediate in size. Textural classes of soil are determined by the percentage of the three particle types (Figure 3.6). **Loam** refers to a soil texture that is a balance of sand, silt, and clay. It has favorable characteristics for most plant growth. Loam is not, however, an equal mix of sand, silt, and clay; it contains relatively less clay.

Within each texture classification, soil particles can vary in size. These size ranges can affect the physical properties of the soil. For example, a soil containing fine sand will drain more slowly than coarse sand.

Soil texture affects a soil's ability to hold water and provide oxygen to the roots, so it has a profound influence on the soil's chemical and biological properties. Texture plays an important role in determining which species of trees will do well in a given site. Although soil texture is an important factor in determining a soil's physical properties, soil structure also plays a key role.

Soil Structure and Pore Space

Chemical and physical changes and the activity of soil organisms cause soil particles (sand, silt, and clay) to be grouped or clumped together. These secondary groups, or clumps, are known as soil **aggregates**. The shape, size, strength, and arrangement of soil aggregates form the **soil structure**. Root growth, freezing and thawing, and burrowing insects and other organisms contribute to changes in soil structure over time.

About 50 percent of uncompacted soil volume is pore space between soil particles. A complex

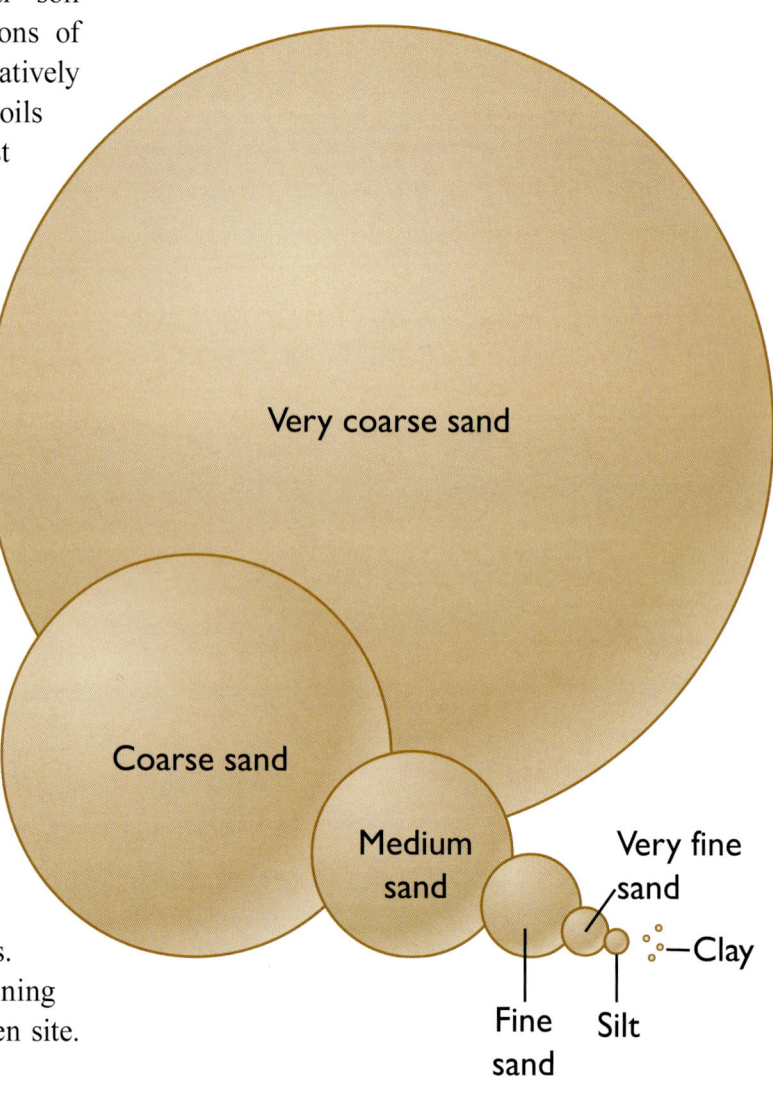

Figure 3.5 Soil texture refers to the relative fineness or coarseness of the inorganic, mineral soil particles—specifically the proportions of sand, silt, and clay. Sand particles are relatively large, resulting in coarser-textured soils. Soils that have a high percentage of clay, the smallest soil particles, are considered fine textured. Silt particles are intermediate in size.

Chapter 3: Soil Science

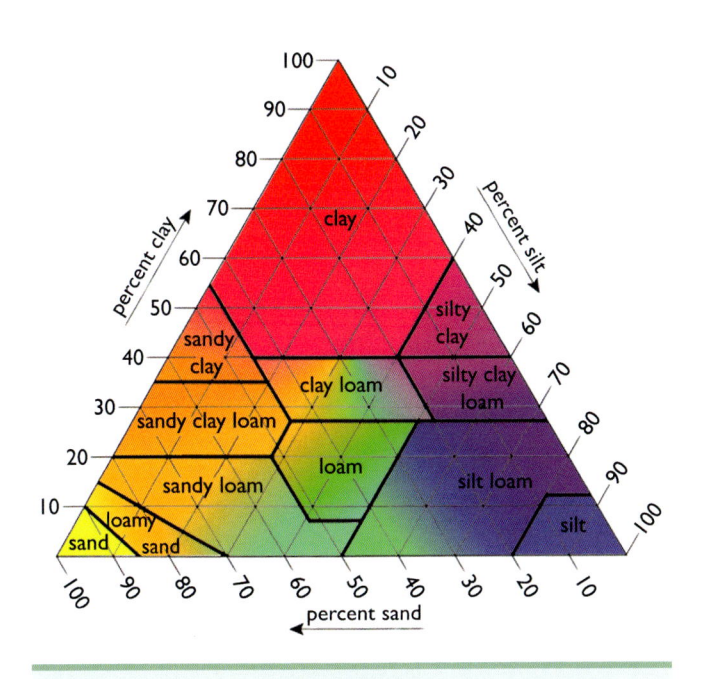

Figure 3.6 The soil texture triangle shows soil types based on the percentage of sand, silt, and clay.

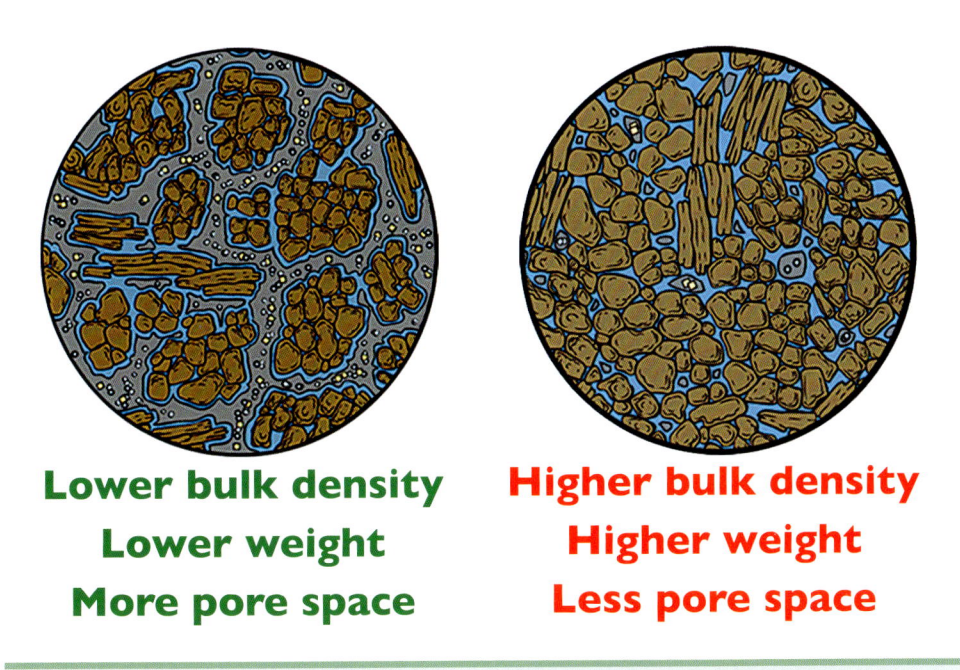

Lower bulk density
Lower weight
More pore space

Higher bulk density
Higher weight
Less pore space

Figure 3.7 The amount of pore space is related to the bulk density of the soil. Bulk density is the mass (weight) of dried soil per unit of soil volume. Bulk density can be used to assess whether adequate pore space exists. It can also be used as a metric to determine the extent of compaction in the soil profile. Lower density (left) is better for tree root growth and health.

network of pores of varying sizes occurs both within and between soil aggregates. The relatively large spaces (mainly between aggregates) are known as **macropores**, while the relatively small spaces (mainly between soil particles) are known as **micropores**. Macropores are too large to hold water against the force of gravity and become filled with air as water drains from the soil. Micropores retain water and are the source of available water to plants between rainfalls. The size and shape of soil aggregates determine the amount of macropore and micropore space in a soil.

Soil texture influences pore space, as well, with coarse soils (greater proportion of sand) being dominated by macropores, and fine soils (greater proportion of clay) having a higher percentage of micropores. For trees and other plants, a well-structured soil has sufficient macropore space for air movement (aeration) and micropore space for water retention. Soils with poor structure have a limited capacity to meet the air and water needs of plants.

Depending on how soil aggregates fit together, there may be more or larger pore spaces for air and water, which can modify the influence of soil texture on drainage and gas exchange. For example, a well-aggregated structure with little compaction facilitates water movement and aeration. Conversely, compacted soil with poor structure is associated with poor water infiltration and percolation.

Bulk density is the mass (weight) of dried soil per unit of soil volume. Bulk density can be used to assess whether adequate pore space exists.

It can also be used as a metric to determine the extent of compaction in the soil profile. **Soil compaction** is defined as an increase in bulk density and a decrease in total pore space. Soil aggregates, and accompanying pore spaces, are easily disrupted by soil compaction. When soil aggregates are destroyed or compressed by vehicles, foot traffic, or other actions, pore space is greatly reduced—especially macropores, which tend to be filled with air. Soils vary in their susceptibility to compaction. Those with a high percentage of fine particles (clays) are easily compacted, while soils with a high percentage of sand are less prone to compaction. Soils with moderate moisture content are generally more easily compacted than dry soils or saturated soils.

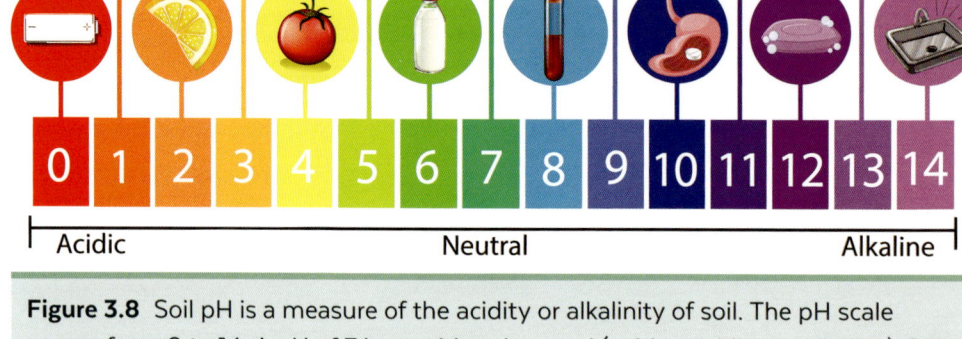

Excessive soil compaction is a major stressor for trees in the urban environment and often a major contributor to decline and dieback. Compacted soils restrict root growth, reduce water infiltration and availability, and limit the movement of oxygen and carbon dioxide in the root zone. Compacted soils can also limit the biological component of soil by reducing the large pores required by larger organisms to move through the soil. Sometimes compacted soils also develop surface crusts. Crusts are created when aggregates are destroyed and the fine soil particles at the surface are oriented like shingles on a roof, inhibiting water infiltration and gas exchange with the air above.

Chemical Properties

Attributes of a soil's chemistry—including pH and cation exchange capacity—can have a significant impact on nutrient availability for trees. Soil **pH** is a measure of the acidity or alkalinity of soil. The pH scale ranges from 0 to 14. A pH of 7 is considered neutral (neither acidic nor alkaline). Soils with a pH less than 7 are acidic, and those with a pH greater than 7 are alkaline. Because pH is a logarithmic function, a pH of 6 is 10 times more acidic than a pH of 7, and a pH of 5 is 100 times more acidic than a pH of 7. Although quite variable for different species, a pH in the range of 6.0 to 6.5 is generally favorable for most trees.

Soil pH has many effects on the ecology and chemistry of the soil. It can affect which species will grow and which soil organisms are present. One of the most important effects of pH on tree growth is the availability of mineral nutrients. Some nutrients are available for uptake within a relatively narrow pH range. For example, in highly acidic soils with a pH of 5.5 or below, phosphorus may be unavailable, while other elements (such as manganese and copper) may become toxic due to high soluble levels. In alkaline soils, iron, zinc, and manganese may be unavailable. When micronutrients such as iron and manganese are deficient in plants, their leaves may show characteristic

Figure 3.8 Soil pH is a measure of the acidity or alkalinity of soil. The pH scale ranges from 0 to 14. A pH of 7 is considered neutral (neither acidic nor alkaline). Soils with a pH less than 7 are acidic, and those with a pH greater than 7 are alkaline.

symptoms of interveinal chlorosis (yellowing of the spaces between leaf veins).

It is possible, though often not practical, to alter soil pH to achieve a more desirable growing medium. Sulfur or other acidifying compounds may be added to temporarily lower the pH, or lime may be added to raise the pH. Altering the pH of the superficial layers of the soil is relatively easy; however, it is more difficult to alter subsurface pH. This is particularly true in urban environments where contamination from alkaline building materials is common. Additionally, many soils have a high **buffering capacity**, or resistance to changes in pH, especially soils high in clay or organic matter.

Minerals required for tree growth (essential elements) dissolve in water, making them available for absorption by tree roots. In solution, these elements are charged particles called **ions**. Negatively charged ions are called **anions**; positively charged ions are called **cations**. The **cation exchange capacity (CEC)** is a measure of the soil's capacity to attract, retain, and exchange positively charged cations. In essence, CEC may be used as a gauge of soil fertility. Soils normally have both negatively and positively charged sites, but negatively charged sites outnumber positively charged sites. Thus, soils have a net negative charge.

Organic matter and clay particles normally have a high negative charge density. This negative charge attracts and holds cations, giving soils high in clay and organic matter a high CEC. Fine-textured soils that contain a relatively high proportion of clay and/or organic matter will tend to be more fertile than coarse-textured soils.

The attraction between cations and soil particles minimizes the tendency of positively charged ions to **leach**, or wash through the soil. Clay soils have a higher CEC, which means they can readily attract, adsorb, and exchange positively charged minerals. Conversely, sand particles have little exchange capacity, and so they are more prone to leaching.

Under some conditions (high evaporation rates or excess fertilization), ions can accumulate to harmful levels. Soils with excess levels of ions in the form of soluble salts are called **saline soils**. Certain types of soils have a tendency to accumulate soil salts in excessive amounts and must be monitored and managed. The recommended treatment for excessive salts in the soil is to leach them through the soil solution with low-salinity water.

Sodic soils are soils in which the cation sodium (Na+) occupies an unusually high percentage of the CEC. Sodic soils have a tendency to crust and are associated with high pH, nutrient deficiencies or imbalances, poor soil structure, and, sometimes, levels of sodium that are toxic to sensitive plants. Correction is sometimes possible with the continued use of low-sodium irrigation water. More intensive remedies involve displacing the sodium from the soil with calcium using a calcium-based soil amendment such as gypsum.

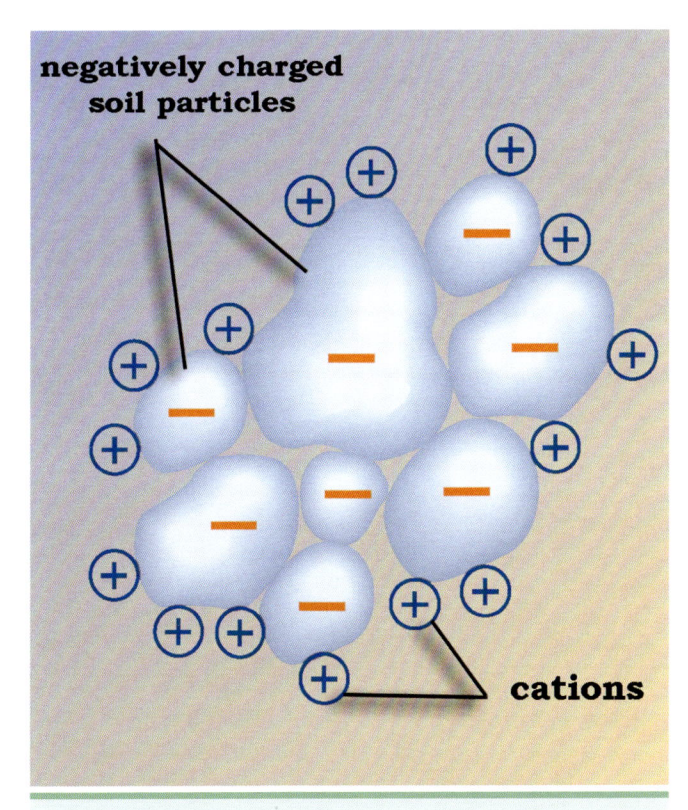

Figure 3.9 Cations are attracted to and held by negatively charged soil particles. The cation exchange capacity (CEC) is a measure of the soil's capacity to attract, retain, and exchange positively charged cations. CEC may be used as a gauge of soil fertility.

Biological Properties

Soil is an ecosystem containing billions of organisms, the vast majority of which are directly or indirectly beneficial to trees. Insects, mites, millipedes, earthworms, fungi, and bacteria all inhabit the soil and litter layer and play a role in developing and maintaining soil structure, organic matter decomposition, and nutrient mineralization.

In addition to small mammals, insects, and worms, additional organisms found in the soil ecosystem include bacteria, fungi, and other microorganisms. Most of these are essential, helping to decompose organic matter or aid in nutrient uptake. Very few cause diseases in plants.

The diversity of organisms living, moving, and interacting in the soil is referred to as the **soil food web**. A healthy, diverse soil food web is a critical component of the health of the overall ecosystem, including trees.

The soil environment immediately surrounding and directly influenced by plant roots is known as the **rhizosphere**. The rhizosphere is an area of intense biological activity in the soil near actively elongating roots where many organisms flourish. As roots extend through the soil, the root caps and external layers are sloughed off, and sugar and other materials (**root exudates**) from the roots are released into the soil. This is a source of organic matter upon which microorganisms feed. Chemical properties within the rhizosphere can be very different from the surrounding soil. For example, pH can be 1 or 2 units higher or lower in the rhizosphere compared with the bulk soil pH and can make more ions available for uptake.

Mycorrhizae, literally meaning "fungus roots," are specialized root structures created when mycorrhizal fungi infect roots of a suitable host plant. Most plants have associated mycorrhizae. Mycorrhizal fungi live in a **symbiotic** relationship with the roots, meaning the fungi and roots both benefit. The roots

Figure 3.10 The diversity of organisms living, moving, and interacting in the soil is referred to as the soil food web. A healthy, diverse soil food web is a critical component to the health of the overall ecosystem, including trees.

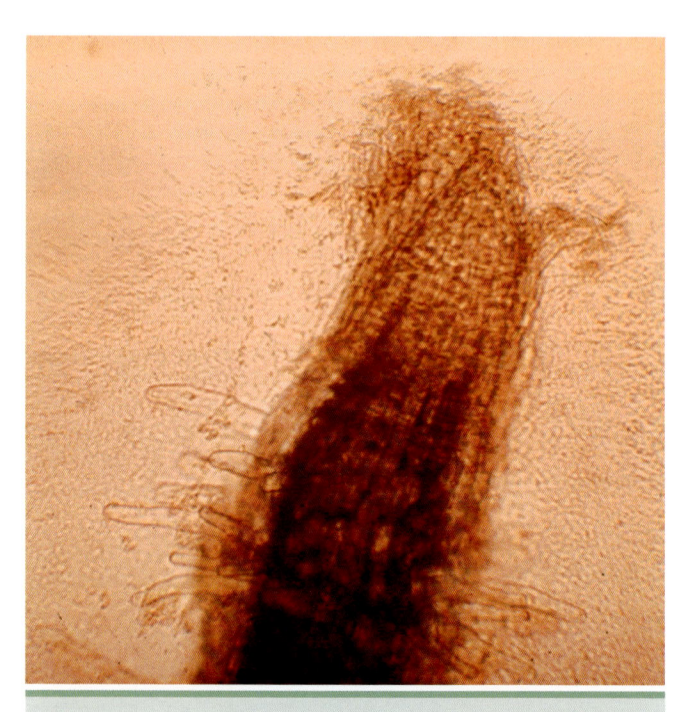

Figure 3.11 The rhizosphere is an area of intense biological activity in the soil near actively elongating roots where many organisms flourish. As roots extend through the soil, the root caps and external layers are sloughed off, and sugar and other materials (root exudates) from the roots are released into the soil. This is a source of organic matter on which microorganisms feed.

Figure 3.12 Mycorrhizae are specialized root structures created when mycorrhizal fungi infect roots. They increase the uptake of water and minerals.

Decomposition releases the nutrients that were bound in the organic material back into the soil, where they become available once again to plant roots.

The bodies of soil microorganisms serve as a storage bank for plant nutrients. Although the nutrients are unavailable to plants while the organisms are alive, as the organisms feed on each other and excrete, nutrients are constantly released back into the soil for plant uptake. The process in which organically bound plant nutrients are converted into inorganic, plant-available forms is called **mineralization**. This storage bank also reduces the loss of nutrients that would otherwise be leached through the soil.

provide a place for the fungi to live and also provide food (sugar). Fungi increase the capacity of roots to absorb water and essential elements, especially phosphorus, to help protect against certain disease-causing fungi and help the tree survive stressful conditions.

Another type of symbiotic relationship exists with plants in the legume family (Fabaceae) and certain species of nitrogen-fixing bacteria. These bacteria form colonies in nodules on the roots and "fix," or convert, nitrogen from the air into forms the plants can use. Examples of leguminous trees (mostly pod-producing) include redbud (*Cercis* spp.), black locust (*Robinia* spp.), golden chain tree (*Laburnum* spp.), and many species that grow in arid climates.

Nutrient cycling is especially important in natural plant systems. As a plant grows, roots absorb essential mineral elements from the soil solution and produce new plant tissue. As seasons pass, plants or plant parts die and are returned to the soil, where they are broken down and eventually decomposed by soil organisms and weathering processes.

Soil Moisture and Plant Growth

Soil and water exist in a dynamic equilibrium that makes it almost impossible to discuss one without mentioning the other. The physical, chemical, and biological properties of soil are all influenced by

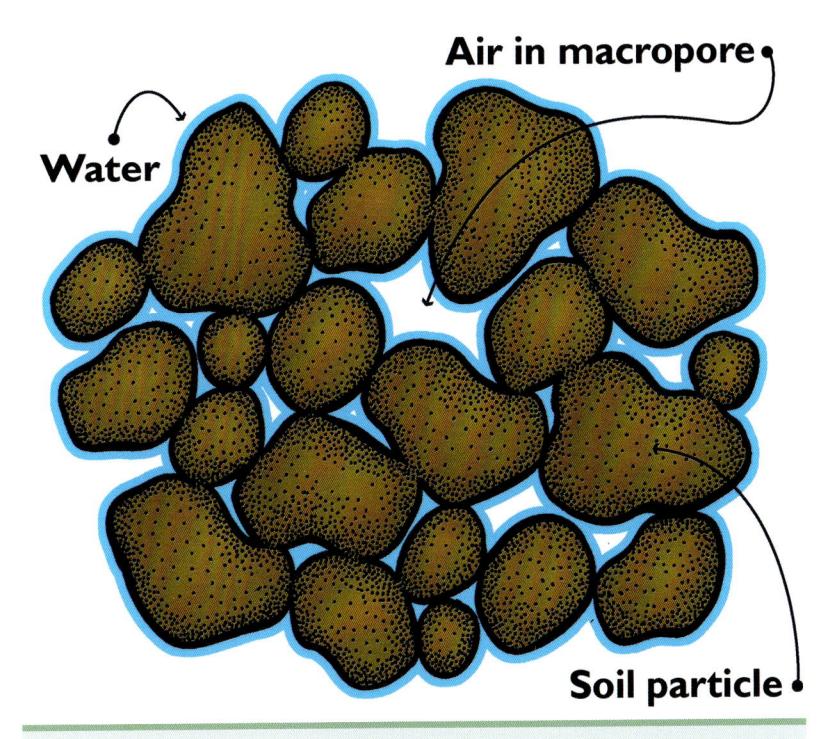

Figure 3.13 Macropores are air-filled at field capacity, but water remains in the micropores, adhering to soil particles.

soil moisture. Arborists must understand this relationship to conduct an effective water management program that will improve and/or maintain tree health.

The volume and size of soil pores and the total surface area of the particles determine the amount of water that a soil can hold. Soils with a large percentage of micropores have a high **water-holding capacity**. Conversely, a soil with a greater amount of macropore space than micropore space has a lower water-holding capacity. For example, clay soils have a high water-holding capacity because pore space is dominated by micropores, while sandy soils with low micropore space have a lower water-holding capacity.

Water that drains from the macropores under the force of gravity is called **gravitational water**. A soil is said to be at **field capacity** immediately after gravitational water has drained away. Water that remains is held in the micropores and is called **capillary water**. Once a soil is at field capacity, water is classified as being available or unavailable. Available water can be absorbed by plant roots, while unavailable water is held so tightly within small pores that very little can be extracted by plant roots. When available water is depleted, plants may experience water stress. When water stress is prolonged and severe, plants may reach a point where they cannot recover, sometimes referred to as the **permanent wilting point**.

Tree roots need both air and water to function and grow. A delicate balance of water and gases (mostly oxygen, nitrogen, and carbon dioxide) exists in the soil pore spaces. Tree roots require oxygen, but they also give off carbon dioxide as respiration takes place. For roots to survive, gas exchange between the soil and the atmosphere must occur. Much of this exchange is by diffusion through the soil surface. Oxygen levels tend to be higher near the soil surface.

If there is insufficient gas exchange from the atmosphere through the soil to roots, an oxygen deficit may occur. Similarly, if the concentration of carbon dioxide increases above critical levels and becomes toxic, root injury can occur. Gas exchange is often limited in saturated and/or compacted soils. In many urban situations, pavement and other hardscape elements can severely limit gas exchange with the atmosphere.

Soil texture plays a major role in water infiltration and percolation. Layers of varying soil texture are common in urban soils and can lead to runoff as well as poor drainage and aeration. If a layer of coarse soil (sand) is on top of a fine-textured soil (clay), water will accumulate in the upper layer as it slowly infiltrates into the lower layer. Conversely, if the coarse layer is below the fine layer, the water will not drain into the lower layer until the upper layer is completely saturated.

If a tree is planted in clay soil and the backfill soil is a coarser soil, the planting hole may act as a bowl and retain too much water (the "teacup effect"). This effect is especially pronounced if nearby surfaces drain into the planting hole. If irrigation and/or precipitation water is held within the root-zone soil for a long time, roots will likely be damaged, although species tolerances for temporary soil saturation vary. Placing gravel in the bottom of

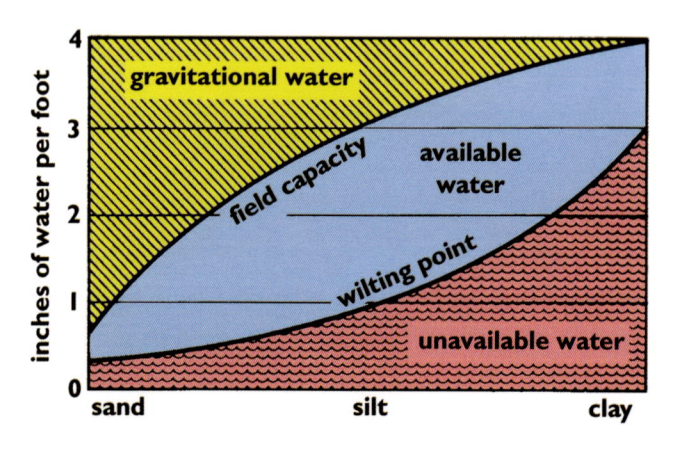

Figure 3.14 Soil texture influences the amount of water available to a tree. According to this chart, if 4 inches of water were applied to silt soil and the gravitational water drains away, 2 inches per foot would remain available (blue), and 1 inch would be unavailable (red). (One inch of water per foot is equivalent to 8 centimeters of water per meter.)

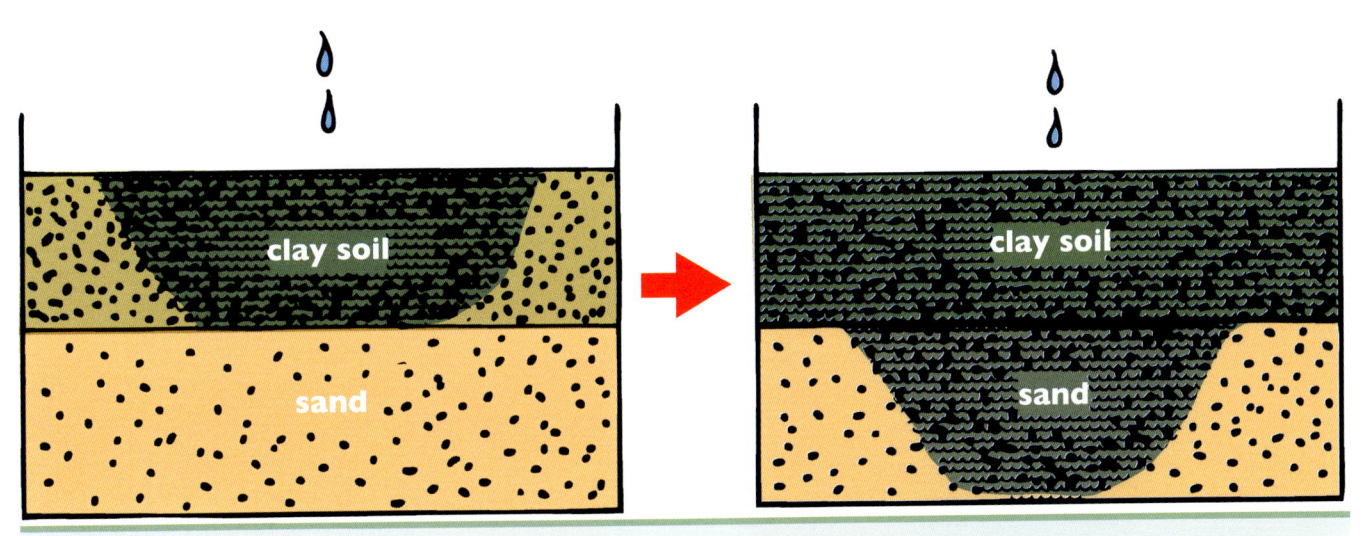

Figure 3.15 This diagram illustrates water infiltration through a layer of fine-textured soil into sand. The upper layer becomes completely saturated before water drains into the lower layer.

planting holes, pits, or containers does not improve drainage, as commonly thought, because the water remains in the finer-textured soil above it until saturation is reached.

Urban Soils

Urban soils are often altered in such a way as to become unfavorable for tree growth and development. For example, urban soils often lack an organic layer. They may be compacted or crusted, inhibiting the flow of water or the exchange of gases. They may have a disrupted soil profile or altered drainage; elevated pH or chemical contamination, causing nutrient deficiencies or toxic damage to plants; or subsurface barriers as a result of building foundations, roads, or underground utilities. The biological makeup of urban soils is typically very different from soils found in natural woodlands. These factors may impact root growth and tree health, eventually leading to decline over a shorter period than would be found in a forest.

Turf, bare ground, and/or pavement replaces the organic layer in many urban soils. Paving may impair aeration and water infiltration. Organic matter reduction decreases biological activity, hampers soil structure development, and interrupts nutrient cycling. Furthermore, the absence of an insulating forest organic layer and understory vegetation can contribute to greater fluctuations in temperature extremes and water availability.

Compaction is one of the biggest problems in urban soils. Foot traffic, vehicle traffic, construction activities, and poor management practices can all lead to compaction. Compaction damages soil aggregates, reduces macropore (air-filled) space, increases bulk density, limits gas exchange, reduces water infiltration, and alters soil organism populations.

Figure 3.16 Urban soils are often altered in such a way as to become unfavorable for tree growth and development.

Figure 3.17 Trees tend to suffer due to soil compaction in urban areas that receive a great deal of pedestrian traffic.

The processes involved with urban development disrupt the ecological balance with trees and soil, creating growing conditions that are often unfavorable to impossible. A large percentage of tree decline situations can be attributed to soil stress.

Structural Soils and Soil Cells

While soils serve important functions for trees, engineers also rely on the structural properties of soils to support roads and buildings. This is one reason severe compaction of soil is common in urban development. Compaction increases the load-bearing capacity of soils, providing a stable base for pavement and other urban infrastructure. Unfortunately, managing soils for buildings and pavement is typically in direct conflict with the needs of trees. There has been considerable effort devoted to finding ways to improve the rooting environment of city trees while maintaining the compression requirements of urban pavements and streets.

Structural soils are soil mixes that can be compacted to meet engineering requirements and still allow root growth and development. Ranging from sand mixes to various combinations of gravel, clay loam, and sometimes hydrogels (used as a stabilizing agent), structural soils are designed to create a load-bearing surface with sufficient pore space to support root growth and establishment of trees. The use of structural soils is not without some concerns and limitations. Proper handling and mixing is critical to achieving the desired mix when placed in the root zone. The expense of using structural soils can be significant. Finally, trees that are planted must be compatible with intermittent water deficits because the water-holding capacity of these soils tends to be lower than that of other soils.

Another approach to providing adequate root space under urban pavement is to create **suspended pavement**, where the pavement rests on a structure rather than on the soil. Thus, the soil beneath the pavement does not need to be compacted and can instead be a high-quality planting soil. In recent years, this is most commonly achieved by using **structural cells**. The concept is to build a three-dimensional grid of cell-like modular units that can be filled with soil to support root growth and development. The cells incorporate a top deck, which is covered with geotextile fabric. A gravel base course is placed next, and then the paving is installed on top. The deck supports the load of vehicles and pedestrians and transfers it into the support grid rather than the soil. While suspended pavement is advantageous for certain projects, its cost can be significant.

Figure 3.18 One approach to providing adequate root space without compaction under urban pavement is to create suspended (supported) pavement where the pavement rests on a structure rather than on the soil. This is sometimes achieved by using structural cells.

Research with different soil types and various configurations of structural cells continues, and collaboration with engineers and urban planners is showing promising results.

Soil Improvement

Although it is possible to correct or improve a number of poor soil conditions associated with urban sites, some are more easily remedied than others. For instance, drainage can be improved and compaction reduced, but these steps are most easily accomplished on sites where trees have not yet been planted. There are more options available for improving soil conditions before trees are planted than after they become established. Soil disturbance within the root zones of established trees has the potential to damage fibrous absorbing roots. Potential benefits and detriments to existing trees should be weighed before performing any soil remediation.

Compaction

Excessive soil compaction is also a leading cause of tree decline in urban areas. In addition to inhibiting drainage and air exchange, compaction also impedes root growth. Compaction problems are best managed by avoidance, which includes protecting future planting sites during construction and controlling foot and vehicular traffic around established trees. Once a soil is compacted, it is difficult to correct. The first step in remediating compaction is physically breaking up the compacted soil. Depending upon the situation, this can be accomplished in a variety of ways. Compaction problems around established trees can be improved using an **air-excavation device** that breaks up soil with high-pressure compressed air while leaving the tree root system more or less intact. These tools should be used with caution because they can, in some cases, cause root damage that may not be easily detected.

Compacted surface soil can sometimes be remediated through tillage with a rototiller or disc harrow. Tilling breaks down valuable soil aggregates, however, and is ineffective at addressing deeper layers of soil compaction. When soil has been built up in lifts or repeatedly graded, soil compaction may extend deep into the soil. In these cases, the affected layers must be broken up with a backhoe or subsoiler. Regardless of approach, breaking up compacted soil is not sufficient on its own. Organic matter should be incorporated into the soil to further improve soil conditions and extend decompaction effects by supporting the formation of soil aggregates over time.

Drainage

Poor drainage is a common problem in urban soils. Options to correct this problem depend on the nature of the drainage problem and whether the goal is to correct problems prior to planting or around established trees. In many cases, poor drainage is caused by soil compaction during construction. In the absence of established trees, pipes, or other underground utilities, it may be possible to improve drainage by breaking up the compacted layer or using other compaction remediation techniques.

In other instances, poor drainage is the result of surface and subsurface flow collecting at the base of

Figure 3.19 Surface water can be directed with grading prior to planting. Options to correct surface drainage problems are more limited in established landscapes.

slopes. One way to mitigate poor subsurface flow is the installation of a French drain that intercepts flow and diverts it away from the area. Where drainage problems are severe, drainage pipes can be installed under the poorly drained area. Attention must be paid to drainage between improved upper soil levels and subsurface layers, which may be compacted.

Figure 3.20 Organic amendments should be well composted and dark in color.

Organic Amendments

Amending soil with organic matter can provide multiple benefits, such as improved water infiltration, increased water-holding capacity, improved aeration, and enhanced CEC. Additionally, the activity of soil organisms responsible for creating and maintaining soil structure is increased by organic additions. Organic amendments should be well composted, black or dark brown in color, and have an earthy odor. The most common organic amendments are composted materials of various types. **Compost** can enrich the soil, reduce the need for chemical fertilizers, help retain soil moisture, and enhance soil biological activity.

Generally, soils are amended with compost at a rate of no more than 15 to 20 percent by volume. Although larger volumes are sometimes used, they can be difficult to incorporate and can contribute to excessive soil settling. Incorporated composts should have a carbon:nitrogen (C/N) ratio of 25:1 or below to avoid risks of short-term nitrogen deficiencies in the soil.

Mulching

Mulching the soil surface with an organic mulch has many beneficial effects. Mulch can reduce surface compaction and crusting, thereby improving water infiltration. Soil erosion is reduced, weed competition is decreased, and temperature fluctuations are moderated. Mulches act as an insulating layer, so they keep soil temperatures cool in the hot summer months, and they reduce evaporative water loss in hot, dry climates. Over time, soil structure can be improved and compaction reduced. Mulching may also alter surface soil pH in some cases. Surface mulching is the least invasive method of improving soil conditions around established trees.

Modifying Chemical Properties

The most common reasons to modify soil chemical conditions are adverse pH or essential element deficiencies. Soil tests and foliage analysis are commonly used to determine whether nutrient deficiencies or excesses exist. Many soil-testing laboratories provide recommendations for nitrogen, phosphorus, potassium, and other elements based on these test results. Cation exchange capacity and pH can be tested for in soils as well. Excess salt in the root zone can be identified in a soil test. If detected, high salt levels can often be corrected by flushing with water.

An issue that is often discussed in relation to soil modifications is the sustainability of any beneficial effects. Some modifications can be considered long lasting, provided that the soil is protected and good management practices are instituted. The costs for these improvements and their longevity vary and must be weighed against the benefits before deciding what actions to take.

Soil improvement limitations and costs vary from site to site. There are more options on a new development site than on a site with mature, established trees. The smaller the volume of the soil in question, the less expensive any modifications will be. In limited cases, soil replacement might be an option, and the soil mix (texture, organic matter, pH, etc.) can be specified. More often, however, working with the existing soil is the only feasible option.

CHAPTER 3 WORKBOOK

1. The majority of the fine, absorbing roots of a tree are in the _____ horizon.

2. Soils are generally described in terms of their _____, _____, and _____ properties.

3. Soil _____ is defined as an increase in bulk density and a decrease in total pore space.

4. True/False—Negatively charged clay particles hold cations near their surface.

5. True/False—A large percentage of tree decline situations can be attributed to soil stress, especially with urban soils.

6. True/False—Soil can hold water so tightly in micropores that the ability of tree roots to absorb the water is restricted.

7. Soil texture refers to the relative coarseness or fineness of a soil. Rank the following from the finest texture (1) to the coarsest (3).

    _____ silt   _____ clay   _____ sand

8. On the pH scale, less than 7 is _____, 7 is _____, and more than 7 is _____.

9. True/False—Over time, mulching can improve soil structure, reduce compaction, and add organic matter to the soil.

10. A pH of 5 is _____ times more acidic than a pH of 7.

11. The process in which ions of essential elements wash down through the soil profile and are lost is called _____.

12. Many essential elements are dissolved in soil water in the form of positively charged particles called _____.

13. Soils with excess levels of soluble salts are called _____ soils. _____ soils are soils in which the cation sodium (Na+) occupies an unusually high percentage of the CEC.

14. The buffering capacity is the resistance of a soil to changes in pH. Clay soils and soils high in organic matter usually have a _____ buffering capacity.

15. The _____ is the zone of intense biological activity near the actively elongating roots.

16. The diversity of organisms living, moving, and interacting in the soil is often referred to as the _____ _____ _____ .

17. Water that drains from the macropores is called _____ water. Following drainage, the soil is said to be at _____ _____ .

18. True/False—Most soil organisms cause disease or decay in tree roots.

19. True/False—Many tree roots exist in a symbiotic relationship with fungi that assist the tree in water and mineral absorption.

20. Compaction problems around established trees can be improved using an _____ _____ _____ that breaks up soil with high-pressure compressed air.

Matching

____ sand
____ buffering capacity
____ field capacity
____ rhizosphere
____ macropores
____ mycorrhizae
____ CEC
____ clay
____ pH
____ micropores
____ gravitational water

A. "fungus roots"
B. measure of acidity or alkalinity
C. fine-textured soil particles
D. water that drains from the macropores
E. ability of a soil to attract and hold cations
F. coarse-textured soil particles
G. tend to be air filled
H. soil after gravitational water has drained
I. soil zone immediately surrounding roots
J. resistance to change in pH
K. tend to be water filled

Challenge Questions

1. How do soil texture and organic matter influence the buffering capacity of a soil?
2. Why are arid soils generally alkaline and high in salts?
3. Explain how the bodies of soil microorganisms serve as a storage bank for plant nutrients.
4. Explain why a planting container filled completely with soil drains better than a container with gravel in the bottom.

5. If, during construction, most of the A horizon and organic layer is stripped away, leaving the B horizon as the uppermost soil, what would be the effects on newly planted trees in the landscape?

6. What are some of the ways that urban soils can differ from forest soils, and how do these differences impact trees?

Sample Test Questions

1. The primary reason that most fine, absorbing roots are typically found near the soil surface is that

    a. roots need both air and water
    b. roots need UV light to drive respiration
    c. the pH of the soil is generally higher near the surface
    d. phosphorus and potassium are more available

2. Microorganisms tend to congregate in the rhizosphere, in part, because

    a. sugar exudates from root tips are a source of food
    b. mycorrhizae fix nitrogen and make it available
    c. root hairs tend to collect cations essential to microorganism growth
    d. bacteria preferentially feed on the meristem tissue at the root tips

3. If a planting hole in a clay soil site is backfilled with sandy soil,

    a. drainage will be improved, helping the tree to establish
    b. nutrients will be more available to the newly established roots within the planting hole
    c. water will drain very slowly out of the planting hole
    d. the improved texture of the backfill will reduce the chances of girdling roots forming later

4. When soil is compacted,

    a. micropores combine to form macropores
    b. soil particles are broken up, giving the soil a finer texture
    c. a high water content will reduce the damaging effects
    d. total pore space and the percentage of macropores are reduced

5. A characteristic of sandy soils in arid regions is that they

    a. tend to become alkaline, and salts build up due to the lack of heavy rainfall
    b. tend to become acidic because basic ions leach out
    c. are fine in texture due to the high sand content
    d. have a high water-holding capacity because rainfall is scarce

Recommended Resources

(See Recommended Resources in back of book for detailed information.)

Best Management Practices: *Soil Management for Urban Trees* (Scharenbroch and Smiley 2021)

Introduction to Arboriculture: Soils online course (ISA)

Introduction to Arboriculture: Trees and Water online course (ISA)

Up By Roots (Urban 2008)

CHAPTER 4
WATER MANAGEMENT

▼ OBJECTIVES

- Describe how water influences plant growth and health.

- Explain the principles of water uptake, transpiration, and the movement of water in the soil.

- Identify soil characteristics and how they affect aeration, infiltration, and water-holding capacity.

- Discuss the factors that affect plant water loss and how plants manage that loss.

- Explain the importance of irrigation for urban landscapes, and review the different methods of irrigation and the advantages and disadvantages of each.

- Discuss potential water problems related to salinity and reclaimed water.

- Discuss the advantages of mulching around trees and shrubs.

- Explain why good drainage within the root zone of plants is important.

KEY TERMS

available water
bubblers
drip irrigation
evapotranspiration (ET)
field capacity
hydrozone
infiltration

infiltration rate
minimum irrigation
percolation
percolation rate
salinity
soil moisture reservoir
spray irrigation

tensiometer
turgid
water budget
water-holding capacity
xeriscaping

Introduction

Water is vital to plants. Trees absorb water and minerals (dissolved in water) from the soil. A large tree can absorb hundreds of liters (1 L = 0.26 gal) of water in a day, and as much as 95 percent of that water may be returned to the atmosphere through transpiration. Without sufficient soil moisture, a tree becomes water stressed and stops growing, and overall plant health can decline. Too much water in the root zone can cause tree decline as well. Maintaining optimal amounts of water in the root zone is critical for tree health and growth.

Irrigation is often necessary in urban planting sites, especially when trees are newly planted. However, irrigation must be managed wisely to avoid problems with volume, frequency, or water quality. Effective water management programs need to consider the water requirements for different tree species, weather conditions, soil conditions, how the water will be applied, water quality, water conservation, and drainage.

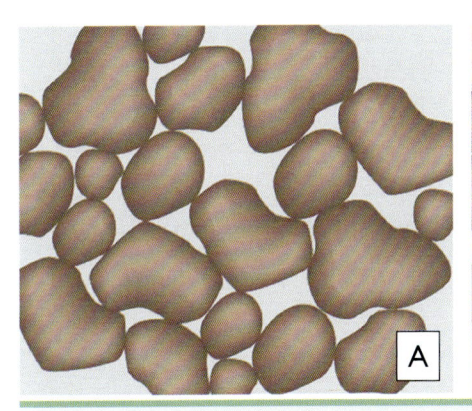

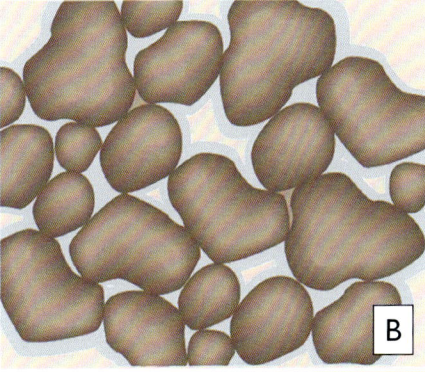

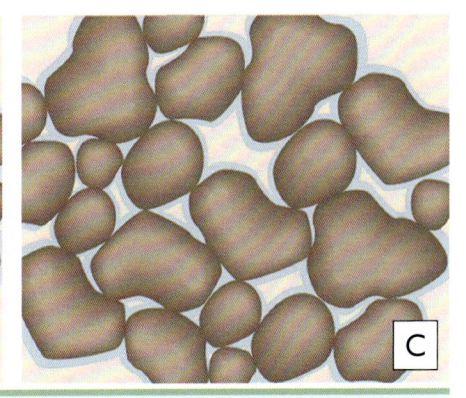

Figure 4.1 Soil moisture conditions. A: Saturation—water (blue) fills both the macropores and micropores. B: Field capacity—water is held by soil particles after surplus has drained by gravity. Air is available in macropores. C: Permanent wilting point—water is held tightly by soil particles and is unavailable to plants.

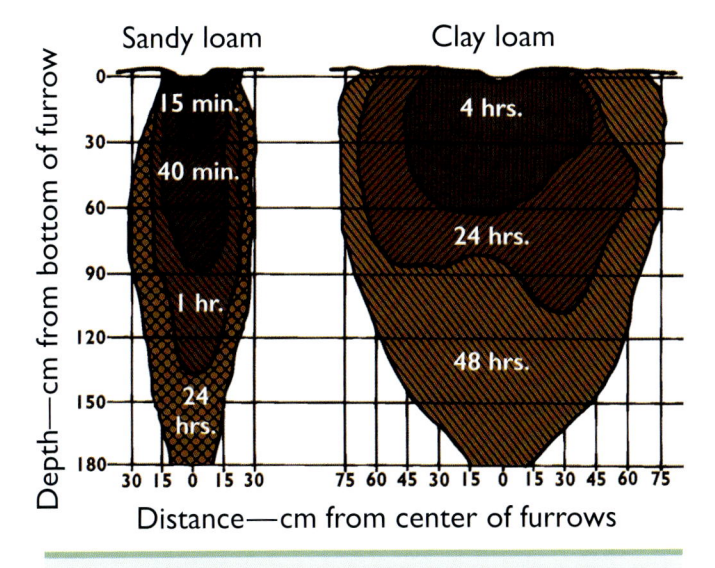

Figure 4.2 Difference in wetting patterns for sandy (left) and clay loam (right) soils.

Trees and Water

Trees require water to grow and survive. Too much or too little water, however, can slow growth and even kill trees. The amount of water needed by a particular species varies with the natural precipitation and soil moisture conditions. In general, looking to a plant's native environment can provide insight to its water requirements. For instance, trees from rain forest areas typically require a consistent supply of water, while those from deserts can survive long periods without water.

Two extremes of water supply are excessively dry soil due to drought and excessively wet (saturated) soil. Drought is an extended period of water deficit that is characteristic of desert and Mediterranean climates where rainfall is infrequent. Lack of water can cause

Copyright © 2022 International Society of Arboriculture

65

Soil Water

Various terms, many of which were introduced in Chapter 3, Soil Science, are used to describe both water retention and water loss. A brief review is useful for this chapter.

Water that drains through the soil is called gravitational water, while that which is retained is either capillary or hygroscopic water. Capillary water, which is also called **available water**, is generally available for plant uptake. Hygroscopic water forms a film on soil particles and is held too tightly by the soil for uptake by plants; it is known as unavailable water. For many soils, approximately one half of the water held after drainage is available for plant uptake, and the other half is unavailable.

After gravitational water has drained, the soil is at **field capacity**, containing both available and unavailable water. When the available water in the soil decreases to a point at which a plant cannot take up enough to replace the water lost through transpiration, the plant will wilt. If soil water diminishes to the point where it is no longer available to the plant, then the permanent wilting point has been reached. The soil moisture content at which the permanent wilting point is reached is highly variable and depends on how an individual plant interacts with the soil, the degree to which it is able to extract water and restrict its water loss, and the atmospheric conditions.

Water-holding capacity is the total amount of water held by a soil after drainage occurs (when it is at field capacity). Water-holding capacity varies with soil texture (sand, silt, and clay content): fine-textured soils (clays) have a greater water-holding capacity than coarse-textured soils (sands). As a result, sandy soils will need to be irrigated more frequently than clay soils.

Water moves predominantly downward but also laterally and sometimes upward in soils. **Infiltration** is the term used to describe water movement from the soil surface into the soil, while **percolation** describes water movement through the soil. Both infiltration and percolation are affected by soil texture: clay soils usually have much slower **infiltration rates** and **percolation rates** than sandy soils. Other factors beyond soil texture that influence percolation and infiltration rates include soil structure (the amount of clumping of soil particles) and bulk density (the mass per unit volume of soil). Compacted soils and soils with poor structure have slower infiltration and percolation rates than soils (of similar texture) with good structure. In situations where the infiltration rate is low, water should be applied slowly and over extended periods of time to wet the soil throughout a tree's root system.

---

root loss, marginal leaf scorch, leaf drop, twig dieback, and plant death. Saturated soils due to flooding or high water tables can lead to root death due to lack of oxygen. This may also result in leaf yellowing, defoliation, and crown dieback. Possible adverse effects include root disease and shallow root development.

Plants can have a variety of adaptations to survive extremes of drought and flooding. Some tree species tolerate short periods of flooding, especially if they are dormant and the water is flowing and retains dissolved oxygen. Trees that are tolerant of prolonged flooding during the growing season may have specialized root and stem tissues that transport oxygen internally. Trees may tolerate sites with high water tables by establishing shallow roots that exchange oxygen near the soil surface.

Figure 4.3 Water loss through stomata in the leaves is called transpiration. Leaves can have adaptations to limit water loss such as a waxy cuticle or sunken stomata.

Figure 4.4 This coast live oak has small, thick, leathery leaves to reduce water loss. The leaves have tiny hairs underneath, which can also reduce water loss.

At the opposite extreme, some plants develop extensive root systems to survive in arid regions. Such plants develop deep, spreading roots that allow them to make use of soil moisture over a large volume and to tap into water sources found deeper underground. By accessing groundwater, trees and shrubs can survive extended periods of little or no rainfall.

Some species have a different strategy for surviving periods of low soil moisture content. Some plants avoid drought effects by shedding their leaves. Other plants become virtually dormant during extended dry periods.

Other plant strategies for controlling water loss involve leaf adaptations in size, shape, or surface. Small leaves have less surface area through which water can be lost. A waxy outer layer (cuticle) may further limit moisture loss to the air. Tiny leaf hairs (pubescence) prevent water loss by trapping an insulating layer of moist air near the leaf surface. Similarly, sunken stomata are protected by a small pocket of air that collects in the depression and serves as a barrier for water loss.

Drought-tolerant plants are not only adapted to survive with minimal water but also can be adversely affected by supplemental irrigation water. Possible adverse effects include root disease and shallow root development. Shallow rooting limits the volume of soil a tree utilizes for water uptake, exposes the plant to fluctuating soil moisture conditions, and can increase the potential for uprooting in strong winds.

Trees that are accustomed to fairly regular rainfall throughout the summer may begin to show stress after several weeks without rain. For some tree species, wilting is common during hot summer afternoons. Leaves may wilt during the day and recover completely overnight. But if the leaves are not **turgid** (fully hydrated) again the next morning, the tree may be seriously stressed. If water stress continues, leaves may begin to exhibit chlorosis, turn brown, or drop completely. Without rain or irrigation, small roots will begin to dry out and die, and the tree will decline.

Trees growing where soil volume is limited, such as small parking lot islands, sidewalk tree pits, and planters, are often the first to show water stress symptoms as their root systems can deplete

Figure 4.5 Leaves may wilt during the day and recover completely overnight. But if the leaves are not turgid (fully hydrated) again the next morning, the tree may be seriously stressed.

the limited soil moisture made available to them. Trees growing in hot, dry, or windy locations may develop leaf scorch even if soil moisture is maintained at optimal levels. Beyond the availability of water in the soil, anything that interferes with the vascular transport system can lead to drought-like symptoms. Girdling roots, trunk injury or cankers, vascular wilt disease, lightning-strike damage, and similar disorders can all affect the uptake and transport of water throughout a tree's vascular system.

A key factor in water management is the rate at which water is lost through transpiration by plants and evaporation from soil. The combined loss is called **evapotranspiration (ET)**. If plant transpiration exceeds precipitation and the soil moisture reaches the wilting point, it may be necessary to provide supplemental irrigation to maintain plant growth and health.

How much water is available to trees depends on the volume of soil occupied by plant roots and the water-holding capacity of the soil. This volume is called the **soil moisture reservoir**. Sandy soils hold less water than clay soils, and plants with small root systems have access to less water than those with large root systems. It is important to estimate the volume of the soil moisture reservoir when determining when and how much water to apply during irrigation.

Irrigation

Most trees grow adequately with natural precipitation. However, trees that are not adapted to the local precipitation patterns, or trees experiencing periods of drought, may need supplemental irrigation. Irrigation requirements vary with species, tree size, the amount of water held in the soil moisture reservoir, and environmental conditions at the site (sun exposure, wind, and humidity). The need for irrigation can be reduced or even eliminated once a tree is established if the species is selected to match the site's environmental conditions and the soil volume is adequate to meet the tree's water requirements.

Newly planted trees of most species require frequent irrigation because the entire root system is contained within the root ball. As the roots grow into the surrounding soil, irrigation frequency should be reduced or eliminated if rainfall is adequate to sustain the tree.

In soils where water infiltration is slow, steps can be taken to avoid standing water. Irrigation should not be applied faster than water can infiltrate. Slow water application can help prevent puddling and runoff. If the problem is caused by compaction, soil aeration may also be considered. Even with proper care, there are limitations to the effectiveness of maintenance practices in managing

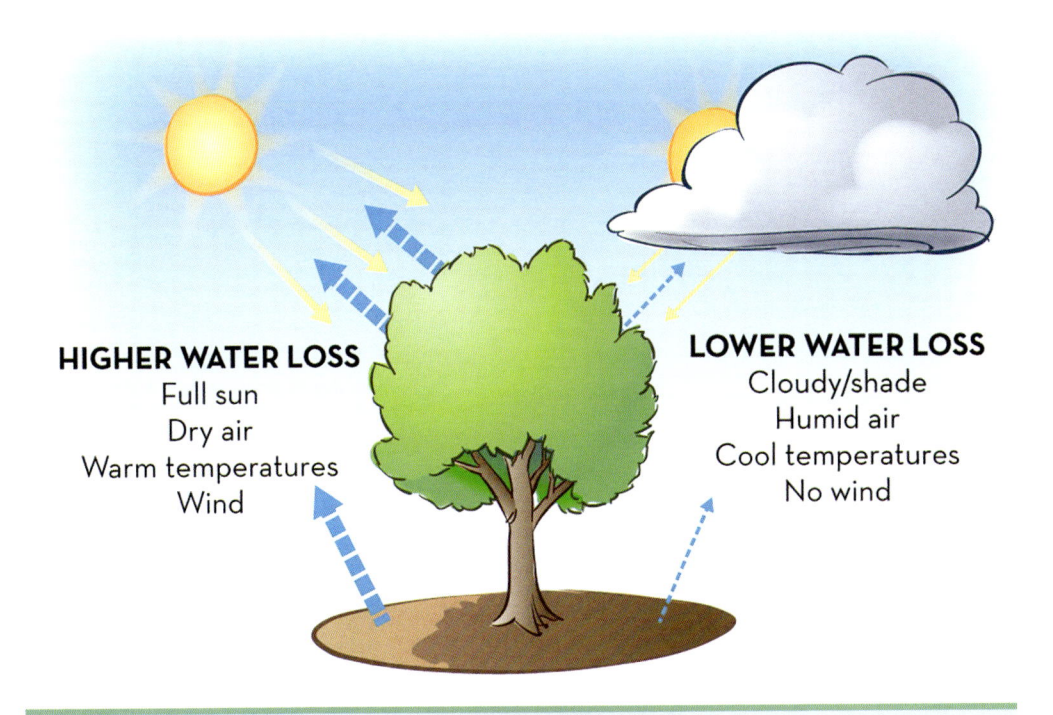

Figure 4.6 A key factor in water management is the rate at which water is lost by transpiration by plants and evaporation from soil. The combined losses are called evapotranspiration (ET), and the rate is affected by weather and site conditions.

Figure 4.7 Sprinkler irrigation is very common in landscapes; most systems are designed for optimal watering of lawns and may not be appropriate for trees. Avoiding direct spray on tree trunks is preferred.

drainage. Careful plant selection is important in choosing trees that will adapt to the soil's drainage conditions.

Irrigation Systems

Although various methods can be used to water trees, sprinkler (spray), bubbler, and drip systems are most common. Each method has specific applications. Routine system monitoring, including nozzle and filter cleaning and repair, is needed to ensure continued effective operation.

Spray irrigation disperses small droplets of water over a wide area. Well-designed and maintained spray systems can apply water effectively and efficiently on relatively level sites and in calm weather. A disadvantage can occur if it applies large quantities over short periods of time, leading to water runoff and erosion. Spray patterns and water distribution can be disrupted in windy conditions, potentially leaving some areas more dry or wet than others. Scheduling irrigation duration (application) times to match infiltration rates and weather conditions, operating during low wind times of the day, and using low-application-rate spray heads can minimize these problems.

Bubblers deliver water in a stream or umbrella pattern at a moderate to fast rate. This is a common irrigation method for newly planted trees. One or two bubblers are installed just outside the root ball. A berm should be created to retain the water until it moves into the soil.

Drip irrigation systems apply water slowly to localized areas of the tree root zone, reducing the potential for runoff, erosion, and water loss by evaporation. Each emitter will wet a relatively small volume of soil. The number of emitters required depends on the size of the area to be

Establishing a Water Budget

A **water budget** is an estimate of how much water is available to a tree (supply) and how much water is required (demand). A water budget is used to schedule when and how much water to apply to sustain trees in the landscape. Water inputs can include rainfall, irrigation, and water table; water outputs are transpiration, evaporation, and runoff. The amount and frequency of irrigation can be estimated by deducting water lost through ET from the amount of water stored in the soil moisture reservoir.

After the amount of water required to establish or maintain a given plant species in a given site is determined, irrigation schedules can be designed to address watering duration, frequency, timing, and distribution.

Irrigation duration. As a general rule, the more water that infiltrates into the soil, the deeper the soil is wetted. Short-duration irrigations, therefore, wet the soil shallowly. Shallow soil moisture then encourages shallow root systems to develop where water is available. In contrast, long-duration irrigations wet the soil deeply and can encourage the development of deeper root systems. Trees with deep root systems tend to be more drought tolerant than trees with shallow roots.

Irrigation frequency. Irrigation frequency is how often water is applied to the soil. The appropriate frequency for a site depends on how much water is held in the soil moisture reservoir and how long it takes for that water to become depleted. The hotter the weather and the more confined the root system, the more often irrigation is needed. Frequent irrigations that do not allow enough time for soils to drain adequately can promote root diseases. Soils and mulch that are high in organic matter should not be allowed to dry completely, because they can have a tendency to resist rewetting, becoming hydrophobic.

Irrigation timing. It is considered best practice to irrigate in the evenings or early morning hours. This is particularly true for sprinkler irrigation systems because less water is lost through evaporation and more water is able to infiltrate the soil. Also, many municipalities have water-use regulations that limit use to evenings and early morning hours. In situations where foliar diseases are a problem, evening irrigation can be detrimental to plants because moisture remains on the leaves for extended periods of time.

Irrigation distribution. Water should be distributed uniformly to as much of the root system as possible. Efforts should be made to keep the root collar dry between irrigations. Wet soil in contact with the lower trunk and root collar can lead to infection by pathogens.

---

irrigated. Emitters should be spread out over the entire root zone and kept away from tree trunks. To apply water to the expanding root system of young trees, emitters should be moved away from the trunk; additional units may be required. It is important to recognize that emitters can become plugged and lines damaged. Sometimes the only sign of a plugged emitter or broken line is a water-stressed plant. Frequent monitoring of drip systems is critical.

Other methods to irrigate trees include high-pressure water injection, soaker hoses, basin irrigation, and temporary, portable watering devices. Portable drip or soaker systems can serve as an effective means of supplying water to newly planted trees, transplanted trees, and established trees during periods of drought. Basin irrigation on level ground distributes water uniformly to the limited root zone of newly planted trees. A basin around a tree is filled with water by a hose, bucket, or other device. Alternatively, when there is no direct water supply, aboveground bags or reservoirs can store and slowly release water for absorption into the soil. High-pressure water injection can deliver water below grade to the root zone, but care should be taken to avoid bypassing absorbing roots near the soil surface. Soaker hoses are portable and can apply water slowly over a longer period of time where runoff may be of concern.

Many irrigation systems are designed and installed primarily to irrigate turf and other landscape plantings. Too often, trees are not properly considered in irrigation management programs.

Overirrigation of trees is becoming a problem in many urban and suburban areas. Disease problems such as root or collar rot are often associated with excessive irrigation. These conditions are particularly troublesome in instances where irrigation is applied directly on the trees' trunks. Tree species that are adapted to dry summers and native to such climates are prone to problems associated with frequent irrigation.

Water Conservation

Water is a very limited resource in many areas, and water conservation needs to be considered as an integral part of water management programs. **Minimum irrigation** provides only the amount of water needed to maintain tree health, growth, and appearance (canopy density and leaf color) using efficient application methods. **Xeriscaping**, or landscaping with drought-tolerant plants, is a popular design alternative for property owners living in areas where rainfall is infrequent and water supplies for irrigation are limited. Although more often practiced in dry regions, the principles of xeriscaping may be applied in a wide range of environments as a means of reducing landscape maintenance, water use, and irrigation expenses. Caution is in order when transitioning to minimum irrigation because plants must adapt to lower soil moisture conditions.

When planning any new landscape, plants with similar water requirements should be grouped together and irrigated on the same schedule (**hydrozone**). This technique usually employs the use of highly efficient irrigation methods. Keep in mind that

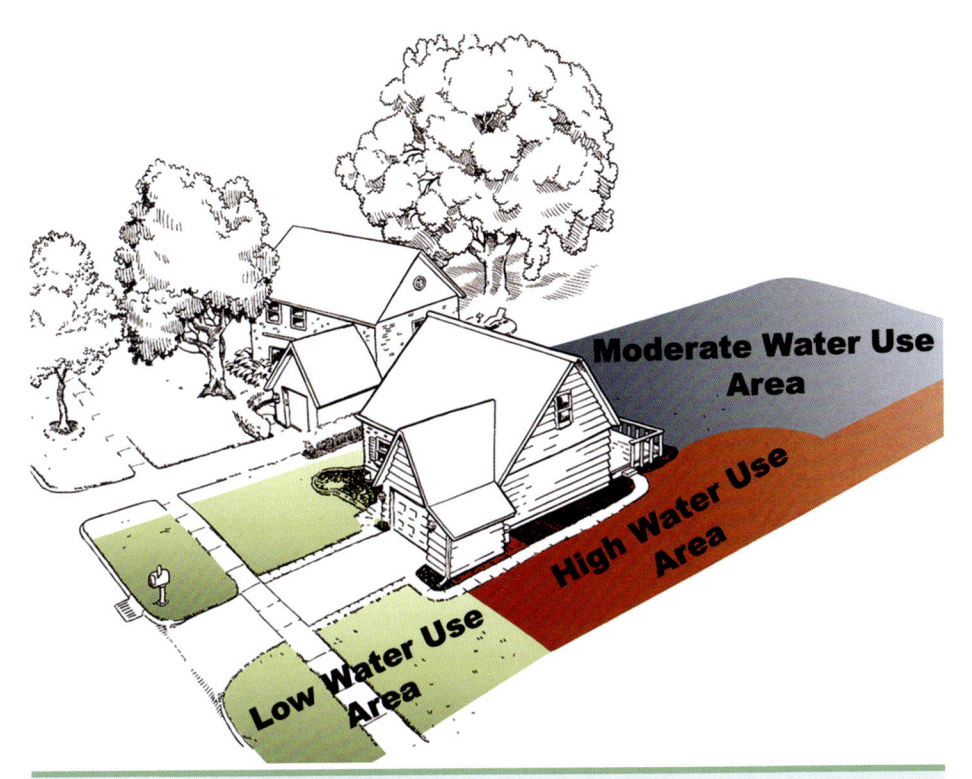

Figure 4.8 When planning a new landscape where minimum irrigation is to be used, plants with similar water requirements should be grouped together and irrigated on the same schedule. This is called hydrozoning.

Figure 4.9 Soil moisture can be monitored with a tensiometer.

Figure 4.10 Under saline conditions, where there are excessive levels of soluble salts, irrigation may need to be provided more frequently. Soil damage can result from excessive sodium in irrigation water.

new plantings require more frequent watering until they develop established root systems.

Another approach to water conservation employs soil moisture monitoring. Devices such as soil probes, **tensiometers**, and electronic moisture sensors are used to measure soil wetness or dryness. Carefully tracking soil moisture before and after irrigations allows the amount of available water to be assessed to adjust irrigation recommendations.

Water Quality

The water source for irrigation affects the quality of the water. The main water sources for trees are groundwater, surface water from rivers and lakes, and, in some locations, recycled (reclaimed) and desalinated seawater.

Water is naturally recycled and purified through the hydrological cycle of evaporation, transpiration, condensation, precipitation, and runoff. As water flows over or through soil and rock, mineral salts dissolve and are carried up by the water. Excess salts in the water can be detrimental or even toxic to plants.

Irrigation water taken from landscape retention ponds can have high levels of sodium and chloride salts. This is especially true where runoff is captured from parking lots where deicing salts have been used. If this water is sprayed on foliage, severe leaf scorch may result. The pattern of scorch on the tree canopy will coincide with the arc of the irrigation spray pattern. *Phytophthora* can also build up in ponds, so this disease should be monitored near bodies of water used to irrigate landscapes.

Recycled or Reclaimed Water

Recycled or reclaimed water is sourced from municipal, industrial, or agricultural wastewater that has undergone some degree of purification or treatment. In regions where rainfall is insufficient to support landscapes, recycled water is becoming an important source for irrigation. Using recycled water for irrigation conserves potable (drinking) water for other uses.

Recycled water may provide essential elements such as nitrogen, phosphorus, and sulfur, which can be beneficial for tree growth. However, recycled water can also be high in salts and other chemicals that can raise the soil pH, cause plant phytotoxicity, and clog irrigation nozzles. Reclaimed water should be tested periodically for soluble salts and other chemicals that could be detrimental to plant health.

It is important to monitor plant health and soil conditions frequently when irrigating with recycled water. A soil **salinity** test should be performed regularly and the salinity values recorded. If soil salinity is increasing, it may be necessary to leach out the salts using a large volume of water. Sometimes periodic rainfall events may be enough to accomplish this, making supplemental leaching unnecessary. Efforts to select salt-tolerant plants can reduce the impact of saline water sources on the overall health of the landscape.

Other Water Conservation Methods

Water conservation techniques include the application of mulch around trees and other landscape plants. Mulching can improve soil structure and infiltration, reduce soil moisture evaporation, moderate soil temperature, limit weed competition (which further reduces evapotranspiration and soil water loss), and reduce soil compaction and erosion. Organic mulches have the added benefit of increasing soil organic matter as they decompose, which enhances soil biology, fertility, and structure. Aside from improving soil and moisture conditions, mulch can also provide a landscape with a well-kept appearance and can reduce trimmer and lawn mower damage to trees. Some commonly used organic mulches include bark chips, pine needles, and wood chips.

In addition to mulching, amendments are sometimes added to soils in an attempt to increase water-holding capacity or improve drainage. Incorporating organic matter into soil may be beneficial, but the effects are often short lived because the organic matter decomposes. Similarly, soil additives such as hydrogels have been used to increase water-holding capacity and serve as an additional source of nutrients for trees, but research shows they are only effective in the landscape for relatively short periods of time.

Landscape water usage can be greatly reduced by limiting the amount of turf areas on a property. Mowed turfgrasses have shallow root systems and quickly lose their ability to access plant-available water as it drains deeper into the soil. As mentioned earlier, proper plant selection can greatly reduce irrigation demands.

Flooding and Drainage

When soil moisture content is above field capacity, water moves downward through macropores in response to gravity (infiltration). This movement can be relatively fast or slow, depending on the soil's physical characteristics (texture, structure, bulk density, depth, and layers). Soils with relatively slow rates

Figure 4.11 When applying mulch, wider is better, but deeper usually is not.

Figure 4.12 Poor drainage can cause root damage to trees. Excess water in the root zone reduces oxygen in the soil, which suffocates roots, can lead to buildup of toxic compounds, and can harm the soil ecosystem.

Figure 4.13 Poorly drained, clay soils sometimes are grayish or blue-gray in color and may have a foul odor.

Flooding can be very damaging to trees, especially in warmer weather conditions. The lack of oxygen in the soil can deprive roots, may change the chemical composition of essential elements, can lead to fermentation in root cells and buildup of toxic compounds, and can create mineral toxicities. Lack of oxygen can also harm the living components of the soil ecosystem.

For some species, just a few hours under flooded conditions will impact photosynthesis and transpiration rates. After several days, most species will experience some root loss. If conditions persist, some species will decline and die. Tree species adapted to floodplains can tolerate longer periods of flooding compared with upland forest species. For species that survive flooding, there can still be long-term consequences. Some will be predisposed to stress factors, including drought and secondary pests (such as root disease, cankers, and borers), from which they may not be able to adequately defend.

Trees that have experienced flood conditions are prone to toppling due to soil failure and/or root loss, and they may also have a higher incidence of root or collar rot. Additionally, grade changes due to erosion and soil deposition can lead to long-term stress. Trees that survive flood conditions should be inspected for hazards and structural integrity of the root system.

of water movement are referred to as being poorly drained. In poorly drained soils, water can occupy macropore space for extended periods, restricting air movement.

Changes in grade should be corrected if possible, and measures to improve soil aeration should be considered.

Improving Drainage

It is important to consider the grade and drainage flow pattern during construction or landscape development. Poor drainage can cause root damage to trees. Drainage problems may occur under various site conditions, including a high water table, shallow soil over bedrock, a layer of impervious compacted soil that limits infiltration or percolation, and soil layers of different textures.

If the site is not already developed, adjusting the grade is the first choice to avoid drainage problems. Attempts should be made to avoid the creation of low spots where water could accumulate or drain slowly. Trees can be planted on mounds or at levels higher than natural grade.

If a developed site is known to have poor drainage, drain tiles can be installed prior to planting to prevent drainage-related problems. Drain tiles are made of clay, concrete, or plastic, although plastic is most commonly used for landscape purposes. The depth and spacing of the tiles depend on soil and planting conditions. Drain tiles remove only gravitational water from the soil. Heavy clay soils or improper irrigation practices can still create situations where the soil remains too wet, even after the gravitational water has drained.

Figure 4.14 If a developed site is known to have poor drainage, drain tiles can be installed prior to planting to prevent drainage-related problems.

Drainage problems in urban landscape sites are often not recognized until after planting is complete. If drain tiles cannot be installed, surface drainage may be improved by changing the grade or trenching. Care must be taken to avoid damaging the root systems of existing trees.

Soils can also be amended to improve drainage, typically with organic materials. With the selection of appropriate materials, macropore space can be increased. However, drainage improvement may require the addition of substantial amounts of amendments and should be done prior to landscape development.

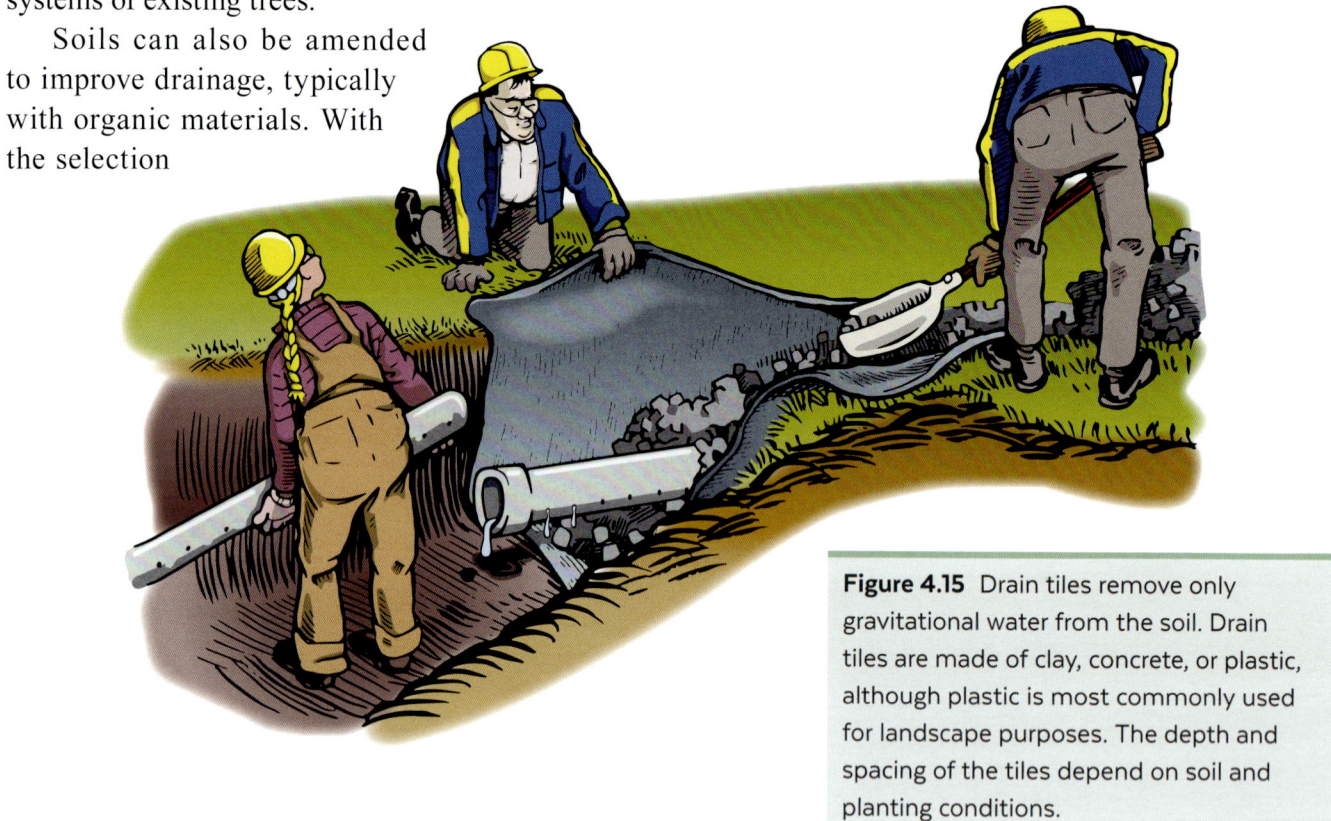

Figure 4.15 Drain tiles remove only gravitational water from the soil. Drain tiles are made of clay, concrete, or plastic, although plastic is most commonly used for landscape purposes. The depth and spacing of the tiles depend on soil and planting conditions.

CHAPTER 4 WORKBOOK

1. True/False—Infrequent, deep soakings of established trees and shrubs are preferable to frequent, shallow watering.

2. The _____ rate is the rate at which water soaks into the soil.

3. Clay soils generally have a greater _____ - _____ _____ than sandy soils, but water percolates through clay soils more slowly.

4. A _____ _____ is an estimate of how much water is available to a tree (supply) and how much water is required (demand).

5. Four irrigation factors that affect a water budget are _____ , _____ , _____ , and _____ .

6. True/False—In most landscapes, it is preferable to minimize irrigation methods that apply water directly to the foliage of plants.

7. List three plant adaptations for surviving low water availability.

    a.

    b.

    c.

8. Name two advantages and two disadvantages of drip irrigation.

    Advantages   Disadvantages

    a.   a.

    b.   b.

9. _____ _____ is designed to maintain plants during periods of reduced rainfall, supplying only enough water to maintain a desired plant quality.

10. In regions where rainfall is insufficient to support landscapes, _____ water is becoming an important source for irrigation.

11. A measure of the rate of water use by plants and evaporation from soil is known as the _____ rate.

12. List five benefits of using mulches around trees.

   a.

   b.

   c.

   d.

   e.

13. Soil probes, _____ , and electronic moisture sensors are tools used to monitor soil wetness or dryness.

14. True/False—If a site is not already developed, adjusting the grade is the first choice to avoid drainage problems.

15. Name three tree health problems that may be associated with flooding.

   a.

   b.

   c.

Challenge Questions

1. What are some of the methods commonly used to estimate how often and how much to irrigate?

2. How can high soil salinity occur? What are the irrigation practices that should be used to minimize high soil salinity?

3. Compare and contrast the effects and symptoms of drought conditions and flood conditions.

4. Describe strategies that can be employed to conserve water in the landscape.

Sample Test Questions

1. When irrigating trees,
   a. infrequent, deep soakings are preferable to frequent, shallow waterings
   b. the most beneficial and efficient time to water is midafternoon at peak sunlight
   c. the foliage should be kept wet at night to reduce transpiration
   d. keeping the soil moist at the root flare reduces girdling root formation

2. Sandy soils

    a. have a greater water-holding capacity than clay soils
    b. have higher infiltration rates than clay soils
    c. do not ever reach field capacity, because drainage is good
    d. have lower percolation rates than clay soils

3. A soil is at field capacity when

    a. it is completely saturated
    b. the permanent wilting point has been reached
    c. gravitational water has drained away
    d. there is no water available to the roots

4. Which of the following is true about irrigating with recycled (reclaimed) water?

    a. Recycled water rarely contains essential minerals.
    b. Recycled water typically lowers the soil pH.
    c. Recycled water is typically high in salts.
    d. Recycled water is typically chemical free.

5. Which of the following is a problem associated with flooding?

    a. lack of oxygen in the soil leading to root suffocation
    b. changes the chemical composition of essential elements
    c. fermentation in root cells and buildup of toxic compounds
    d. all of the above

Recommended Resources

(See Recommended Resources in back of book for detailed information.)

Abiotic Disorders of Landscape Plants (Costello et al. 2003)

Introduction to Arboriculture: Trees and Water online course (ISA)

CHAPTER 5

TREE NUTRITION AND FERTILIZATION

▼ OBJECTIVES

- Describe the essential elements that a tree requires and how these elements are absorbed.
- Discuss the problems that can be associated with excessive fertilization.
- Explain why determining nutritional requirements and availability is the first step before making any fertilization recommendations.
- Discuss the reasons for fertilizing urban trees.
- Explain the various types of fertilizer and the advantages and disadvantages of each.
- Describe the different methods of fertilizer application and the advantages and disadvantages of each.
- Discuss the efficacy and limitations of common soil additives.

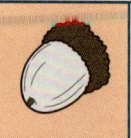

KEY TERMS

- atmospheric deposition
- cation exchange capacity (CEC)
- chelates
- chlorosis
- complete fertilizer
- controlled-release nitrogen (CRN)
- decomposition
- drill-hole fertilization
- essential elements
- fertilizer analysis
- fertilizer burn
- fertilizer ratio
- foliar analysis
- foliar application
- implant
- injection
- inorganic
- internal cycling
- leaching
- liquid injection fertilization
- macronutrient
- micronutrient
- nitrogen fixation
- nutrient deficiency
- nutrient limitation
- organic
- prescription fertilization
- slow-release fertilizer
- soil analysis
- subsurface application
- surface application
- volatilization
- water-insoluble nitrogen (WIN)

Introduction

Plants require certain nutrients to function and grow. A nutrient is an **essential element** that is involved in tree metabolism and plant processes or is necessary for a plant to complete its life cycle. For trees growing in a forest or other natural setting, these elements are normally present in sufficient quantities from the weathering of minerals in the soil, the **decomposition** of organic litter, and **atmospheric deposition**, such as from rainfall and dust particles.

Landscape and urban trees, however, often grow in severely altered, compacted, and fragmented soils, some of which do not contain sufficient amounts of essential elements. In addition, the physical and chemical properties of a disturbed site may reduce the availability of nutrients. For example, landscape maintenance practices such as raking and collecting leaves will reduce the amount of organic matter returned to the soil and, in turn, the quantity of elements released through nutrient cycling. As a result, it may be necessary to supply additional nutrients through fertilization to meet management and performance objectives.

Fertilizing a tree can increase growth, improve flowering and fruiting, and, under certain circumstances, help slow the decline in health resulting from nutrient deficiencies. However, if the fertilizer is neither needed nor applied appropriately, it may not benefit the tree at all—and may even adversely affect the tree and the environment. Fertilization can increase susceptibility to certain pests and can accelerate decline. Research shows that nitrogen fertilization can trigger a tree to allocate more energy to new growth rather than to defense. Trees with satisfactory growth and vitality that do not show symptoms of nutrient deficiency do not require fertilization. It is important to recognize when a tree would benefit from fertilization, which elements are needed, and how and when they should be applied, if at all.

ISA's Best Management Practices: *Tree and Shrub Fertilization* offers detailed information on fertilization principles and practices. Specifications for tree fertilization should clearly define the objectives and identify the type of fertilizer, along with its application rate and timing, method, and location.

Figure 5.1 Landscape and urban trees often grow in severely altered, compacted, and fragmented soils, some of which do not contain sufficient amounts of essential elements. Deficiency symptoms may include reduced growth, small leaves, and chlorosis (yellowing).

Essential Elements

Depending on species, plants require 16 to 19 essential elements to support growth and normal physiological processes. Carbon (C), hydrogen (H), and oxygen (O), which are obtained from the atmosphere and water, account for more than 90 percent of the mass of the tree. The remaining essential elements are absorbed from the soil and compose less than 10 percent of a tree's mass. The annual demand for essential elements to support growth and plant functions is met through three primary sources:

- Uptake from the soil
- Internal cycling
- Wet (rainfall) and dry (dust) atmospheric deposition

Table 5.1
Examples of pests promoted by nitrogen fertilization.

adelgids
ambrosia beetles
aphids
browsing mammals
caterpillars including gypsy moth, spruce budworm, eastern tent caterpillar, white-marked tussock moth
Dioryctria borers
Diplodia tip blight
Dothistroma needle blight
fireblight
Japanese beetle
lacebugs
leaf beetles
leafhoppers
leafminers
pine pitch canker
pine tip moths
plant bugs
psyllids
sawflies
scales
spider mites
thrips
whiteflies

Generally, young, rapidly growing trees have a high demand for essential elements obtained from the soil. As trees grow and age, the contributions from the various sources change, and more of the annual demand is met through reusing elements taken up from the soil during the previous year(s), a process known as **internal cycling**.

Table 5.2
Nutrient demand profiles. Characteristics of high- and low-demand species and conditions.

High Demand	Low Demand
Deciduous	Evergreen
Young	Old
Indeterminate growth	Determinate growth
Fast growing	Slow growing
Healthy	Unhealthy (except for nutrient deficiencies)
Full sun	Shaded

Trees take up essential elements dissolved in water through their roots. Each of the essential elements has a specific role in a plant and cannot be replaced by another element. Trees require certain elements, known as **macronutrients**, in relatively large quantities. The three primary macronutrients are nitrogen (N), phosphorus (P), and potassium (K). They are not necessarily of greater importance but are needed in greater quantities. Secondary macronutrients include sulfur (S), magnesium (Mg), and calcium (Ca). Relative to the primary macronutrients, these elements are required in moderate quantities but are still considered macronutrients.

In most landscapes, nitrogen is often the element that limits annual growth and is required in the greatest quantity. The high demand for nitrogen stems from the fact that it is a constituent of chlorophyll and proteins, making it critical to photosynthesis and other plant processes. In nature, the primary source of nitrogen is the decomposition of organic matter in the soil, a process driven by microorganisms (including fungi and bacteria). Other natural sources of nitrogen include atmospheric deposition and **nitrogen fixation** (conversion to a plant-usable form of atmospheric nitrogen).

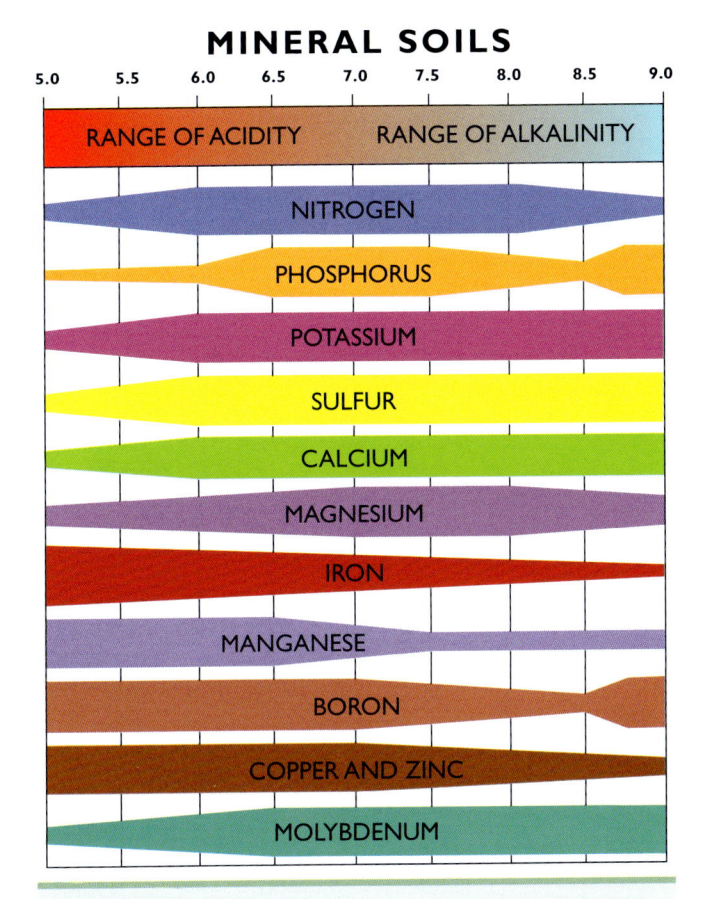

Figure 5.2 Soil pH affects the availability of nutrients for plant uptake. For example, when the soil pH is greater than 7, iron tends to become unavailable for plant uptake. This chart shows the relative availability of essential elements at different pH levels.

Figure 5.3 Magnesium deficiency on palm.

Nutrient deficiencies adversely affect a tree's ability to function properly. If left unchecked, deficiencies stress the tree, making it susceptible to other harmful agents, and can even lead to premature death. For example, nitrogen deficiency in trees is expressed as reduced growth, small size, and **chlorosis** (yellowing or whitening of the leaves). In this instance, chlorosis is a symptom indicating the tree is not functioning properly.

Initial symptoms of nitrogen deficiency are typically observed in older leaves because nitrogen is mobile within plants, meaning it moves out of older leaves and into new growth. If there has been a deficiency for a moderate length of time, the symptoms may be present in both young and old leaves. The availability or quantity of nutrients in the soil may not be the sole reason behind a tree displaying deficiency symptoms. Problems that affect root uptake (such as waterlogged soils or construction damage) or movement in stems (e.g., wilt disease, cankers, borers) can initiate the expression of symptoms.

In forests, an organic layer consisting of leaves and woody debris in various states of decomposition naturally accumulates on top of the mineral soil. This organic layer moderates soil temperature and moisture content and is a source of nutrients. Nitrogen released by microbial decomposition of soil organic matter is available for uptake by trees. Because soil nitrogen comes largely from decomposed organic matter, removing leaf litter and other natural sources of nitrogen can affect the quantity of nitrogen and other essential elements available to plants.

Nitrogen can also be lost due to leaching or **volatilization** (loss in gaseous form into the atmosphere). Leaching occurs more frequently with well-drained, sandy soils with low organic matter content. Nitrogen deficiencies are most common in sandy soils low in organic matter, especially if irrigation is heavy.

The remaining essential elements, known as **micronutrients**, are required in lesser quantities. Regardless of the level of annual demand, all of these nutrients are capable of limiting growth and/or

Figure 5.4 Yellowing of the foliage (chlorosis) on red maple (*Acer rubrum*) likely caused by a deficiency of manganese in the foliage. Iron and manganese can be unavailable in high pH soils.

causing a deficiency. It is important to recognize, however, that **nutrient limitation** does not equate to **nutrient deficiency**. Nutrient deficiency affects tree health; limitation affects the rate of growth. It is beyond the scope of this introduction to explore the symptomology of nutrient deficiencies. There are numerous, region-specific agencies and resources to assist with diagnoses.

Soil Testing and Nutrient Analysis

The process of setting an objective for plant health, conducting analyses, determining the need for supplemental nutrition, and, if warranted, selecting a fertilizer to achieve the objective is known as **prescription fertilization**. It is matching the tree's annual demand for nutrients relative to the supply of nutrients. The site's ability to supply nutrients to support growth and functioning can be determined through soil and/or tissue sampling and analysis and an assessment of the site in which the tree is growing.

Typically, a **soil analysis** provides estimates of the quantities of plant-available nutrients, soil organic matter content, pH, and **cation exchange capacity (CEC)**. The CEC is the ability of the soil to hold onto and release positively charged ions (essential elements). It is strongly influenced by soil organic matter content, the amount of clay in the soil, and soil pH. Cation exchange capacity is high in clay soils and soils with a high organic matter content. Conversely, sandy soils and soils with low organic matter have lower CEC.

Soil pH affects the availability of nutrients for plant uptake. For example, iron (Fe) is most available when the soil pH is less than 6. When the soil pH is greater than 7, iron tends to become unavailable for plant uptake.

Soil organic matter provides an estimate of the long-term nitrogen supplying power of the soil. Soils having 2 percent to 5 percent or more organic matter content will supply enough plant-available nitrogen to meet the annual demand of most trees. Plant-available nitrogen may limit annual growth in soils having less than 2 percent organic matter content.

Foliar analysis (analysis of nutrient content in the leaves) reflects the tree's ability to acquire and use nutrients from the soil under current conditions. As such, an examination of the growing location, including available rooting area, soil depth, bulk density, damage to the root system, and changes to the soil grade, typically accompanies both soil and foliar analyses. Because the concentration of elements in a tree can change during the growing season, it is important to have a consistent sampling period in which to make evaluations. Also, data identifying critical foliar concentrations is limited

Figure 5.5 A soil test is only as good as the sampling procedure. A typical procedure is to collect six to ten cores from representative locations in the root zone and mix them together.

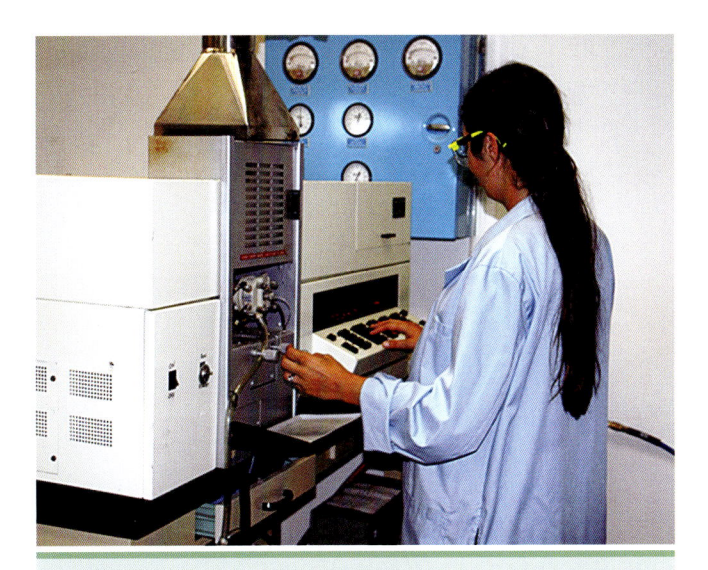

Figure 5.6 Soil and foliar analyses are available from commercial and public laboratories.

for most tree species. Confident interpretations of foliar analysis for a tree with symptoms can be made by comparing the results against a healthy, vigorous tree of the same species.

Matching the supplemental nutrition to include only those elements identified to be either limiting or deficient ensures the correct nutrients are applied to the landscape, preventing unneeded chemicals from being added to the environment. The purpose of analysis is to correlate the probability of the desired plant response (improved vitality, appearance, growth) to the results of the soil and/or foliar analysis and the assessment of the site in which the tree is growing. For example, if the analysis result for a given element is very low, there is a high probability of positive plant response to fertilization

Collecting Soil Samples for Nutrient Analysis

A soil analysis is only as good as the sampling procedure, so it is important to collect a representative sample of the site, avoiding areas where nutrient levels may be very high or very low.

A typical procedure is to collect six to ten cores from representative locations in the root zone of the tree or shrub bed. Sampling depth is 10 to 20 cm (4 to 8 in), depending on where the majority of the fine roots are concentrated. These cores should be mixed together in a clean, nonmetallic container or in a soil sample bag. Management areas that are not representative of the overall landscape should be sampled individually.

---

with that element. Conversely, if the analysis result is high, there is a low probability of response to treatment.

A soil analysis has greater value if done in conjunction with a foliar analysis. Results provide information about nutrients that have been absorbed and assimilated by the plant. Foliar nutrient analyses are the most reliable means for determining deficiencies. However, they do not provide information about *why* the elements are deficient. Assessing the site along with soil and foliar analyses will provide opportunities to correct the problems.

Soil and foliar analyses are available through commercial and public laboratories. A list of laboratories may be available from local agricultural extension offices or universities. Before sampling, arborists should contact the lab for advice on sampling methods and quantities, packaging and transport, and other important protocols.

Fertilizer

Fertilizers are available in many forms and combinations. A **complete fertilizer** is defined as one that contains nitrogen (N), phosphorus (P), and potassium (K). The **fertilizer analysis**, listed on the container, expresses the proportion of the fertilizer as a percentage by weight of total nitrogen (N), available phosphate (P_2O_5), and soluble potash (K_2O), always listed in the same order. For example, a fertilizer with an analysis of 10-2-4 contains 10 percent nitrogen, 2 percent phosphate, and 4 percent potash. A 23 kg (50 lb) bag of this fertilizer would contain 2.3 kg (5 lb) of nitrogen, 0.5 kg (1 lb) of phosphate, and 1 kg (2 lb) of soluble potash. **Fertilizer ratio** is calculated by dividing each of the analysis numbers by the lowest number in the group. A 9-3-3 analysis therefore has a 3:1:1 fertilizer ratio.

Fertilizers are available in either **organic** or **inorganic** forms. Organic fertilizers are derived from plants or animals and provide the benefit of a carbon (food) source for soil organisms. Inorganic fertilizers are mineral based and typically release their elements relatively quickly when dissolved in water. Available nutrients in the form of inorganic ions are absorbed by oppositely charged sites on the root membrane. These same ions, when in excess concentration, are also responsible for plant "**fertilizer burn**" by drawing water out of the roots. An advantage of inorganic fertilizers is that solubility in water is less affected by temperature.

Organic fertilizers also release inorganic ions but do so more slowly because the molecules require time to be broken down by soil microorganisms. Organic fertilizers are composed of carbon-based molecules and can be either synthetic or natural. One advantage of organic fertilizers is that they are not leached as readily from the soil.

Slow- or **controlled-release nitrogen (CRN)** fertilizers are preferred when fertilizing trees. These types of fertilizer provide a long period of time for uptake by roots, reducing the amount of nutrients that may be leached and reducing salt or fertilizer burn. The percentage of **water-insoluble nitrogen (WIN)** or controlled-release nitrogen on the label will determine if a fertilizer is slow release. The preferred level of WIN for trees and shrubs is 50 percent or more of the total nitrogen.

Phosphorus and potassium are typically found in sufficient amounts in most soil environments. When deficiencies of these elements are confirmed, they can be included in a fertilizer mix. Application rates should be guided by soil and/or foliar nutrient analysis.

Excess potassium can limit utilization of calcium, magnesium, and nitrogen. Overapplication of phosphorus has been linked to eutrophication (excessive nutrient enrichment) of fresh water and reduction of some mycorrhizal fungi populations. Excess phosphorus can also affect the ability of roots to take up zinc, manganese, and iron even if it is present in the soil. Therefore, phosphorus is often excluded from fertilizers applied near water and where sufficient natural sources in the soil exist. Unnecessary phosphorus applications have been banned in many jurisdictions.

The Secondary Macronutrients: Sulfur, Calcium, and Magnesium

Sulfur deficiency is more commonly associated with crop plants than landscape plants, although sulfur is sometimes used to lower soil pH. Tree and shrub growth in low pH (acidic) and/or sandy soils may be limited by calcium and magnesium. The common treatment when both calcium and magnesium are deficient is the application of dolomitic limestone. If calcium is low and magnesium is sufficient in the soil, calcitic limestone is the preferred treatment. Application rates are determined by soil and/or foliar nutrient analysis and by soil pH. Both dolomitic and calcitic limestone increase soil pH. When using either as a fertilizer to correct nutrient limitations, apply at rates that will not increase soil pH outside of the optimal range for the plant species.

The Micronutrients: Iron, Manganese, Zinc, Boron, and Copper

Micronutrient deficiencies are often seen in high pH (alkaline) soils, sandy soils, and soils with naturally low levels of these elements. These deficiencies are species specific—often affecting certain tree species while not affecting different species of trees nearby. Although these elements are not required in large amounts, a deficiency may result in effects on the health of a tree. For example, iron chlorosis is a condition that results when a tree is not absorbing sufficient quantities of iron (Fe). This condition is usually associated with high pH soil. Young leaves are small and chlorotic between the leaf veins, while older leaves tend to be darker green. Iron deficiency can eventually contribute to the death of the tree. Manganese (Mn) and zinc (Zn) are micronutrients that are commonly deficient in trees, resulting in symptoms similar to iron deficiency. Manganese

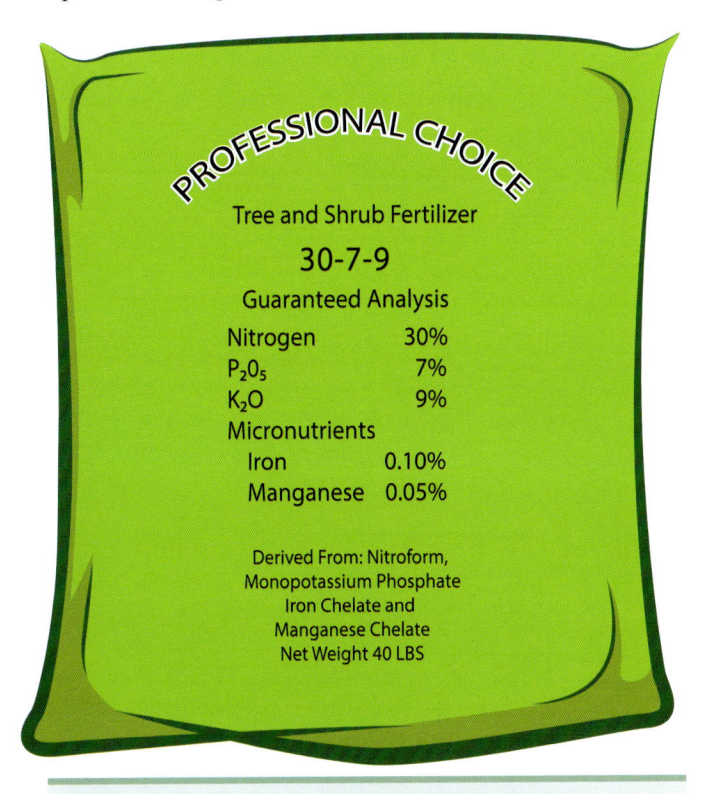

Figure 5.7 The fertilizer analysis, listed on the container, expresses the proportion of the fertilizer as a percentage by weight of total nitrogen (N), available phosphate (P_2O_5), and soluble potash (K_2O), always listed in the same order.

symptoms similar to iron deficiency. Manganese deficiency in palms can lead to mortality. Molybdenum (Mo), copper (Cu), chlorine (Cl), boron (B), and nickel (Ni) are micronutrients that are much less likely to be deficient.

The availability of many essential elements is pH dependent. The fertilizer chosen for a particular site should consider soil pH. When fertilizing a site where the soil is slightly alkaline for the species, choosing an acidic fertilizer may increase fertilizer effectiveness. Selecting a chelated form of a micronutrient can improve availability in alkaline conditions. **Chelates** are chemical compounds that bind the micronutrients with organic molecules. Micronutrients can be applied to the soil or foliage, or they can be injected into the xylem.

Fertilizer Application

Rates

The amount of fertilizer to apply should be determined based on the objectives, such as promoting growth or, if a deficiency exists, improving the health and vigor of the tree. Whenever possible, soil and/or foliar analysis results should be used to provide information on nutrient availability. Climate, weather, species, and other site and plant factors must also be considered when determining application rates.

Leaching is the washing of chemicals through the soil from rainfall or irrigation. Normally, nitrogen and other water-soluble elements are continually being leached from the root zone. Not only does this make nutrients unavailable to the tree, but the leached nutrients can also pollute groundwater, lakes, and streams. To reduce the potential for leaching, apply the minimum amount of fertilizer to achieve management goals using organic or slow-release forms, and avoid overirrigating sandy soils.

Because nitrogen is the element required in the greatest amount, it is the basis upon which fertilizer application rates are determined and quantified. Best practices recommend that application rates do not exceed 1.5 to 2.9 kg of slow-release nitrogen per 100 m^2 (3 to 6 lb/1,000 ft^2) or 0.5 to 1 kg of quick-release nitrogen per 100 m^2 (1 to 2 lb/1,000 ft^2). Quick-release fertilizers are not typically recommended. They can readily leach from the root zone, have a higher potential to cause fertilizer burn, and should be used only when the objective of fertilization cannot be met with **slow-release fertilizer**.

Because leaching and runoff from nutrient application can have a negative impact on water quality, fertilization has become a controversial subject, and research recommends lower rates than what have previously been used. If supplemental nutrition is warranted through analysis and assessment, a low application rate is advised and must be accompanied by a follow-up evaluation to determine if the prescription met the desired management objective(s).

Timing

Regardless of when fertilizer is applied, it may not be readily absorbed or utilized until root growth begins. Uptake and metabolic demand are low in the dormant season, and some of the more soluble forms of nitrogen may leach from the soil before being utilized. Nitrogen uptake peaks during the growing season when metabolic need is greatest.

One of the factors limiting fertilizer uptake is water availability. Studies on mature trees show that response to fertilization is greatest when moisture levels are adequate.

Application Techniques

In **surface application**, the fertilizer is applied over the soil surface using a spreader (dry formulations) or a sprayer (liquid formulations) calibrated to apply the desired amount of nitrogen. Depending on the fertilizer type, the area may need to be thoroughly watered following application to dissolve the fertilizer and to wash it off the grass and into the soil.

The advantages to surface application of fertilizer to trees include ease of application, relatively low cost, and speed of application. Disadvantages include possible contact with nontarget organisms and the potential for volatilization or runoff into nearby rivers and streams. Volatilization occurs

Chapter 5: Tree Nutrition and Fertilization

Figure 5.8 Liquid injection fertilization. Fertilizer dissolved or suspended in water is injected under hydraulic pressure into the soil using a soil injector.

most frequently on high pH soils and when temperatures are high. Avoid surface application where runoff is likely.

Where trees are growing in a lawn, turfgrass will compete significantly for available nitrogen. Although tree roots may get some benefit, turf roots will outcompete tree roots in absorbing essential minerals. **Subsurface application** is designed to place fertilizer below the majority of the turfgrass roots. Keep in mind, however, that the majority of the fine, absorbing tree roots are located in the upper portion of the soil.

One subsurface application method is **drill-hole fertilization**, although its use is decreasing. This method places granular fertilizer in holes drilled into the soil throughout the main root area. Although this application method can partially aerate the soil, applicators must avoid drilling into roots or placing fertilizer in direct contact with roots.

A more common subsurface application technique is **liquid injection fertilization**. Fertilizer dissolved or suspended in water is injected under hydraulic pressure into the soil using a soil injector. The advantages of this method are better distribution and uptake of fertilizer because water is injected into the root zone. This method applies fertilizer in the upper centimeters (inches) of soil where most of the absorbing roots are growing.

Some minerals are more mobile in the soil than others. Nitrogen (N) in its water-soluble form, for example, is highly mobile and will readily leach out of the soil in a short time. However, it is easily applied to the soil from the surface. Sulfur (S), in contrast, is highly immobile. A surface application of elemental sulfur will not distribute the mineral throughout the soil profile, so a soil amendment of sulfur must be incorporated by physically mixing it into the soil. Amending soil with immobile minerals can be done before trees are installed, but it is not feasible to do once trees are established without risking mechanical injury to existing roots during the incorporation process.

Most minerals are absorbed through the root systems of plants. Occasionally, there are circumstances in which more rapid nutrient uptake is desirable, such as when it is not possible to physically access the root system. In these cases, foliar application or tree (xylem) injection may be an option.

Foliar application is sometimes employed as a short-term treatment to correct nutrient

Figure 5.9 Liquid injection and drill-hole application techniques can leave dark spots in turf, where the grass benefits from the fertilizer.

Copyright © 2022 International Society of Arboriculture

91

Figure 5.10 Foliar application is sometimes employed as a short-term treatment to correct nutrient deficiencies. For example, chelated iron sprays may be used to provide a rapid, although temporary, treatment of iron chlorosis.

Figure 5.11 Microinjection application is employed to introduce nutrients directly into the xylem of trees. It relies on transpirational pull to move materials systemically upward in the xylem.

deficiencies. For example, chelated iron sprays may be used to provide a rapid, although temporary, treatment of iron chlorosis. Micronutrient spray applications are most effective when made just before a period of active growth. Response time varies, but one or two applications per year will usually be effective. Not all plant species respond to foliar treatments, though, and it may not be appropriate for species with thick cuticles. Arborists should also consider that some fertilizer will drip from the foliage. Because of the possibility of staining, care must be taken around hardscape and other finished surfaces.

Implants and **injections** are techniques employed to introduce nutrients directly into the xylem of trees. Implants, such as capsules, and some injections rely on transpirational pull to move materials systemically upward in the xylem. Some injection techniques involve supplemental pressure to reduce injection time. Implants and injections have been used successfully to introduce essential elements into trees to treat micronutrient deficiencies, but they are usually not practical for supplementing macronutrients because of the volume of material required.

Uptake and distribution are best achieved when the tree is actively transpiring and moving water upward into the crown. For this reason, do not use implant or injection techniques on drought-stressed trees. Stem injection may also cause damage to the water-conducting ability of the xylem during drought conditions.

Tree injections create wounds, and multiple wounds caused by injection holes in the same area of the tree may lead to wood discoloration and decay. To minimize these injuries, trunk injections should not be performed more than once per year on the same tree. Additionally, to minimize injection wounds and facilitate uniform distribution within the crown, drill small, clean holes close to the base of the tree at the root flare. To reduce chances of a blockage that inhibits uptake, make the insertions as soon as possible after drilling the holes.

Soil Additives

Compost and some organic mulches can improve plant health and growth and may decrease the need for fertilization. As organic materials decompose, they release essential elements into the soil. Long-term benefits provided by compost and organic mulches can include improved soil structure, increased water- and nutrient-holding capacity, and improved aeration and water infiltration. However, the addition of excess organic matter can lead to other concerns. Incorporation of organic matter should be based on soil analysis.

Many products are available and marketed as soil additives. Soil additives range from various forms of organic matter to vitamins, compost teas, and seaweed. Some additives, such as biochar, may improve soil biology and structure, providing indirect benefits for the health of the tree. Unfortunately, research is lacking on the efficacy of many soil additives, and, in some cases, there is limited evidence of their value. Worse, some may be contaminated with heavy metals or other harmful substances.

One popular group of additives is mycorrhizal soil inoculants. Recognizing the important role of mycorrhizae in water and mineral absorption, researchers have sought to encourage mycorrhizal development in tree roots. Mycorrhizae are the result of a symbiotic relationship between tree roots and specific fungi. Mycorrhizae cannot be added directly to the soil, but the soil can be inoculated with spores (and hyphae) of mycorrhizal fungi. The goal is to introduce the fungal inoculants, which might in turn form mycorrhizae. Mycorrhizae are influenced by soil pH, minerals in the soil and plant, and other competing mycorrhizae and soil microbes; therefore, inoculation should be adapted to match the plant species and site as much as possible. Although inoculation has been very successful in poor soils where fertility and microbe populations are low, fewer benefits have been shown on more favorable soils.

Figure 5.12 Incorporation of compost mulches can improve plant health and growth and may decrease the need for fertilization.

CHAPTER 5 WORKBOOK

1. Trees take up essential elements, dissolved in _____ , through the roots.

2. _____ are elements required by trees in relatively large quantities.

3. The macronutrient _____ is a constituent of chlorophyll and, if deficient, can cause reduced growth and chlorosis.

4. _____ fertilizers release nutrients over an extended period of time and can be a source of carbon.

5. The _____ _____ , listed on the container, gives the relative percentage of nitrogen, phosphorus, and potassium.

6. A 23 kg (50 lb) bag of 20-10-5 fertilizer contains _____ kg (or _____ lb) of actual nitrogen.

7. If fertilizer "burn" or leaching are potential problems, it may be desirable to use a _____ - _____ fertilizer.

8. The most important factor for good uptake of fertilizer elements is adequate _____ .

9. True/False—Surface application of fertilizer is relatively inexpensive and makes the fertilizer available in the upper few centimeters (inches) of soil.

10. True/False—Fertilization can increase susceptibility to certain pests.

11. True/False—An advantage to foliar application of fertilizer is that it usually lasts up to five years.

12. Name two limitations to trunk implants and microinjections.

    a.

    b.

13. _____ is the washing out of chemicals down through the soil.

14. Fertilization recommendations should be based on _____ and _____ analyses and assessment of the tree and site.

15. Studies show that nitrogen fertilization can trigger a tree's energy allocation toward growth, sometimes at the expense of _____ .

Matching

___ micronutrients A. element needed in the largest quantity

___ 10-0-4 B. provides pH, CEC, nutrient information

___ leaching C. nitrogen, phosphorus, potassium

___ macronutrients D. iron, manganese, boron

___ nitrogen E. washing down through the soil

___ soil analysis F. fertilizer analysis

Challenge Questions

1. Explain how a nutrient can be plentiful according to a soil analysis and still be deficient in a tree.

2. Discuss the potential drawbacks to supplemental fertilization in the landscape, and describe how they are minimized.

3. What are the advantages to using slow-release forms of nitrogen?

4. Compare and contrast the various methods of fertilizer application, including surface broadcast, drill-hole method, liquid injection, foliar spray, and implants or injections.

Sample Test Questions

1. A 36 kg (80 lb) bag of 10-6-4 fertilizer contains how many kilograms (pounds) of actual nitrogen?

    a. 1.8 kg (4 lb)
    b. 2.7 kg (6 lb)
    c. 3.6 kg (8 lb)
    d. 4.5 kg (10 lb)

2. A complete fertilizer contains

    a. all 16 essential elements
    b. nitrogen, phosphorus, and potassium
    c. organic and inorganic nitrogen
    d. equal amounts of N, P, and K

3. A tree may not respond immediately to fertilizer application if

   a. a slow-release fertilizer was applied
   b. there is inadequate soil moisture
   c. the tree is not actively growing
   d. all of the above

4. A soil test may not identify a nutrient deficiency problem in a plant because

   a. the tests are not reliable
   b. the nutrient content can change after collecting
   c. the soil may contain adequate nutrients, but something may be inhibiting uptake
   d. no one knows which levels of nutrients in soils are adequate

5. Which of the following statements about soil additives is true?

   a. Addition of organic matter is always beneficial to plant growth and health.
   b. Mycorrhizal inoculants generally show beneficial effects when added to most established landscape soils.
   c. Biochar has not been shown to have any beneficial effects for plants.
   d. Some composts and other sources of organic matter can contain heavy metals.

Recommended Resources

(See Recommended Resources in back of book for detailed information.)

Best Management Practices: *Tree and Shrub Fertilization* (Smiley et al. 2020)

Applied Tree Biology (Hirons and Thomas 2018)

Introduction to Arboriculture: Fertilization online course (ISA)

ANSI A300 (Part 2) *Soil Management a. Assessment, b. Modification, c. Fertilization, and d. Drainage* (American National Standards Institute 2018)

CHAPTER 6

TREE SELECTION

▼ OBJECTIVES

- Discuss the benefits of trees to the landscape and the environment.
- Describe the site characteristics that should be considered prior to tree selection.
- Explain the characteristics of a tree that must be considered when selecting a species.
- Explain what to look for in selecting healthy, vigorous planting stock.

KEY TERMS

acclimation adaptability exfoliating fastigiate functional goals	habit hardiness introduced species invasive species microclimate	native species naturalized species pest resistance site analysis susceptibility

Introduction

Trees are the largest, longest-living organisms in the urban landscape. They beautify and enrich our lives throughout the year with a variety of forms, flowers, fruits, seasonal leaf colors, and interesting bark. They are an important component of the urban ecosystem, sequestering carbon, providing oxygen and cooling shade, and filtering out many pollutants from the air. Trees deflect noise and wind, block unwanted views, and protect the soil from erosion. They can also reduce rain runoff by capturing and holding rainwater on the leaves and branches before it reaches the ground. Trees have even been found to have positive health and societal effects such as reducing stress, reducing crime rates, and reducing recovery time for hospital patients.

Too often, however, people take trees for granted and do not think about the benefits they provide. Arborists understand that the benefits and contributions of trees are largely dependent on careful planning and selection. In fact, if a tree is not properly matched to the landscape site, it can become more of a liability than an asset. Choosing the right tree for a particular site is an important decision to ensure long-term benefits, beauty, and satisfaction. A poorly planned tree installation will eventually require more frequent and costly maintenance. The goal for any tree planting is to maximize the benefits and minimize the inputs necessary for it to be sustainable.

Figure 6.1 Tree health and sustainability in the landscape are largely dependent on appropriate tree selection before planting.

Matching Tree and Site

The first part of tree selection is matching the tree with the site. Different tree species can have varying environmental requirements for light, water, soil conditions, growing space, and other conditions. The environmental characteristics of a particular planting site determine which plants will thrive. In addition, most trees are planted to fulfill a particular function in the landscape, such as providing shade for a backyard patio. So, it is important to choose a tree species or cultivar that will fulfill the desired function and be capable of growing in the cultural and environmental conditions at the site. Selecting the right tree for the right site increases the tree's chances of survival; reduces the ongoing maintenance costs of pruning, modifying soil, fertilization, and pesticide application; and allows the long-term benefits of the tree to be realized. A well-selected and planned planting means the tree could be around for many generations of people and arborists to enjoy and admire.

It is necessary to set priorities when selecting a tree because there are many site conditions and plant characteristics to consider. The highest priorities are those affecting the survival of the tree, such as cold and heat tolerance, moisture requirements, pH adaptability, and light needs. Sometimes the **functional goals** cannot be met entirely because of site limitations. For example, an evergreen species might be selected to provide year-round screening. If shade is a limiting factor, a broad-leaved evergreen species might be preferred because few needled evergreens grow well in shade.

Trees respond in various ways to environmental site conditions, especially temperatures. These adjustments can be stated as adaptability and acclimation.

Adaptability is the genetic ability of a species to adjust to different environmental conditions. Some species are more adaptable than others. For example, red maples (*Acer rubrum*) tend to grow well across a large range of environments, while most palms will grow only in relatively warm climates.

Acclimation is the gradual process by which a given tree adapts to changes in its environment. This usually involves physiological or morphological changes. For example, some trees acclimate to shade by developing larger, thinner leaves. Trees with variegated or colored foliage tend to lose variegation or color when grown in the shade, maximizing their photosynthetic capacity.

While it is possible for some trees to acclimate to certain site conditions such as shade and low moisture, a tree will have a much greater chance of survival if planted in a site that mimics its natural growing conditions. Some references provide lists of trees adaptable to certain site conditions such as shade, wet soils, and urban conditions. Although such lists may be very helpful, it is important to keep in mind that many factors influence survival. Good cultural practices in the nursery are important, especially for first season survival after transplanting into a landscape setting. Proper planting technique, especially planting depth, is critical. After that, the tree's genetics become the long-term determining factor in survival. Returning to the red maple example, although the species has a wide tolerance range, individual trees selected from a subtropical seed source may not do well in a temperate climate, and vice versa.

Figure 6.2 Functional uses of plants. A: Groups of plants can be used architecturally to form walls, canopies, and floors. B: Plants can have an engineering use to reduce the glare of lights. C: Deciduous plants can be used to screen the hot sun in summer and let the sun through in winter. D: Plants can form living sculpture.

Site Considerations

Competent professionals do a **site analysis** before a landscape plan is designed. The site analysis evaluates pertinent existing site conditions to determine factors that will affect plant selection. In addition, the functional goals of the design are outlined. This helps the designer to select plants that are appropriate for the site. Many site characteristics must be taken into consideration in plant selection.

The climatic or environmental conditions of the region must be considered. This includes not only the climate of the geographic zone but also the **microclimate** of the planting site, which can be affected by buildings, topography, pavement, and other surroundings. For example, a protected urban courtyard may be notably warmer than a rural subdivision, which could affect the survivability of some trees over time.

Plants may be limited by growing space, water availability and drainage, soil texture and pH, light levels, weather, and site use or function. Trees can suffer if one or more of these factors are not appropriate. They cannot be expected to thrive, or even survive, if their basic biological needs are not met.

The amount of growing space available is important when selecting the type of tree to plant in a particular site. The area above and below ground must be large enough to allow the tree to reach its potential mature height, branch spread, trunk diameter, and root volume without interfering with surrounding objects. It is easy to forget that many trees can grow more than 30 m (100 ft) in height if conditions are favorable. Many trees grow wider than taller, and roots have the potential to extend well beyond the farthest-reaching branches.

Large trees can be wonderful assets, especially for shade, as long as they have enough room to grow, both above and below ground. For example, tall-growing trees should not be planted beneath or near utility wires. Likewise, trees without sufficient soil growing space may grow poorly and can interfere with sidewalks and curbs. Similarly, by knowing the average crown spread at maturity, one can avoid planting a tree too close to a building where it may cause damage to the roof or siding, or too near an intersection where it may block visibility.

Figure 6.3 Tall-growing trees should not be planted beneath or near utility wires.

Light levels may significantly affect plant growth and survival. Excessive shade can be a problem for some trees because many trees are not adapted for low light conditions. Their branches grow long and spindly, and they may drop leaves. Similarly, excessive light may cause additional stress to trees. Bright or reflected light and heat from buildings and paved areas can also cause severe problems if a tree cannot adapt to those conditions. Some trees

Figure 6.4 Ample space for root growth is essential but is often overlooked.

will develop scorched foliage and may wilt in high light and temperature conditions. Typically, trees that are adapted to bright light and heat have small, thick foliage. Some species have sunken stomata to reduce transpiration.

The soil conditions of the site must also be considered. It is advisable to get a soil analysis to determine the texture, pH, soluble salts, and nutrient levels of the soil. Another important soil quality is its water-holding capacity and drainage. Trees planted in a site that is too wet or too dry will often die within the first year. If the soil is compacted, tree growth may be greatly reduced because of insufficient oxygen in the root zone. Tree managers must be alert to soil contaminants such as road salt or herbicides in the planting site that could be toxic.

Table 6.1
Site Considerations

Climate
Hardiness zone
Heat/cold
Rainfall/snowfall
Sunlight/other lighting
Prevailing winds
Exposure
Soil
pH
Drainage
Soil texture
Bulk density
Cation exchange capacity (CEC)
Nutrient analysis
Soil volume
Planting site
Buildings and other structures
Paved surfaces
Clearance needs below branches
Plans for future development
Overhead and underground utilities
Intended use of site
Intended function of plant
Other plantings
Trees and shrubs
Planting beds
Groundcovers/flowers
Turfgrass
Maintenance to be provided
Irrigation
Postplanting care
Ongoing maintenance
Other requirements

Once the site has been analyzed, the next step is to consider the design criteria. Design criteria are based on the functions the tree is expected to serve. These functions include engineering, architectural, and personal safety considerations such as controlling pedestrian traffic or hiding unsightly building features. Climate controls, such as wind screens, or shade for an east or west exposure may be desired. Even aesthetic considerations such as colored foliage, attractive bark, or showy flowers can be design criteria. An ornamental tree may be planted as the focal point of a landscape. Client needs and goals are major factors in plant selection.

Urban sites, especially street tree plantings, pose some of the biggest challenges. Usually the underground space is limiting, and the original soil has been greatly disturbed. Soil compaction may be extreme, the pH may be inappropriate, and moisture levels are likely to be suboptimal. Add to this the stresses of high reflected heat and light, deicing salt injury, and the threat of vandalism, and it becomes clear why municipal arborists struggle to find trees that will survive. Selecting trees that can also meet the design criteria may be next to impossible. Some urban sites are simply not suited for trees.

Tree Considerations

Many factors must be considered when selecting a tree for a given site. These include cold hardiness, heat tolerance, drought tolerance, size at maturity, form, insect and disease resistance, nuisance potential, and maintenance requirements.

Specific plant characteristics may make a tree more desirable. These characteristics might include **exfoliating** bark, features that attract birds, or an interesting branching habit. Some trees are admired for their flowers or leaf color change in the autumn. Other plant characteristics, such as excessive leaf, fruit, or twig drop, may preclude the use of certain trees adjacent to a sidewalk, patio, or parking lot.

Arborists should also consider future maintenance inputs when selecting tree species. Some plants require more water than others, and supplemental irrigation or drainage may be necessary in some landscape situations. Ensuring adequate water is especially important in the first and second growing seasons following planting. Be very cautious in choosing plants that will need long-term irrigation for success. Even though irrigation might be installed and promised, future maintenance may be lacking, or regulations may even disallow. It is best to install trees that will not require irrigation after the first two seasons. Some trees may need to be pruned regularly for health or size maintenance.

Table 6.2
Tree Considerations

Hardiness
Heat tolerance
Growth habit/form

Size
  Height
  Spread
  Root zone requirements

Aesthetic attributes
  Showy flowers
  Attractive fruit
  Attractive bark
  Interesting foliage
  Exceptional fall color

Resistance to insects and diseases
Drought tolerance
Tolerance of drainage problems
pH requirements
Salt tolerance
Light requirements

Known issues
  Common pests and disorders
  Poor structure/weak wood
  Propensity to form surface roots
  Issues with dropping branches/flowers/fruit/leaves
  Thorns

Maintenance requirements

Figure 6.5 Attractive foliage can be a desirable tree characteristic.

Figure 6.6 Some trees are selected for their showy flowers.

Figure 6.7 Fruit can be ornamental and can be attractive and beneficial to wildlife.

Select plants that realistically match the maintenance inputs for a healthy, functional tree in the landscape.

Fast-growing trees are often tolerant of poor soil conditions. They may quickly reach a size where they provide functional benefits such as shade and screening. However, such trees often have weak, brittle wood and develop multiple branches with included bark that may break easily in storms. They usually do not live as long as trees with a more moderate growth rate.

Selecting a tree with a particular growth form may offer a solution to a growing-space problem. For example, a large tree with a narrow, upright growth **habit** (form) could be planted closer to a building than one with a spreading form. Cultivars of some tree species are available with different growth forms such as **fastigiate** (upright), pyramidal, or weeping. Trees with different growth forms also provide a variety of architectural effects in the landscape.

Figure 6.8 Some trees have attractive bark.

Figure 6.9 It is important to also consider characteristics such as thorns or dropping fruit, which can create a nuisance in some settings.

The root system of a tree is an important consideration during the selection process and for locating a tree on a site. The rooting space available can restrict the choices of trees that can be planted. Some trees have a strong tendency to form surface roots that can damage pavement and cause problems for lawn mowers.

Hardiness is a plant's ability to survive winter growing conditions and often refers specifically to low temperatures. Several plant authorities in the United States and other countries publish maps of hardiness zones that label each geographic area. References list either the lowest (coldest) zone in which a tree will thrive or a range of zones where the plant will grow (for example, zones 5 to 8). However, plants that are considered hardy in a given zone may still decline or die because of low temperatures if their roots are above ground in containers or if the microclimate around the plant is unusually cold.

Just as some plants are not suited for severely cold temperatures, others may be stressed by high temperatures or dry conditions. For example, many species of spruce (*Picea* spp.) are native to cool-weather locations. If planted in warmer climates, these trees are stressed by summer heat and may be subject to mite infestation and diseases. Sometimes deodar cedar (*Cedrus deodara*) will grow well in Mediterranean-like climates but will do poorly in more humid locations in the same hardiness zones. Heat zone maps, which are similar in use to hardiness zone maps, help in the selection of plants that will grow in warmer climates.

With arborists taking a more holistic approach to tree health, more consideration is given to tree selection. Of particular concern is **pest resistance** or **susceptibility** to pests such as insects and

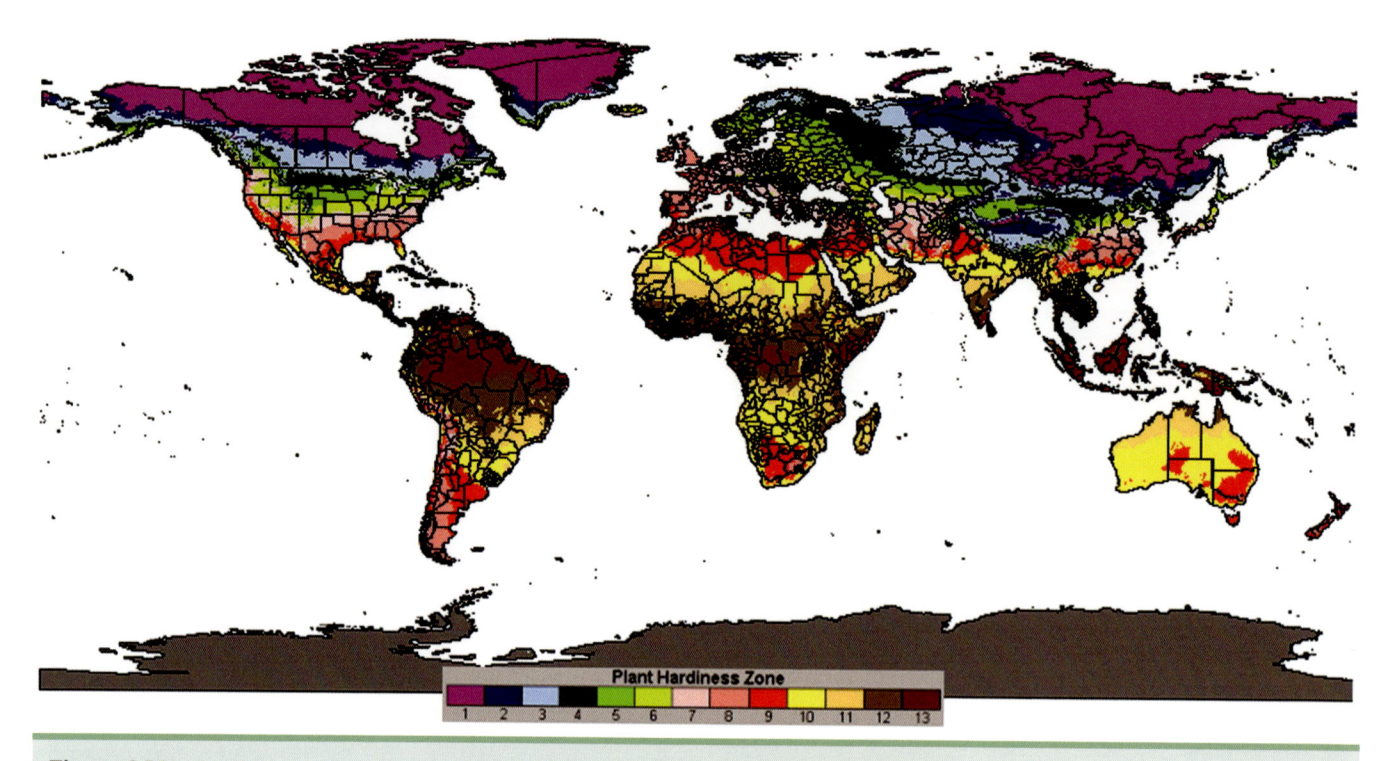

Figure 6.10 Hardiness zone maps are available for various geographic regions of the world.

diseases. It is important to select a tree that will be able to withstand stress on a given site. Cultivars selected for their resistance to certain insects or pathogens or their relative tolerance of environmental stresses are available. For example, many crabapple (*Malus* spp.) cultivars are resistant to apple scab, a disfiguring disease of crabapples in temperate regions.

There is considerable debate about the use of **native species** versus **introduced** (nonnative or exotic) **species**. Native species usually grow well in their natural range; however, the urban environment is not always representative of the natural range. For several hundred years, species of trees and other plants have been introduced to our landscapes from different parts of the world. Some introduced species have been very successful, and some have become **naturalized species**, reproducing and thriving in their new settings for decades or even centuries. There are other instances, however, where introduced species are creating many problems. Some have been very invasive, virtually taking over landscapes and nearby natural areas. An **invasive species** is defined as a species that is considered nonnative to the ecosystem and whose introduction causes or is likely to cause economic or environmental harm. Invasive species should not be planted.

An arborist should select plants that will perform well in a given site. In areas where new plantings are limited to native species by legislation, arborists sometimes have difficulty finding nurseries that grow appropriate species for urban plantings. Urban sites typically have altered soils and microclimates that do not match the native conditions of any species. In addition, urban plantings adapt to the cultural and maintenance practices of the site, as well as factors involved with human interaction. Landscape professionals should choose from a list of plants that best match site conditions as indicated in a site analysis, which includes considerations such as compaction, drainage, long-term maintenance needs, and other environmental challenges.

Figure 6.11 Palms have a wide variety of growth forms that can be incorporated into landscape designs.

Climate Change

Climate conditions around the world are changing over time. In many places, temperatures are rising, storm frequency and severity are increasing, and rainfall patterns are changing. Arborists should select trees capable of growing in a site's current climate and those that will tolerate expected changes in climate. For example, as local summer peak temperatures increase, trees that naturally grow at lower elevations and latitudes may become better long-term choices. In some cases, even native trees may be less desirable to select for long-term plantings.

Selecting Trees at the Nursery

Once a tree species has been chosen, the next step is selecting high-quality trees from a reputable source. For a tree to be successful when transplanted into a landscape, it is important to begin with a healthy plant.

Look for a vigorous plant with good shoot growth. Healthy, vigorous plants establish quickly in the landscape. Plants in poor health can attract pests and require more maintenance. Examine the leaves and branches. Choose trees with healthy foliage (if present), and check for the presence of insects or disease.

Trees with good vertical and radial branch spacing are more likely to develop strong structure as they grow. Look for good trunk taper from the roots up to the branches. Most young trees should have a single, dominant leader and good radial distribution of branches. Scaffold branches should be free of included bark and should be less than half the diameter of the parent stem. Avoid trees with many upright branches unless an upright or columnar form is desired. Foliage should be evenly distributed on the upper two-thirds of the tree and not concentrated at the top. Check the plant for mechanical damage. Do not purchase a tree that has an injury to the trunk. Also, ensure that the trunk is not loose in the soil ball or container, which could indicate improper handling and disruption of the root-soil contact.

Examine the root ball of the tree. In the field, on trees large enough to have a trunk flare (root flare), it should be visible. If not, excavate the soil to trunk flare depth prior to digging. On smaller-diameter trees, the trunk flare may only show as a slight swelling at ground level. In general, some structural roots should be found near the soil surface. If the tree is already balled and burlapped, it should have a solid ball that has been kept moist and protected from drying. If the plant is in a small container, remove the container to check the root system. Roots that are brown or black or have a foul odor indicate a health problem. The trunk flare should be visible at or just below the surface level. You may have to excavate some soil to verify this condition. The trunk flare and large roots should be free of circling or kinked roots. Soil removal near the root collar may be necessary for inspection. The first lateral roots should be near the soil surface.

Selection of the particular tree from the nursery can be as important as selection of an appropriate tree species. Selecting a healthy plant with good form can help ensure that the tree will be an asset in the landscape.

Figure 6.12 Always inspect the trunk flare and roots when selecting a tree at the nursery.

Figure 6.13 Roots can circle inside of containers and may later become girdling roots.

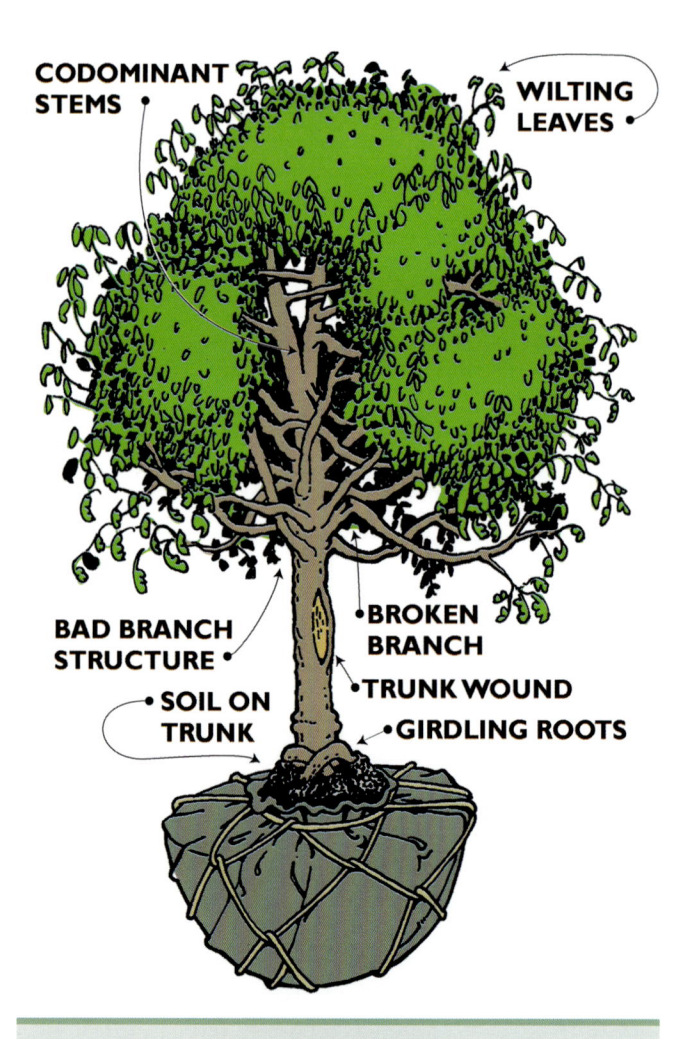

Figure 6.14 A few minutes of inspection at the time of selection can avoid long-term problems for a tree.

CHAPTER 6 WORKBOOK

1. Name five tree species that would *not* be appropriate for planting under utility wires.

    a.

    b.

    c.

    d.

    e.

2. _____ is the ability of a tree to withstand low temperatures and winter stresses in a given site.

3. True/False—Although a tree may be considered hardy in a given area, it may decline or die if the roots are unprotected.

4. Name three site characteristics that must be considered in tree selection.

    a.

    b.

    c.

5. Upright, pyramidal, and weeping are three examples of tree _____ that are important in selection.

6. If a particular disease is known to be a problem, a tree species or cultivar should be selected that has _____ to that disease.

7. Name three plant characteristics that may make a tree aesthetically desirable.

    a.

    b.

    c.

8. _____ is the gradual process by which a tree adapts to changes in its environment.

9. True/False—A tree listed as adaptable to wet soil conditions will always thrive if planted in those conditions.

10. Name five characteristics to consider when selecting a tree in the nursery.

   a.

   b.

   c.

   d.

   e.

Challenge Questions

1. Give examples of the various functions that trees serve in the landscape.

2. Name five tree species in your area that are adaptable to each of the following site conditions: wet soils, shade, full sun, dry soils.

3. Name five plant attributes that may make a tree desirable for a given area.

4. Why is the geographic origin of a tree important to consider?

5. Discuss the merits and limitations of using only native species in new plantings as opposed to naturalized or introduced species.

Sample Test Questions

1. The primary climatic factor that determines hardiness zones is

   a. north-south location
   b. temperature, rainfall, and winds
   c. east-west location
   d. low temperature extremes

2. Trees to be planted under utility lines should be

   a. tolerant of heavy top pruning
   b. low growing to remain below the lines
   c. vase shaped or overarching to clear conductors
   d. all of the above

3. Some trees acclimate to shade conditions by

   a. developing larger, thinner leaves
   b. developing smaller, thicker leaves
   c. producing stomata mostly on the upper leaf surfaces
   d. developing variegated foliage

4. Fastigiate trees have a growth form that is

   a. upright
   b. weeping
   c. overarching
   d. vase shaped

5. Which of the following is a true statement?

   a. Floodplain species will always grow well in wet soils.
   b. Trees adapted to grow in full shade will perform well in full sun.
   c. Some tree species are adapted to hot, dry, or bright light conditions with small, thick foliage and sunken stomata.
   d. Most evergreen conifers are very shade tolerant and tend to scorch in full sunlight.

Recommended Resources

(See Recommended Resources in back of book for detailed information.)

Dirr's Encyclopedia of Trees and Shrubs (Dirr 2011)

Introduction to Arboriculture: Selection online course (ISA)

The Practical Science of Planting Trees (Watson and Himelick 2013)

CHAPTER 7
INSTALLATION AND ESTABLISHMENT

▼ OBJECTIVES

- Describe the techniques and procedures used to plant and transplant trees.

- Explain how using recommended planting techniques can improve tree establishment and survival chances and prevent longer-term issues associated with root health and arrangement.

- Identify the various types of planting stock, proper techniques for planting each, and the advantages and limitations associated with them.

- Identify early care requirements of newly transplanted trees as well as the potential pitfalls associated with improper care practices.

- Explain the special considerations and procedures for transplanting palms.

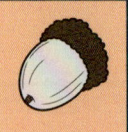

KEY TERMS

backfill
balled and burlapped (B&B)
bare root
container-grown
containerized
drum lace
girdling root
guying

hardened off
inorganic mulch
organic mulch
planting specifications
portable watering device
root ball
root pruning
staking

substrate
transplant shock
tree guard
tree spade
tree wrap
wire basket

Introduction

The key to successful establishment of nursery stock is to transplant high-quality plants using good planting procedures. The transplanting process can be quite disruptive, and improper planting can lead to increased stress and long-term physiological disorders. However, tree planting techniques and transplanting processes are among the most studied aspects of arboriculture. A well-informed arborist should be aware of the current best management practices associated with tree installation and establishment. ISA's Best Management Practices: *Tree Planting* provides detailed information on planting and transplanting procedures.

Figure 7.1 The key to successful establishment of nursery stock is to transplant high-quality plants using good planting procedures. Inspect all trees before accepting delivery. Look for damage, pests, and poor structure. Ensure that there is a good root system without girdling roots and locate the trunk flare.

Stock Types

Landscape trees are produced and transplanted as one of four main stock types: **bare root**, **containerized**, **container-grown**, and **balled and burlapped (B&B)**. Each type has its advantages and disadvantages. Cost, site requirements, species, regional production techniques, and **planting specifications** (job requirements) often determine stock type availability and selection.

Bare Root

Bare-root trees are usually small, easy to transplant, and less costly than other stock types. They arrive with many of their larger roots fully intact, but they lose fine roots when harvested from the field nursery in which they were grown. Because there is no soil moved with the root system, they are lightweight, which reduces shipping costs and facilitates handling at the jobsite. When planting bare-root trees, it is vital that the roots be kept moist. Bare-root trees are normally planted during the dormant season before roots and buds begin to grow. If not planted immediately, temperate-zone bare-root trees should be stored cold, but not frozen, with moist packing

Figure 7.2 Bare-root nursery stock. Because there is no soil moved with the root system, they are lightweight, which reduces shipping costs and facilitates handling at the jobsite. When planting bare-root trees, it is vital that the roots be kept moist.

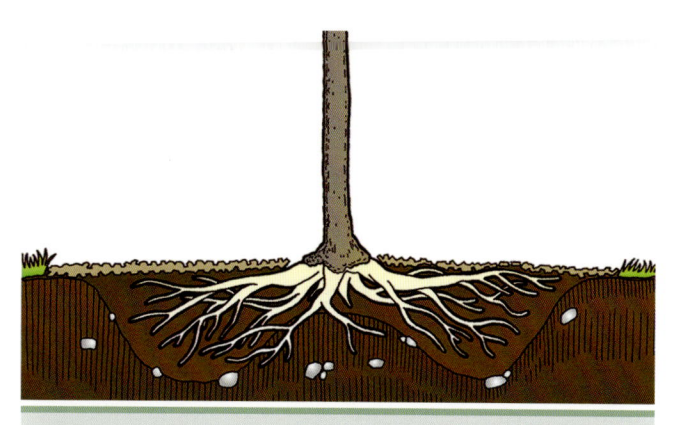

Figure 7.3 When planting bare root, spread and distribute the roots to prevent kinking or circling. Exposure of the roots to air can lead to desiccation; therefore, be sure to place the backfill around the roots to minimize air pockets.

material around the roots. Usually, only deciduous trees or conifer seedlings are handled bare root.

Plant bare-root trees with the main structural roots near the soil surface. Do not dig planting holes so deeply that loose **backfill** (soil used to fill the hole) is required to elevate the tree to the correct height. Spread and distribute the roots in a spoke-like manner to prevent kinking or circling. Exposure of the roots to air can lead to drying out (desiccation); therefore, be sure to place the backfill firmly (but do not compact it) around the roots to minimize air pockets. Water gently while gradually backfilling the planting hole. Bare-root trees may require staking to remain upright once they leaf out.

Bare-root trees that were dug in autumn and overwintered in cold storage may benefit from a preplanting procedure known as sweating. This will result in breaking dormancy and growth starting. Because there is no gradual change from winter to spring in a storage cooler, sweating is substituted as an artificial and accelerated spring warming period. Species that often require sweating include common hackberry (*Celtis occidentalis*), birch (*Betula* spp.), honeylocust (*Gleditsia triacanthos*), sugar maple (*Acer saccharum*), and most oaks (*Quercus* spp.). In the sweating procedure, the root system of a bare-root tree is surrounded by a moisture-holding material such as straw or burlap, and the entire tree is wrapped in plastic to retain a high level of humidity. The trees are then placed in a warm, shaded area until the buds start to swell and bud scales separate. The tree is ready for planting as soon as the bud swelling begins. Failure to plant before or immediately after budbreak may lead to plant death.

Containerized and Container-Grown

Many trees are sold in nurseries in containers. The root systems of container-grown trees are completely intact and in a soil or a **substrate** (growing medium) that is held within a container. If properly watered and maintained, container-grown trees can be planted any time of the year that the ground is not frozen. If planted after leaf drop, roots can begin establishment before the next growing season.

Not all trees sold in containers are container-grown; often, bare-root trees are grown in the nursery fields, planted in containers, and then made available for sale. If they have recently been potted, there will not be an established root system in the container.

Remove containers before planting. Removal of biodegradable pots, such as natural peat pots (generally limited to smaller-sized nursery stock or seedlings), is optional. It may be preferable to remove a biodegradable container unless the root system will not hold together without it. When selecting containerized trees, check the root system. Sometimes the roots will have grown in circles and will be matted along the wall of the container. These circling roots should be pulled apart or removed. Slicing or shaving off the outer 2.5 cm (1 in) of the **root ball** sides and bottom (if needed) has been shown to be particularly effective in promoting outward root growth after planting. Roots should regenerate quickly, and if soil moisture is maintained, the mild stress is preferable to the development of circling roots. Excessively pot-bound trees (e.g., trees with woody roots circling within the container) should be rejected. As container-grown trees may be in multiple containers during production (going from smaller to larger), it is important to look for signs of circling throughout the root ball and not just at the outer edge of the current root ball. Circling roots can often be detected by gently probing the upper surface of the root ball.

If circling roots are allowed to grow across or around a stem, they can become **girdling roots** over time as the stem and roots expand in diameter. Stem-girdling roots can choke off vascular tissues in the tree, leading to decline and death of branches or even the entire tree. Girdling roots are often associated with planting too deeply or with compacted soil beyond the root ball.

Balled and Burlapped

Some trees are balled and burlapped (B&B) in the nursery. In this process, field-grown trees are dug with a portion of the root system remaining intact and wrapped with burlap. Although some roots are preserved within the root ball, it is estimated that as much as 90 percent of the fine, absorbing roots can be lost in digging. As a result of the potentially high amount of root loss, it is critically important to keep root balls moist throughout the handling and transplanting process. The burlap and twine used to wrap the root ball keep the surrounding soil in place. This minimizes root exposure to desiccation and reduces the risk of root breakage.

Natural-fiber burlap is biodegradable, but some nurseries use treated or synthetic burlap. Synthetic burlap may act as a barrier to root growth and should be removed to the extent possible. Removal of the top and upper sides of treated burlap is advised. The same may be done for roots in biodegradable burlap if desired. Peeling back and removing the burlap from the surface of the root ball may reduce wicking of moisture from the exposed burlap. The burlap is often held in place with twine. To avoid girdling the tree, remove all twine, whether natural or synthetic, that is tied around the trunk.

Larger balled-and-burlapped trees often come in **wire baskets** to maintain the structure of the ball during handling. In poor soils, wire baskets can last decades and may partially girdle roots before being grown over by root expansion. The impacts of this are debatable, and research has not determined differences in growth or long-term tree stability when trees are planted with or without wire baskets intact. If girdling remains a concern, the top portion of the wire basket can be removed; however, staking may be necessary to stabilize the tree. This should be done once the tree is placed in the planting hole and partially stabilized with backfill soil to avoid damage to the root ball. Removing the basket can

Figure 7.4 Some larger balled-and-burlapped trees come with their root balls enclosed in wire baskets to maintain the integrity of the ball during handling.

Figure 7.5 Cut all twine around the trunk to prevent girdling. Because burlap may act as a barrier to root growth, remove the top and upper side portions, even if biodegradable burlap was used.

Figure 7.6 If a traditional basket is used, remove as much of the top portion of the wire as is practical. Low-profile baskets do not interfere with root growth and should not create maintenance conflicts.

Figure 7.7 A saucer-shaped planting hole two to three times the width of the root ball at the soil surface, sloping down to about the width of the root ball at the base, is ideal. In very compacted clay landscape soils, the soil can be loosened to a wider area to facilitate root growth.

also minimize damage to machinery if the stump needs to be ground out in the future.

Planting Techniques

The most vigorous root growth occurs near the soil surface. Root growth from the lower portions of the ball is often reduced due to poor soil aeration or poor drainage, especially in heavy or shallow soils. This should be taken into consideration when digging the planting hole. A saucer-shaped planting hole two to three times the width of the root ball at the soil surface, sloping down to about the width of the root ball at the base, is ideal. In very compacted clay landscape soils, the soil can be loosened to a wider area to facilitate root growth. Because the planting hole may act as a basin and hold water, especially in compacted or clay soils, oxygen levels in the bottom of such holes are low and not conducive to rapid root growth. To the extent that it is practical, wider, saucer-shaped planting holes are recommended.

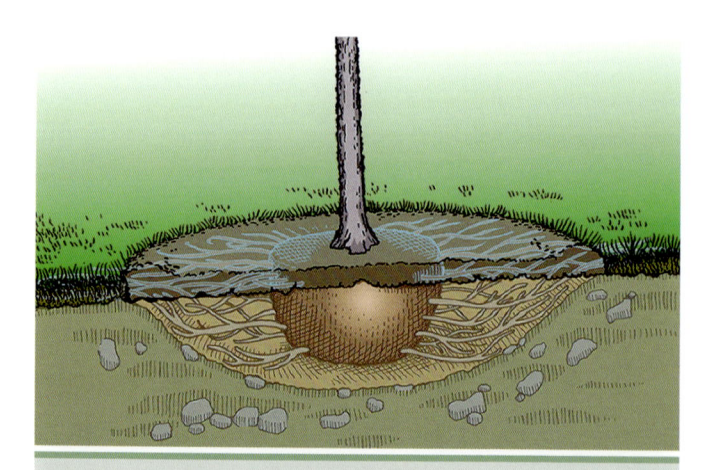

Figure 7.8 The most vigorous root growth occurs near the soil surface. Root growth from the lower portions of the ball is often reduced due to reduced soil aeration or poor drainage, especially in heavy or shallow soils.

A common planting mistake is planting too deeply. The planting hole should never be deeper than the distance from the trunk flare to the bottom of the root ball. Planting too deeply can

Chapter 7: Installation and Establishment

Call Before You Dig

Always locate any underground utilities before digging. This is not only an essential safety consideration—in some areas, it is the law. Many countries around the world have utility-location services. In the United States, every state has rules and regulations governing digging, some more strict than others. Most states or regions have one-call services that allow the caller to arrange for all utility types to be marked by calling one number.

occur unintentionally because containerized and balled-and-burlapped trees can arrive from the nursery with too much soil or substrate covering the structural roots. Because the trunk flare is not always visible on young trees, it is very important to locate the primary structural roots. At least two primary roots must be located within 2.5 to 7.5 cm (1 to 3 in) below the soil surface. Measure root depth using a probe or stiff piece of wire at a distance of 10 cm (4 in) from the trunk. It is often necessary to remove some excess soil so that the trunk flare will not be buried. If there are no roots in the upper portion of the root ball, the root system may be undersized, and the tree should be rejected and returned to the nursery.

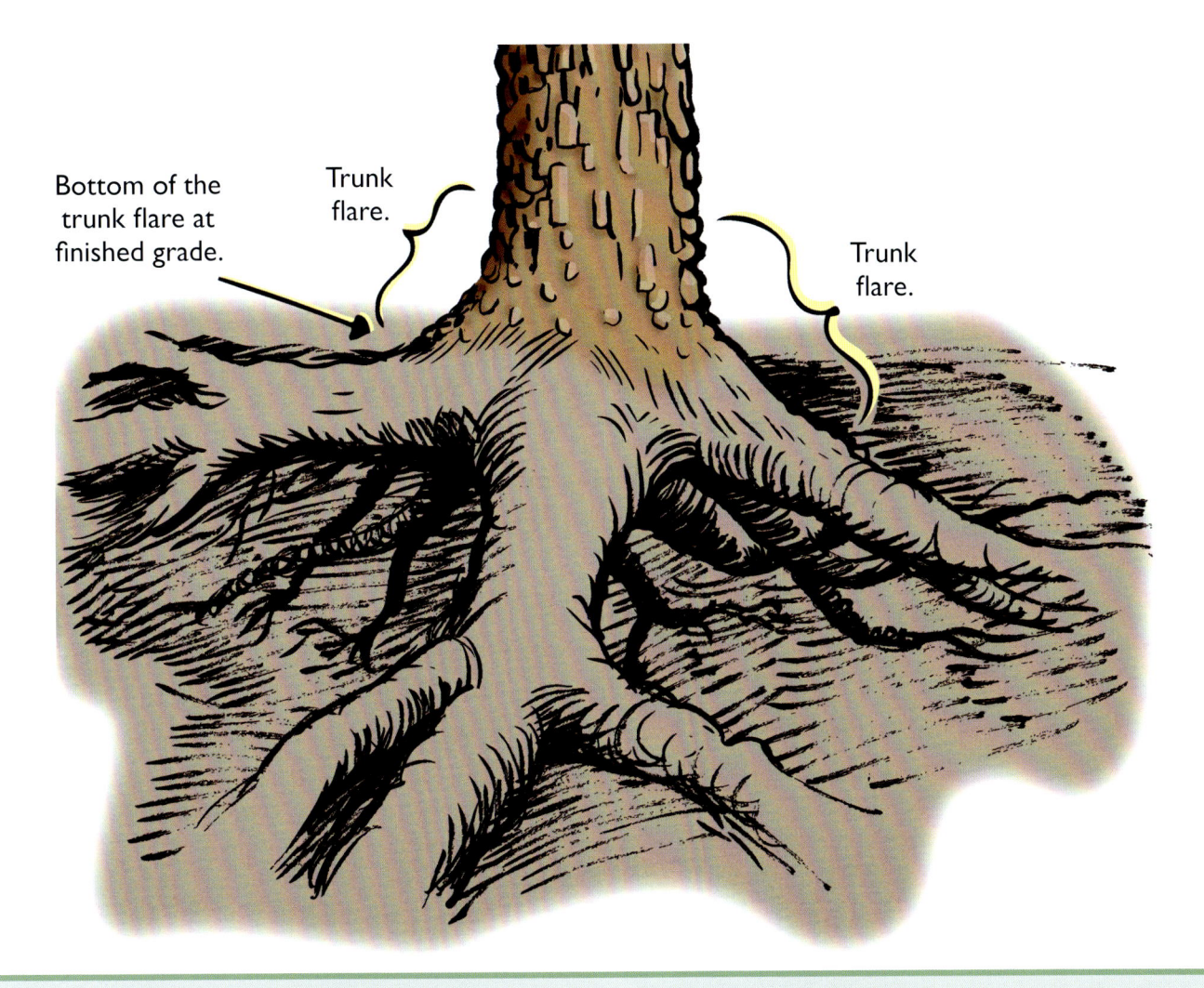

Figure 7.9 The bottom of the trunk flare should be at finished grade and at least two primary roots should be near the soil surface.

Figure 7.10 Planting too deeply can occur unintentionally because containerized and balled-and-burlapped trees can arrive from the nursery with too much soil or substrate covering the structural roots. Because the trunk flare is not always visible on young trees, it is important to locate the primary structural roots.

Once the actual root depth is determined, determine the appropriate planting depth, avoiding digging the hole too deep. For instance, if the root ball measures 40 cm (16 in) deep, yet there are 10 cm (4 in) of soil over the structural roots, dig the planting hole to a depth of 30 cm (12 in). Planting a tree too deeply can stress, drown, or suffocate roots; it may also enable soilborne insects or pathogens to enter the trunk.

Do not add fill to the bottom of the hole. The root ball will settle in the days and weeks after planting, resulting in a final depth that is deeper than originally planted. In areas where the surrounding soil is dense clay, plant trees slightly shallower, with 5 to 8 cm (2 to 3 in) of the root ball higher than the original grade. Soil may be added gradually to taper out from the top of the ball to the original grade, but it should not be placed directly over the root

Figure 7.11 Measure the depth of the root ball before digging the hole.

Figure 7.12 Slicing the outer roots of a containerized plant (left) before planting can help reduce the occurrence of circling roots. Shaving the outer edge (right) has been shown to be even more effective.

Figure 7.13 Drainage can be a problem in heavy clay soil. One option is to plant trees slightly shallower, with 5 to 8 cm (2 to 3 in) of the root ball higher than the original grade. Soil may be added gradually to taper out from the top of the ball to the original grade, but it should not be placed directly over the root system.

system. The exposed ball and surrounding area can be covered with 5 to 10 cm (2 to 4 in) of mulch.

Never lift a tree with a soil ball by the stem. Because small, fibrous, absorbing roots are easily broken, handle the root ball carefully. Place the tree in the planting hole gently and check to see that the trunk flare will be no deeper than the soil grade in well-drained soils, or 5 to 8 cm (2 to 3 in) above grade in poorly drained soils.

Backfill around the tree using the same soil that was removed from the planting hole, if possible. Research has shown that modifying the backfill of soil with amendments does not assist in tree establishment and growth unless the soil at the site significantly restricts root growth. Abrupt texture changes within the soil profile can have a profound effect on water movement due to the possible creation of an interface between the site soil and the fill around the root ball. If the planting site's original soil is not conducive for root growth, implement extensive soil improvement procedures (as discussed in Chapter 3, Soil Science) to provide a gradual transition and make the soil texture around the root ball as uniform as practical.

While backfilling, gently tamp the soil around the ball and add water to minimize air pockets. Large pockets of air can allow the roots to dry out. Firm the soil around the bottom of the root ball so that the tree is vertical and adequately supported. Slightly tamp down the remainder of the backfill as the hole is filled. Water thoroughly and slowly after backfilling.

A regular and effective irrigation schedule is crucial for new root development and establishment. Watering containerized trees can be especially important because they are often in soilless substrate, which dries out much faster than the soil of a B&B plant. On sloped sites, mound the remaining backfill soil into a dike or berm beyond the outer edge of the root ball to collect water over the root

Figure 7.14 Forming a berm over the root ball can help collect water where it is needed.

zone. If the tree is planted in a location where it will not be closely monitored for irrigation needs, such as in a park, it may be a good idea to use a **portable watering device**. A range of devices are available that provide the tree with a reservoir of water that is dispensed slowly over the root ball. Such devices must also be monitored, however, and should not become a reason for neglecting a newly planted tree.

Transplanting

Transplanting a tree involves the additional procedures of digging and preparation for moving. Some species are easier to transplant than others. Because digging a tree for transplanting can remove as much as 90 percent of its root system, species that have a low transplant tolerance should be moved when conditions are optimal.

In general, the best time to transplant most tree species, in temperate climates, is in the early spring or autumn while the tree is dormant. In moderate climates, many deciduous plants can be moved just after autumn leaf drop, when the moisture level in the soil is relatively high and the soil is still warm. Transplanting at this time gives the roots a chance to grow and begin to establish before the ground freezes. Some plants are more easily transplanted in the spring before budbreak. Transplanting dormant trees reduces demand on soil moisture because transpiration is minimal. Generally, evergreen trees are also more easily transplanted while dormant.

In colder climates, very large trees are sometimes moved in the winter when the ground is frozen and heavy equipment used around the tree will do minimal damage to the surrounding area. If the outer 10 to 15 cm (4 to 6 in) of the root ball are frozen, it will be easier to move without damage. Frozen root balls require less wrapping because they are more cohesive and easier to handle. Sometimes the tree can be pre-dug and thoroughly mulched before the ground freezes. The hole at the new location should also be prepared before freezing.

If extra care is taken, some trees can even be moved during the growing season. When trees are dug from a nursery field, they may be stored in a protected holding area to be **hardened off** (a process that acclimates balled-and-burlapped trees to water stress when dug with foliage). New root generation may begin if stored for a long period. Also, it is not uncommon for some leaf drop to occur due to reduced water availability because of the reduced fine root system. This is usually a temporary situation, and the tree will often produce new foliage if kept well watered until the fine root system can reestablish.

The techniques used in digging vary little with species, size, and soil type. The procedures described here are for digging and transplanting large trees. The techniques are similar for smaller trees, which are easier to transport.

Transplanting should be planned in advance to significantly improve the chances of survival and increase the rate of establishment. One tactic that must be planned in advance is **root pruning**. Root pruning is the process of predigging around a root ball to increase the density of root development within the final ball. The digging process severs existing roots and stimulates root regeneration. Root pruning may be repeated multiple times before the tree is actually moved. Each successive cut is made several centimeters (inches) out from where the roots were previously severed. Although

Chapter 7: Installation and Establishment

Figure 7.15 Root pruning severs existing roots to stimulate new root development within what will be the future root ball of a transplant tree.

root pruning encourages growth below ground, top growth may be temporarily reduced.

Before moving the tree, tie together as many of the branches as possible to prevent them from breaking during transport. When securing the branches, take care to avoid damage to the bark. Potential tissue compression or branch breakage may result from tying the branches too tightly.

When a tree is dug for transplanting, the size of the root ball is usually based on tree caliper. The width of the ball is determined based on tree caliper; the depth of the root ball may vary with species, soil texture and other conditions, irrigation practices, and the age and size of the tree. Larger trees need deeper root balls to encompass more roots to ensure adequate regrowth, as well as anchorage and stability. Depth can be less for smaller trees or for trees that are not deeply rooted.

More detailed guidance is available in ISA's Best Management Practices: *Tree Planting*.

Make the first cuts around the perimeter of the root ball with a sharp spade or shovel. A clean cut is necessary to avoid tearing or breaking roots. If the ball is being dug with machinery such as a backhoe or trencher, dig the initial ball several centimeters (inches) larger than the final ball size. Perform the shaping and final cuts by hand. As larger roots are located, cut them with loppers or a saw. Some recommendations call for sterilizing tools with alcohol or bleach before cutting to minimize the risk of transferring disease. While digging the trench, avoid standing on the root ball; the edge of the ball could break down and damage the roots. Once the ball has been dug to the desired depth, it can be shaped. The ball should taper on the sides, slanting inward toward the base, and it should stand on a pedestal of soil for shaping and burlapping before it is undercut.

Place burlap on the sides and across the top of the ball. Pin the burlap together using nails, taking tucks to pull the burlap snug. The burlap should cover the full circumference of the root ball, with a bottom skirt of burlap hanging over the pedestal. Pin the skirt of

Figure 7.16 The weight of the root ball is supported, and the trunk is protected, when lifting this large tree for transplanting.

Copyright © 2022 International Society of Arboriculture

burlap to the ball after the tree is taken out of the hole. For additional support of large soil balls, **drum lace** them with manila rope or sisal. If the tree is very large, consider boxing it on-site before moving. Undercut the tree once the root ball has been laced and secured. For a large tree, use a crane or other mechanical device to remove it from the hole. Place chains or large slings around the ball and attach them to the crane hook. Do not lift the tree by the trunk. Doing so can cause trunk injury and serious damage to the root ball. If additional support and balance are necessary, attach cables to the trunk. Once the tree is out of the hole, fasten the burlap to the bottom of the ball.

When transporting a tree to the planting site, take measures to protect it. Pad the trunk well to protect it from injury. Loosely wrap or cover the crown of the tree with burlap to minimize drying and/or wind damage. Special permits may be necessary to transport large trees on public roads.

Mechanical Tree Spades

A **tree spade** is a mechanical device used to transplant trees. It encircles a tree and forces several large blades downward into the ground at an angle, forming a root ball. The blades can make clean cuts through roots unless the roots get caught between blades. Damaged roots can be severed cleanly after the tree is removed from the hole. The tree is lifted hydraulically from the hole and can be transported.

Tree spades are available in various sizes. Do not attempt to transplant trees that are larger than the spade's size limitations. Although larger trees can be removed from the ground, transplant survival is not likely, because the ball size must be proportional to the tree size. Centering the tree spade around the tree will provide a more uniform root ball and, in most cases, maximize the amount of root system captured. On sloped sites, the vehicle frame should be supported on the downhill side.

Tree spades can be used to dig planting holes, although planting in a larger, saucer-shaped hole is preferred. Holes dug in moist clay soils with tree spades tend to be glazed on the sides, which can inhibit root growth into the surrounding soil. It is best to eliminate the potential for trouble by using a hand tool, or other cultivation tool, to expand the loosened "saucer" or the planting hole and break up glazed surfaces before planting. This method loosens the soil around the ball and helps eliminate any gaps that could allow roots to dry out. If a tree spade is used to dig the hole, dig no more than six hours before planting, or keep the hole covered and moist. If tree spades are used for transplanting, take care to obtain a vertical root ball. Planting on a slope can be a problem because the ground level will be different from the tree's original site. Also, if practical, place the tree in the new location with the same orientation it had in the original site.

Figure 7.17 Mechanical tree spades are often used to transplant trees. It is important to use a tree spade that is large enough for the tree being transplanted.

Transplanting Palms

In general, palms transplant readily, and often the size that can be moved is limited only by the equipment available. In fact, planting younger, grass-stage palms may be more difficult as palms rely on the water stored in their trunk to survive the root loss associated with transplanting. Depending on the species, palms initiate new root growth from cut roots and/or from the stem, which can affect the recommended width of the root ball. In general, a 30 cm (12 in) radius is sufficient, but the recommended depth varies because many roots initiate vertically along the stem. Palms are often dug from the field, mechanically or manually. It is essential to keep the root ball moist, and a burlap wrap may be needed if replanting is not immediate. Sandy soil has a tendency to fall away from the roots, which could lead to root desiccation, so moisture management is important.

Depending on the species and site conditions, research supports the removal of some of the fronds and tying up of the remaining fronds to protect the terminal bud, reduce water loss, and accelerate establishment after planting. The sabal palm (*Sabal palmetto*) experiences root dieback to its trunk when dug. As such, survival is increased when all of the fronds are removed before transplanting. For other species that do not experience this dieback, removing fewer fronds is recommended. Tie up the remaining center fronds to protect the sole terminal bud (apical meristem) during digging, transport, and planting. If the bud dies, the palm will die.

The procedure for digging the hole for a palm is similar to other species of trees to be planted: the hole should be larger in diameter than the root ball, but it must not be too deep. Misguided practitioners may plant a line of palms at various depths to create a uniform height of the tops. This inevitably stresses the palms and can lead to health problems and often the death of the plants. Use the same backfill that came out of the hole, and be sure that any drainage problems are controlled; some species of palms tend to be susceptible to problems from poor drainage.

Figure 7.18 Large palms can be transplanted with remarkably small root "balls."

Large palms are often planted using a crane due to their height and weight. Slender stems may need to be supported with braces to prevent breaking. It is particularly important to prevent damage or death of the terminal bud, especially in single-trunked specimens, as this is the sole aboveground growing point. Once planted, palms are usually braced with boards. Make sure to protect the stem, and do not nail the braces directly into the trunk. This practice results in permanent damage to the stem.

In tropical areas, the time of year for transplanting palms is usually not critical as long as water availability is adequate. In subtropical areas, it is preferable to transplant palms in the spring and early summer when rainfall is good and soil temperatures are increasing. High-value palms are sometimes given a fungicide drench to reduce disease problems during the transplant period.

Early Care

Proper planting alone does not guarantee that a newly planted tree will survive through its first growing season and until it is established. As was

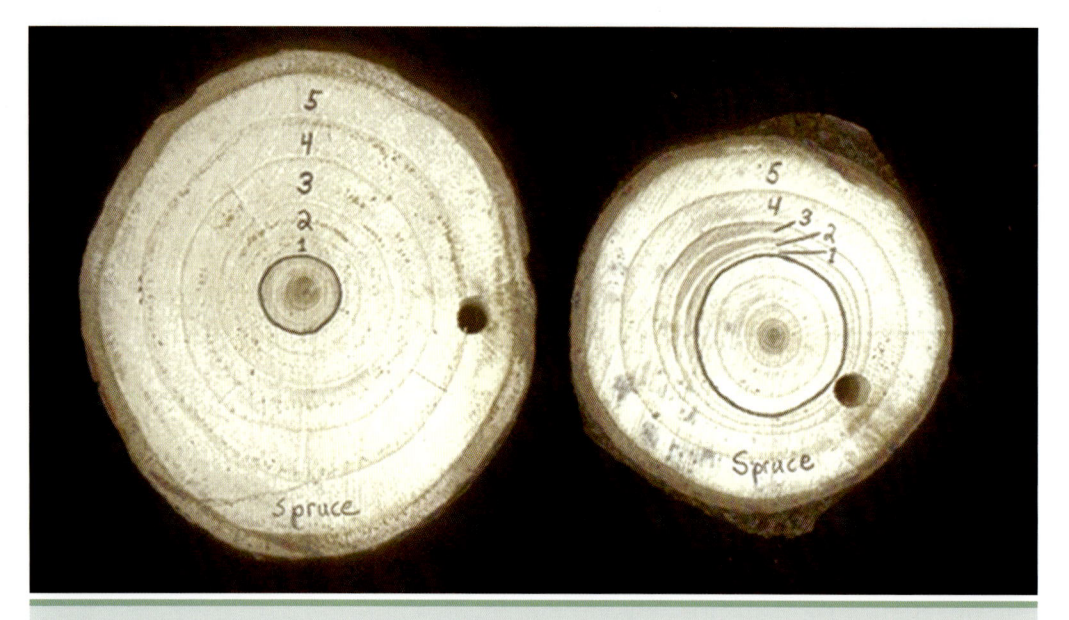

Figure 7.19 Because of recovery and reestablishment rates, it is sometimes possible for smaller-diameter transplants to recover more quickly and outgrow larger transplants.

mentioned in the previous section of this chapter, transplanting can be tremendously stressful to a tree. In an attempt to compensate for root loss and other deficiencies, several establishment practices are commonly used by the arboriculture and landscaping industries. By employing some or all of these early-care techniques after planting, the likelihood that a tree will survive and become fully established is much greater.

Watering and Irrigation

Transplanting is a major operation from which most trees recover slowly. Except for container-grown trees, a large portion of the root system is lost in digging, and a tree must reestablish sufficient roots to sustain itself. Even container-grown trees must establish a normal, spreading root system. A newly planted tree's ability to absorb and transport water and minerals is greatly reduced. The tree experiences postplanting water stress referred to as **transplant shock**.

The rate of recovery and reestablishment after planting or transplanting varies with the species, planting season, soil, site conditions, moisture availability, climate, and tree size and vigor. The general rule of thumb for reestablishment, in temperate climates, is one year for every 2.5 cm (1 in) of tree caliper (diameter). Therefore, it is sometimes possible for smaller-diameter transplants to recover more quickly and outgrow larger transplants. In tropical climates, trees establish faster.

Proper watering is the key to survival of newly planted trees. Water requirements will vary with species, plant size, climate, and soil type. Trees should be watered based on need, not by a calendar or clock. To assess need, check the root ball, backfill, and surrounding site soil manually, and water when the soil dries out. At that time, apply sufficient water to moisten the soil to 0.3 m (1 ft) deep. A slow, gentle soaking of the root ball is the preferred way to water. It is very important to monitor the soil moisture level of the root ball itself. Often, the soil or growing substrate from the nursery will dry out more quickly than the surrounding soil. Also, the ball may actually repel water, especially if it is wrapped in burlap. Many container substrates can be difficult to rewet if left to dry out significantly between watering. If the soil around the root ball dries out too much, root growth will be halted, and the establishment period will be lengthened. If it stays dry for too long, the tree will not survive.

Adequate water is critical for tree survival and new root growth. On the other hand, excess water accumulation in the planting hole (which displaces soil oxygen) is just as detrimental. When monitoring the soil moisture level, always check the root ball and the surrounding soil. Watering must be

Figure 7.20 Adequate watering is the key to survival of newly planted trees. Irrigation bags, one type of portable watering device, are a possible solution to meeting water needs during establishment.

appropriate for tree species, soil type, and drainage. Even trees planted in areas where rainfall is plentiful may need supplemental water during the establishment phase. Ensure there is a plan for water management.

Fertilizing

Given the moisture stress associated with transplanting, fertilization is generally not recommended at the time of planting. Excessive fertilizer salts in the root zone can be damaging and can lead to increased water stress. If fertilizer is used in the first growing season, application of a slow-release nitrogen fertilizer is recommended.

Mulching

Mulches are materials placed over the soil surface to maintain moisture, reduce competition from weeds and turfgrass, and improve soil structure. Use of organic mulches is one of the most beneficial practices for the health of a tree. Mulch can also give landscapes a desired appearance. However, if mulch is too deep or if the wrong material is used, it may cause significant harm to trees, so proper application is important.

Although mulches are available commercially in many forms, they can be classified into two categories: **organic mulch** and **inorganic mulch**. Organic mulches are usually derived from plants or are byproducts of industry, agriculture, or municipal waste. Some of the most common types of organic mulches are shredded bark, hardwood chips, and pine needles.

The plant material present in organic mulches releases essential nutrients and improves soil structure as it decomposes. Although this decomposition is beneficial for soil structure and for the plant, it also creates the need for periodic replenishment. The rate of decomposition depends on particle size, material being used, and the regional climate. Straw, leaves, and lawn clippings will decompose faster than bark or wood chips. In warmer climates, the rate of decomposition will be much faster than it is in colder climates.

Inorganic mulches include stone, lava rock, pulverized rubber, and other materials. These mulches do not decompose and do not need to be replenished as often as organic mulches. This characteristic makes inorganic mulches a popular choice for many contractors because minimal maintenance is required after installation. However, inorganic mulches do not improve soil structure, add organic materials, or provide nutrients. For these reasons, most horticulturists and arborists prefer and recommend organic mulches.

Cover the area around the tree with about 5 to 10 cm (2 to 4 in) of organic mulch. Do not place mulch directly against the stem of the tree. Direct contact may lead to bacterial or fungal infections such as crown rot, adventitious roots, rodent feeding, and insect pests. The broader the mulched area, the more effective it will be at reducing competition from other plants and retaining moisture. The actual size of the area to be mulched depends on the size of the tree. For a 2.5 to 5 cm (1 to 2 in) caliper tree, a circle of mulch at least 1.8 m (6 ft) in diameter is recommended. Do not place black plastic under the mulch; it restricts water movement and oxygen availability to the roots.

A broad mulch circle is desirable; a deep one is not. Excessive amounts of mulch often accumulate over time, largely due to repeated applications of fresh material to restore the color or appearance of an established bed or mulch circle. Although it may be necessary to replace decomposed material, make sure there is not a buildup of organic mulch. "Volcanoes" of mulch around trees can reduce oxygen and water availability to the roots and may lead to decay. Overmulching may also limit nitrogen availability and affect soil pH. Mulch piled high against the trunks of young trees may provide shelter and concealment for rodents that may feed on the lower stems.

Tree Stabilization

Staking or **guying** of newly planted trees is not always necessary. In fact, staking may even have detrimental effects on tree development. When compared to trees that have not been staked or guyed, young trees that have been supported for prolonged periods produce less trunk taper, develop smaller root systems, and are more subject to breaking or tipping after stakes are removed. Stabilized trees may become injured or girdled by the staking or guying materials. However, stakes are frequently included in tree planting specifications and may be

Figure 7.22 Ties should be broad, smooth, and somewhat elastic.

Figure 7.23 All too often, support materials are left in place too long and girdle the tree. Support stakes or guy wires should generally be removed after one growing season.

Figure 7.21 Stabilize the tree as low along the trunk as is practical while still providing the necessary support.

expected by clients. Clear communication can help to set expectations around staking needs.

Some trees, however, cannot stand upright without support. Bare-root trees, trees grown in small containers with a loose potting substrate, containerized trees with large crowns, balled-and-burlapped trees with wire removed, and large conifers may require support while they establish, especially in windy sites. In some instances, stabilization may be recommended to reduce movement of the root ball and subsequent damage to the fine, absorbing roots. In urban sites, stakes are sometimes used to protect young trees from lawn care damage and to reduce vandalism.

Figure 7.24 Root ball anchoring allows for increased trunk movement in wind compared to staking, while still effectively stabilizing the root ball. This method eliminates the risk of trunk damage from staking or guying materials.

Normally, one to three stakes are used to stabilize a tree. If two stakes are used, a single strip of flexible attachment material secured at the top of each stake will be sufficient for support. If possible, avoid driving the stake through the root ball, because it can damage the roots. Stabilize the tree as low along the trunk as practical while still providing the necessary support. If the tree is secured too rigidly, it will develop a less sturdy root system and a smaller-caliper trunk with insufficient taper and may become girdled or break above the point of attachment. The material used to attach the tree to the stake should be broad, smooth, and somewhat elastic. The use of wire—typically passed through a hose in an effort to limit damage—is not advised, because the wire can still girdle the tree's stem.

Trees larger than 10 cm (4 in) in diameter are often stabilized with guy wires. Trees are generally guyed with three or four wires (covered with plastic sleeves) that are secured to the ground using stakes or another anchoring device. Placing guy wires low on the trunk allows the top of a tree to move freely and develop taper while maintaining root ball stability. Attaching the guys above two-thirds the height of the tree is not advised, because the higher the support is, the more upright and inflexible the top of the trunk will be.

Guying materials in contact with the tree trunk should be wide, smooth, nonabrasive, flexible, and, if possible, photodegradable. Wires that are attached to the ground anchors are subsequently attached to the guying materials rather than to the stem of the tree. Any modification to reduce girdling and friction will minimize damage to the tree. On large trees, guy wires can be attached to the tree with eye screws. This method wounds the tree but causes minimal damage to a vigorous tree and eliminates the chance of girdling by guy wires. The size of the eye screw or eye bolt used should be proportional to the size of the tree.

Another option for tree stabilization is the use of root ball anchors.

Check stabilization systems several times during the first year to be sure that they have the right tension and are not injuring the tree. Support stakes or guy wires should generally be removed after one growing season. If support systems are left in place for more than two years, the tree's ability to stand alone may be reduced and the chances of girdling injury increased.

Tree Wraps and Tree Guards

Many early references recommend wrapping the trunks of newly planted trees to protect against temperature extremes, sunscald, boring insects, and drying. Recent research demonstrates that temperature differentials at the bark are greater with **tree wrap** than without. Tree wrap also tends to hold moisture on the bark and can lead to fungal problems. Insects often burrow between the bark and the tree wrap; these insects can be more problematic with the wrap than without it. If tree wrap is used, it should be light in color and allow for air and water to pass through freely. Although paper wrapping is often used, better materials (some photodegradable) are available and recommended if wrapping is necessary.

Sometimes plastic or metal mesh **tree guards** are installed around the trunks of newly planted trees. These guards have the advantage of minimizing damage caused by animal feeding or deer rubbing and from mowers and trimmers. Most guards are loose fitting and may have slots or holes for ventilation. It is important to remove or replace trunk guards as they are outgrown.

Figure 7.25 Once planted, palms are usually braced with lumber. Make sure to protect the stem, and do not nail the braces directly into the trunk.

Root ball anchoring allows for increased trunk movement in wind compared to staking while still effectively stabilizing the root ball. This method eliminates the risk of trunk damage from staking or guying materials. Root ball anchoring is suitable for soil-ball or container-grown trees. Follow the manufacturer's instructions for anchor installation and to avoid damaging roots.

Pruning

Pruning following planting should be limited to corrective pruning. It was once thought that up to one-third of the crown of a newly planted tree should be thinned to compensate for the loss of roots. This practice was known as compensatory pruning. Research shows that trees will grow and establish

Chapter 7: Installation and Establishment

more rapidly if pruning is limited to removing broken or damaged branches.

Although the need for pruning at the time of planting is usually minimal, trees should be pruned while still young to establish good structure. Training young trees can greatly reduce the need for major structural pruning later. Many defects such as weak branch attachments can be corrected early, creating trees that are structurally stronger than untrained trees. Training young trees at planting may be appropriate in instances where the trees are not likely to receive structural pruning in the foreseeable future.

The leader is the central stem of the tree. Generally, there should be only one leader. If competing leaders exist, the strongest, most vertical stem should be maintained while the others are removed or reduced. Follow-up pruning will be needed in subsequent years to remove or reduce newly developed competing stems and to ensure that the new leader is in a vertical position.

Figure 7.26 Metal guards are often used to protect young trees in public places. If left in place too long, they can distort or restrict growth as the tree gets larger.

Figure 7.27 Pruning at the time of planting should be limited to removal of damaged branches and improving structure. Training young trees, which ideally takes place over multiple years, can greatly reduce the need for drastic structural pruning later.

Planting Specifications

Planting specifications are detailed statements of procedures and standards for planting trees or shrubs in a landscape. Arborists often work with contractors from other green-industry professions and must write planting specifications before they bid out work. Other times, arborists must work according to these specifications to fulfill a bid. The amount of detail and breadth of these specifications can vary greatly. When writing a contract, it is important that all of the key terms be clearly defined. What constitutes a "healthy" or "vigorous" plant may be different for each party involved in a bid. Timelines should be established for each step in the installation process and should include deadlines for any follow-up maintenance or inspections. The length and terms of warranties should also be defined and documented within the planting specifications.

Local and national standards provide a system for sizing and describing trees and shrubs. Standards serve as a tool for communication and to ensure quality. Planting specifications should be drafted using the terminology found in Best Management Practices: *Tree Planting* and, in the United States, the ANSI A300 transplanting standard.

Chapter 7: Installation and Establishment

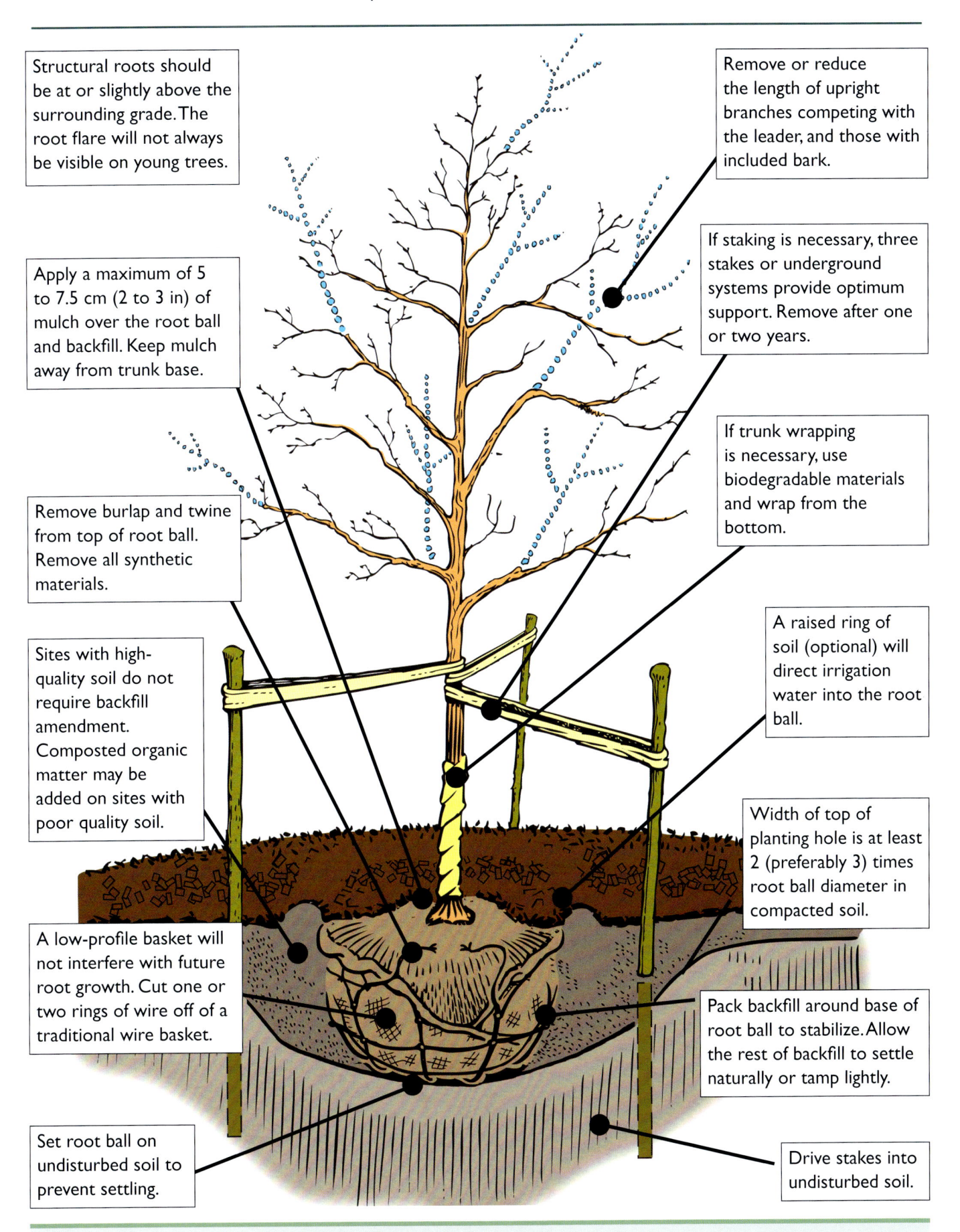

Structural roots should be at or slightly above the surrounding grade. The root flare will not always be visible on young trees.

Remove or reduce the length of upright branches competing with the leader, and those with included bark.

If staking is necessary, three stakes or underground systems provide optimum support. Remove after one or two years.

Apply a maximum of 5 to 7.5 cm (2 to 3 in) of mulch over the root ball and backfill. Keep mulch away from trunk base.

If trunk wrapping is necessary, use biodegradable materials and wrap from the bottom.

Remove burlap and twine from top of root ball. Remove all synthetic materials.

A raised ring of soil (optional) will direct irrigation water into the root ball.

Sites with high-quality soil do not require backfill amendment. Composted organic matter may be added on sites with poor quality soil.

Width of top of planting hole is at least 2 (preferably 3) times root ball diameter in compacted soil.

A low-profile basket will not interfere with future root growth. Cut one or two rings of wire off of a traditional wire basket.

Pack backfill around base of root ball to stabilize. Allow the rest of backfill to settle naturally or tamp lightly.

Set root ball on undisturbed soil to prevent settling.

Drive stakes into undisturbed soil.

Figure 7.28 Balled & burlapped tree planting diagram.

CHAPTER 7 WORKBOOK

1. Trees are generally available from the nursery in one of four forms:

    a.

    b.

    c.

    d.

2. Bare-root trees are normally planted when _____ , before buds begin to grow.

3. _____ roots can become a problem because they can constrict the vascular system in the trunk or in other roots.

4. Planting holes should be dug _____ to _____ times the width of the root ball at the surface, with the sides sloping down to the diameter at the base of the root ball.

5. Trees that are dug in the nursery are often wrapped with _____ to help keep the root ball intact and to reduce exposure of the roots to air.

6. The planting hole should never be _____ than the root ball.

7. True/False—Planting a tree too deeply can stress, drown, or suffocate roots; it may also enable soilborne pathogens to enter the trunk.

8. True/False—Research has shown that soil amendments used in the planting hole generally do not promote establishment and growth.

9. True/False—Digging a tree for transplanting can remove as much as 90 percent of the absorbing roots.

10. In temperate climates, the best times to transplant most trees are _____ and _____ .

11. True/False—When a tree is dug for transplanting, the size of the root ball is usually based on tree caliper.

12. True/False—If trees have wire baskets to help maintain the integrity of the root ball, these baskets should never be removed, nor should the top portion be cut off at planting.

13. True/False—Adding fill to the bottom of the hole can cause the root ball to settle in the days and weeks after planting, resulting in a final depth that is deeper than originally planted.

14. Predigging around a tree to create a more densely rooted ball is called _____ _____.

15. True/False—Staking of newly planted trees is not always necessary.

16. Name three adverse effects of staking or guying trees.

    a.

    b.

    c.

17. True/False—The material used to attach the tree to the stake should be broad, smooth, and flexible.

18. Warm soil temperatures and adequate soil _____ are the optimal conditions for new root growth.

19. True/False—In general, palms are more difficult to transplant than most other kinds of trees.

20. When transplanting palms, the fronds are often tied up to protect the solitary _____ _____.

21. Transplant shock is mainly due to _____ stress from the greatly reduced root system.

22. If fertilizer is applied at planting, it should be a _____ - _____ type to avoid excess salt buildup in the root zone.

23. True/False—There is no advantage to pruning one-third of the tree crown at the time of planting (i.e., compensatory pruning).

24. The most important maintenance factor in the survival of a newly planted tree is proper _____.

25. True/False—Tree roots may suffocate if the tree receives too much water after planting.

Challenge Questions

1. What are the advantages and disadvantages to planting each of the following ways: bare root, balled and burlapped, and container-grown?

2. Why is the use of soil amendments in the backfill of planting holes not generally recommended?

3. What are some of the special considerations for transplanting palms?

4. What are some of the physiological effects of staking on the early growth and establishment of a tree?

5. Pruning of one-third of a tree's crown to compensate for root loss when transplanting has not proven to be beneficial. Why might this be the case?

Sample Test Questions

1. Staking or guying when planting a tree is
    a. done only for bare-root trees
    b. not necessary for trees greater than 15 cm (6 in) in diameter
    c. not always required or necessary
    d. for promoting a larger and stronger root system and better trunk taper

2. When mulching a newly planted tree, care should be taken to ensure that
    a. the mulch extends all the way to the base of the tree trunk
    b. the depth is not excessive to the point where soil aeration is negatively impacted
    c. the mulch is sourced only from quality, inorganic sources
    d. an underlying layer of plastic is laid first to prevent weed growth

3. Planting a row of palms at various depths to create a row of uniform height
    a. will not be harmful as long as the soil is sandy and well drained
    b. can lead to death of the deeply planted trees
    c. is acceptable due to the deep rooting of palms
    d. is recommended to achieve deep rooting in sandy soils

4. When planting a container-grown tree,
    a. slice or shave off the outer roots if they are circling or matted
    b. place soft fill in the bottom of the planting hole to encourage taproot growth
    c. backfill the hole with a soilless growth medium to encourage root growth
    d. plant the tree slightly deeper in the planting hole than ground level

5. The most important reason to prune a tree when transplanting is to
    a. compensate for root loss
    b. invigorate the tree
    c. reduce growth at the tips
    d. remove damaged branches and improve structure

Recommended Resources

(See Recommended Resources in back of book for detailed information.)

Best Management Practices: *Tree Planting* (Watson 2014)

The Practical Science of Planting Trees (Watson and Himelick 2013)

ANSI A300 (Part 6) *Planting and Transplanting* (American National Standards Institute 2018)

Introduction to Arboriculture: Planting online course (ISA)

Introduction to Arboriculture: Early Care online course (ISA)

CHAPTER 8

PRUNING

▼ OBJECTIVES

- Describe the objectives for pruning trees, shrubs, and palms.

- Compare and contrast the various pruning systems and how they differ in use.

- Identify the structures associated with branch attachments, including the function of the branch protection zone.

- Recognize the types of pruning cuts, and explain how to use them to meet objectives.

- Discuss the effects of pruning on trees, including response and recovery and how these are affected by timing and the amount of pruning.

- Contrast the pruning of palms with the pruning of trees and shrubs.

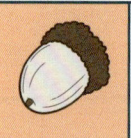

KEY TERMS

ANSI A300
antigibberellin
branch bark ridge
branch collar
branch protection zone
branch removal cut
branch union
clearance pruning
CODIT
codominant stems
compartmentalization
directional pruning
espalier
frond

heading cut
included bark
inflorescence
internodal
lateral
leader
lion tailing
permanent branches
plant growth regulator
pollarding
pruning objectives
pruning system
reduction
reduction cut

restoration
scaffold branches
shearing
structural pruning
subordinate
temporary branches
three-cut method
topiary
topping
watersprout
wound dressing
woundwood

Introduction

Pruning is a common maintenance procedure that is often necessary to reduce risk; provide clearance; improve tree structure, health, and appearance; and meet client needs. Arborists should understand the biology of trees in order to promote tree health and structure during the pruning process.

Pruning can be beneficial or detrimental. Effective, science-based pruning helps trees become sustainable, functional elements in the landscape and provide many benefits; poor pruning practices can shorten the life of a tree and can increase the likelihood of structural failure.

Arborists must understand how the tree will respond and recover from pruning cuts. Each cut may affect the tree's future growth and development, and improper pruning can cause damage that remains for the life of the tree. Removing live branches from a tree removes stored resources in the wood. Removing foliage reduces the tree's photosynthetic capacity and may reduce overall growth. However, a tree may respond to pruning with vigorous new growth.

Many countries now have or are developing standards for pruning. In the United States, specifications for pruning should be written in accordance with the American National Standards Institute (**ANSI**) **A300** standards. ISA's Best Management Practices: *Pruning* provides guidelines for pruning trees and other woody plants.

Figure 8.1 Several consequences can occur when pruning is not performed, including: (1) an accumulation of broken branches; (2) dead branches; (3) branches that interfere with site function; (4) weak, codominant stems; (5) defects such as included bark; and (6) in some species, development of suckers.

Pruning Systems

An important consideration prior to pruning trees or shrubs is the application of a particular **pruning system**. The pruning system is a technique or procedure that is applied to develop the desired long-term form of the plant. In choosing a pruning system, the design of the landscape and the client's goals need to be considered. The landscape design of the site can be formal, informal, or a combination; identifying the intent of the design dictates pruning practice. Pruning systems include natural, topiary, pollard, espalier, pleach, fruit production, and bonsai. Arborists most often use the natural pruning system.

Natural

The natural pruning system is considered an informal style that improves the natural form and shape. Site conditions, such as proximity to a building, may limit the natural appearance of a tree using this system, as compared to a tree growing in an open area such as a park. Pruning within the natural system may involve branch removal and/or reduction to avoid conflict with nearby infrastructure, encourage strong architecture, allow desirable views, and provide clearance.

Specialized Pruning Systems

Pruning may be utilized to create special effects. Two specialized pruning systems that are fairly common with trees are espalier and pollard. **Espalier** is a combination of pruning and training branches that are oriented in one plane, usually supported on a wall, fence, or trellis. The pattern can be simple or complex, formal or informal. This technique has been used for centuries on fruit trees, often where space was extremely limited. Another pruning system that is usually limited to shrubs and small, ornamental trees is **topiary**, in which plants are pruned into specific shapes.

Pruning Systems

Natural
The natural pruning system is an informal style used to retain and promote the characteristic form of the species or cultivar in its current location.

Topiary
Topiary is a formal pruning system in which shrubs, vines, or trees are pruned into a specified shape by shearing and/or pruning.

Pollard
Pollarding is the regular removal of young shoots by pruning them back to a common point on the scaffold, known as a pollard head. Pollarding is used to maintain trees in a certain size range, often with a semiformal appearance.

Espalier
Espalier is a formal system for developing plants to grow in a two-dimensional plane, such as along a wall or a fence.

Pleach
Pleaching is a system that involves interweaving branches horizontally to form an arbor, wall, allée, or arching tunnel.

Fruit Production
There are numerous subsystems used in the commercial production of tree fruit. These systems are specific to the tree species, but the intent of all of them is to maximize fruit production.

Bonsai
Bonsai is a general term for a system that maintains container-grown trees at a small size.

Figure 8.2 Pruning systems may be utilized to create special effects: A. topiary; B. espalier; C. pollarding; D. pleaching.

Pollarding is a training system that involves severe heading cuts the first year followed by sprout removal annually or every few years to maintain a desired size or appearance. The pollarding process should be started when the tree is young. It is not an effective technique on every species of tree. To begin the pollarding process, make **internodal** cuts at specific locations. Do not make any additional internodal cuts after making the initial cuts. Sprouts grow from beneath the pruning cuts, and pollard heads, called knobs or knuckles, will begin to develop at these points. Annually or every few years, remove the sprouts that grow from these pollard heads, usually during the dormant season, taking care not to cut into or below the knobs. Knobs are the key differentiating factor between pollarding and topping. If the knobs are damaged or removed in subsequent pruning, the branches will react as they would on a topped tree.

Pruning Objectives

No plant should be pruned without a clearly defined **pruning objective**. The objective for each tree is determined by the desired goals of the pruning and the design of the landscape. Establishment of objectives must consider the pruning system, plant health, growth habit, plant size, structure, species characteristics, expected growth response, client expectations, location, and site usage. A specification for pruning should be developed that states the appropriate objective(s).

Common objectives for pruning include

- improving structure
- reducing risk
- providing clearance
- managing crown size

Figure 8.3 Examples of pruning objectives include removal of hazards and increasing clearance for pedestrians and vehicles.

- reducing crown density
- restoring structure
- rejuvenating shrubs

These objectives serve as examples and can be expanded or shortened to account for site conditions and expectations. The decision of which stems and branches to remove or reduce is as important as making correct pruning cuts. Even if the physical cutting of each branch is technically correct, pruning live branches will still create wounds and decrease overall photosynthesis. If the wrong branches, too few branches, or too many branches are removed, pruning will not achieve the defined objectives.

Improving Structure

Strong tree and branch structure is important for reducing the likelihood of failure. If young trees are trained—pruned to promote good structure—as they grow, they will likely be more stable and sustainable in the landscape. Training young trees in this way is known as **structural pruning**. Among other things, structural pruning usually includes the selective removal of dead, diseased, broken, or weakly attached branches. It also includes the identification and removal of structural defects (such as weak **branch unions**), creating a dominant leader and appropriate branch spacing. When done correctly, structural pruning should decrease a tree's potential for failure as it grows larger and should decrease the need for future maintenance.

With single-stemmed trees, a strong central leader is usually desired. If two branches develop at the tip of the same stem, they may form **codominant stems**. Each codominant stem is a direct extension of the trunk. Branch unions and codominant stems that have very narrow angles of attachment can sometimes have a higher likelihood of failure at the point of attachment, especially if included bark has developed within that union. **Included bark** is bark that becomes enclosed inside the attachment as the branches grow and develop. This weakens the branch attachment, making it more likely to split apart.

It should be noted that not all codominant branch attachments have the same likelihood of failure. The presence of included bark and decay are of primary concern. However, the formation of the attachment plays a role as well. A codominant configuration can be categorized as V-shaped or U-shaped. V-shaped configurations, with bark of one branch pressed against bark of another, tend to be weaker than U-shaped configurations and can result in bark inclusions. Also, trees sometimes grow wood in such a way to strengthen the unions,

Figure 8.4 On most trees, there should be only one leader, which is usually the strongest, most vertical stem.

Figure 8.5 Trees should be structurally pruned when young to develop well-spaced branches. Clustered branches are often weakly attached.

Figure 8.6 Branch unions with very narrow angles of attachment and codominant stems can sometimes have a higher likelihood of failure at the point of attachment, especially if included bark has developed within that union. Included bark is bark that becomes enclosed between two or more closely growing stems.

which, over time, can significantly reduce the likelihood of failure.

Research has shown that trees sometimes form "natural braces" in the form of crossing branches, intertwined branches, and even natural grafts that can reinforce V-shaped unions. If arborists remove these reinforcements while the tree is young, natural branch movement in the wind will encourage the development of wood to strengthen the union. If, however, a tree matures with natural braces in place,

Training Young Trees

The process of training young trees can be described in five steps.

1. Remove broken, dead, dying, diseased, or damaged branches.
2. Select and establish a desired structure, most often a dominant **leader**. On most young trees, there should be only one leader, which is usually the strongest, most vertical stem. To promote and maintain a single leader, competing stems should be **subordinated** (reduced to become laterals) or removed.
3. Select and establish the lowest **permanent branches**. The height of these branches is determined by the location and intended function of the tree. For example, the lowest permanent branch on a tree located between a street and a sidewalk may need to be higher on its street side than on its sidewalk side because vehicle traffic requires a higher level of clearance than pedestrian traffic. Lower branches can be left to develop on appropriate species and in appropriate locations, such as an arboretum. The diameter of the lowest permanent branch selected should be less than half the diameter of the trunk at the point of attachment.
4. Select and establish **scaffold branches**. These branches should be selected for good attachment, appropriate size, and desired spacing in relation to other branches. Scaffold branches should be well spaced, vertically and radially, on the trunk. As a guideline, vertical spacing should be at least 0.5 m (18 in) for trees that will be large at maturity, and about 0.3 m (12 in) for smaller trees.
5. Select and subordinate **temporary branches** below the lowest permanent branch and among the scaffold branches. These branches should be retained temporarily because they provide carbohydrates through photosynthesis, contribute to trunk taper development, and provide shade to the young trunk tissues. The smaller temporary branches can be left intact; larger ones may need to be subordinated, unless subsequent pruning will be frequent. Temporary branches are intended to be removed in later years. This training process should be spread out over many years, if practical, as the tree develops and matures.

movement and the development of strengthening wood at the union will be minimal. Removal of well-developed natural braces at this stage could lead to failure of the union because it has lost its support.

Recognizing potentially problematic branch unions is important when making pruning decisions. Removing or subordinating one codominant stem while the tree is young can improve the tree's structural stability in the long term.

Reducing the Risk of Failure

Pruning to reduce the likelihood for branch or whole tree failure is a primary objective for many large trees in urban areas. Branch, stem, and whole tree failures are influenced by many factors including species, defects, site conditions, loads, size, and response growth (see Chapter 12, Tree Risk Assessment and Management). The best way to reduce likelihood of failure in the future is to perform structural pruning when trees are young and when needed throughout the life of the tree. This approach will develop strong crown architecture. As trees grow larger and crown architecture is established, risk mitigation has a different emphasis. In some cases, specific branches must be removed to mitigate risk. To reduce the likelihood of trunk and root

Chapter 8: Pruning

Figure 8.7 Only qualified line-clearance tree workers or qualified line-clearance trainees should engage in line-clearance work.

Figure 8.8 New growth following topping tends to be clustered at the stub tips and is weakly attached.

failure, it may be necessary to reduce the height and/or spread of the crown. When pruning trees that produce large fruit, removal of the fruit or developing **inflorescence** (flowers) may be a mitigation option to reduce risk associated with the fruits.

Providing Clearance

The objective of **clearance pruning** is to reduce interference between current and future branches and nearby people, activities, infrastructure, buildings, traffic, lines of sight, desired views, or adjacent plants. Clearance pruning plays an important role in ensuring safe and reliable electrical utility services and, in some cases, regulatory compliance.

Directional pruning is the preferred way to provide clearance, by directing branch growth, often when applied before the tree grows into an area of interference. Branches that currently interfere (or will do so in the future) should be removed or reduced. When reducing these branches, make the reduction cut (described later in this chapter) to a **lateral** branch that is not growing in the direction of the clearance zone when possible. This directs growth away from the specified clearance area and promotes compatible branch structure. Clearance pruning often must be done on a regular cycle because new growth eventually will fill the void created by pruning.

Trees in urban and residential settings may need to have lower branches removed. Another objective is crown elevation, which focuses on removing the lower branches of a tree in order to provide clearance for buildings, signs, vehicles, pedestrians, and sight lines. Avoid excessive removal of lower branches so that development of trunk taper is not affected and structural stability is maintained. Similarly, when performing vista pruning, which is pruning to enhance a specific view, be careful not to jeopardize the health or stability of a tree.

branches. Retaining healthy, vigorous branches will help to maintain the structural integrity and form of the tree and may delay the time until it will need to be pruned again.

Topping trees as a means to reduce height and/or spread is considered an unacceptable practice as it often leads to decay, disease, development of branches with poor attachment to parent branches, and greater likelihood of branch failure. Topping temporarily reduces tree size by cutting live branches and leaders to stubs, without regard to long-term tree health or structural integrity.

Reducing Crown Density

Reducing crown density (previously called thinning) is sometimes performed to increase wind or light penetration, achieve aesthetic objectives, or promote interior foliage development. This objective is accomplished by selective removal (in young trees) or reduction (in mature trees) of the longest and largest branches. The dominant leader is rarely reduced or removed.

In most circumstances, however, the removal of live interior and lower lateral branches should be avoided. Removing too much foliage can have adverse effects on the tree, resulting in production of **watersprouts** on interior branches, which is often

Figure 8.9 If the height of a tree must be reduced, cuts should be made back to strong laterals or to the parent branches.

Size Management

Sometimes the crown of a tree must be reduced in height or spread. Crown **reduction** is used to reduce the size of a tree. Size management is accomplished with reduction cuts or, rarely, heading cuts (both described later in this chapter). The size of cuts should be as small as possible to reduce the likelihood of decay. The number and location of remaining branches should be considered, including the possibility of sunscald on newly exposed

Figure 8.10 Large internodal cuts often fail to close and decay may develop within the stub.

148

Figure 8.11 Lion tailing is an effect caused by removing an excessive number of inner branches and foliage. This displaces weight to the ends of the branches and may result in sunburned bark tissue, watersprout development, reduced branch taper, weakened branch structure, and breakage.

Figure 8.12 If a tree has been topped previously and has sprouted vigorously or has sustained storm damage, crown restoration can improve its structure and appearance.

a response to overpruning. Removal of a majority of lower or interior branches, leaving foliage clustered at branch ends (**lion tailing**), is not an acceptable practice.

Crown Restoration

If a tree has been topped or has sustained storm damage and is now responding by growing an overabundance of weakly attached sprouts, crown **restoration** can improve its structure and appearance. Pruning to restore the crown structure consists of the selective removal of some watersprouts, stubs, and dead branches. Choose one to three well-spaced sprouts from the ends of damaged branches to become permanent branches and to re-form the crown with a more natural appearance. Selected vigorous sprouts may need to be subordinated to control length growth and ensure adequate attachment for the size of the sprout. Managing sprouts growing further down along the damaged branch is also important. Most of them should be retained early during the restoration process, with selective reduction and removal occurring as necessary over subsequent pruning cycles.

In some cases, heading cuts are necessary to initiate new shoot development from a damaged branch as an alternative to removing the branch altogether. Restoration pruning requires patience and pruning over a number of years.

Managing Wildlife Habitat

Pruning can have a positive or negative impact on wildlife that depend on urban trees. Removal of branches can eliminate shelter, food, and nesting sites. For example, many features that arborists consider structural defects, such as cavities, can be used as nesting sites by birds and mammals. Dead, dying, and decaying branches also provide food or habitat

> ## Why Not to Top
>
> 1. **Starvation**: Topping removes a large portion of the crown, reducing the tree's ability to manufacture energy resources to recover from the pruning cuts. It also removes reserve carbohydrates stored in branches.
> 2. **Insects, diseases, and decay**: The large stubs of a topped tree have a difficult time forming woundwood. The terminal location of these cuts, as well as their large diameter, prevent the tree's chemically based natural defense system from doing its job. Large pruning wounds are vulnerable to insect invasion and the spores of wood-decaying organisms.
> 3. **Weak branch attachments**: A new branch that sprouts after a larger branch is shortened is more weakly attached than a branch that develops normally.
> 4. **Rapid new growth**: The objective of topping is size management, usually reduction in height, but this reduction is only temporary. The resulting sprouts (often called watersprouts) are far more numerous than normal new growth, and they can elongate so rapidly that the tree returns to its original height in a very short time—and with a far denser crown.
> 5. **Tree death**: Some trees are more tolerant of topping than others, surviving for a few years. Many trees do not have the ability or resources to recover from this extreme pruning, leading to their decline and eventual death.
> 6. **Ugliness**: A topped tree is a disfigured tree. Even with its regrowth, it never regains the grace and character of its natural form.
>
> Adapted from the Arbor Day Foundation

for wildlife. Pruning activities can also be loud and disruptive to wildlife and can cause them stress and even to abandon their nests. Arborists should take extra precautions when pruning in sensitive habitats and during breeding seasons, when they are mostly likely to encounter wildlife easily affected by disturbance.

Pruning can also be used as a means to maintain and promote wildlife habitat. Practices such as reducing or removing branches can increase the time that a tree can remain in an urban landscape. In cases where the management objective is to promote habitat for wildlife species that use cavities or dead and decaying trees, arborists may intentionally create cavities or install nesting sites.

In addition, arborists should be aware that removing or disturbing active nests, especially of protected species, is illegal in many locations.

Tree Biology, Architecture, and Pruning

Trees are dynamic biological organisms that respond to external actions such as pruning, storm damage, and pests. Arborists need to understand basic biological concepts to be able to predict and interpret how trees will respond to different types of pruning cuts.

Pruning live branches reduces a tree's capacity to photosynthesize and manufacture carbohydrates, at least for a short period. Pruning live branches will cause some dysfunction to sapwood near the pruning cut. Pruning also creates wounds that the tree must expend energy to close and defend. Routine live branch removal does not necessarily improve tree health, and some practices, such as topping, can have adverse effects on both tree health and structure. In addition, excessive branch removal from the interior of the crown (lion tailing) can actually increase the likelihood of branch failure. The costs mentioned above and benefits of pruning, such as improved branch architecture and reduced failure potential, should be assessed when developing pruning objectives and when deciding how much pruning will occur (sometimes referred to as pruning dose).

Compartmentalization of decay (damage) in trees, or **CODIT**, is a model describing the defense strategies that limit the spread of decay based on physical and chemical properties of the wood in living trees. Wound closure occurs when **woundwood** develops around the edges of a cut or wound, eventually covering it with new tissue. Compartmentalization relies on the tree's genetic capacity and available resources to close and seal wounds. In general, the larger the cut and the slower the growth rate, the longer the wound will take to close, potentially affecting decay initiation and movement. Large wounds are frequently associated with decay; smaller wounds will typically close more readily than larger wounds, usually resulting in little or no decay.

Branches are connected to other branches or stems with clearly defined (and usually visible) forms of tissue connections. A supportive growth pattern develops at the branch union where the diameter increases for both the branch and stem each growing season. Lateral meristem around the circumference of each branch and stem continue to generate tissues as they increase in diameter. A branch with good structural development usually has its base deeply embedded in the trunk or stem, forming a knot. This ensures greater strength as the branch collar at the base of the attachment overlaps wood tissue surrounding the branch, creating a strong union and stable attachment.

The size of a branch in relation to the trunk is more important for branch attachment strength than is the attachment angle. The bending strength of a branch union can be affected by the ratio of the diameter of a branch to the stem or the branch from which it emerges. When branches remain small relative to the trunk diameter, a swollen **branch collar** often develops around the base of where the branch connects to the stem (the branch union). A branch collar is formed by overlapping and intermingled branch and trunk wood and is often representative of a strong union.

Inside the branch union on many trees, specialized wood, sometimes termed axillary wood, is formed central to the connection between the trunk and the branch. This wood

Figure 8.13 When branches remain small relative to the trunk diameter, a swollen branch collar often develops around the base of where the branch connects to the stem. Note the location of the branch bark ridge in the center of this branch union.

Figure 8.14 Inside the branch union on many trees, specialized wood, termed axillary wood, is formed central to the connection between the trunk and the branch. This wood is typically much denser than other wood and exhibits twisted and intermingled wood grain, which serves to strengthen the connection.

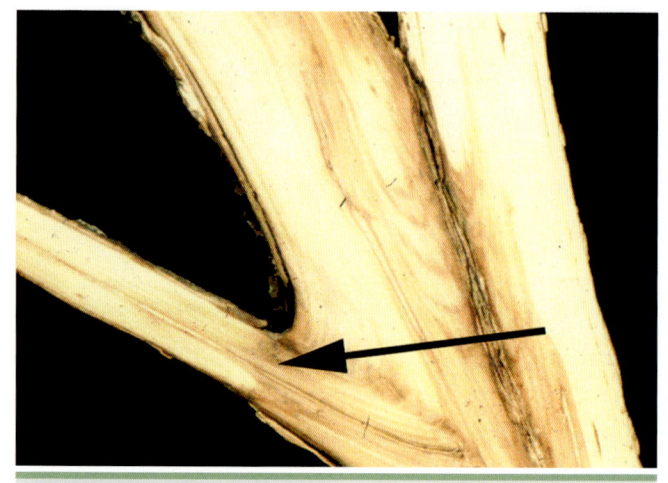

Figure 8.15 The branch protection zone has chemical and physical properties that slow the spread of decay into the trunk.

is typically much denser than other wood and exhibits twisted and intermingled wood grain, which serves to strengthen the connection. An external sign of the development of this specialized wood is the junction's **branch bark ridge**, a raised strip of bark. The axillary wood and chemical deposits in the cells form a cone-shaped area within the branch union called the **branch protection zone**. The effect of the branch protection zone is to slow or block the spread of decay organisms into the trunk. If the collar is removed, as with a flush cut or rip, decay can more readily spread into the wood of the stem or parent branch behind the pruning cut. Not only do flush cuts result in unnecessarily larger wounds, but they can also slow or reduce the ability of the tree to close the resulting wound.

Knowing how branches are attached to stems enables the arborist to make proper pruning cuts. Pruning cuts made in the appropriate locations may promote compartmentalization, the natural process of defense in trees to wall off decay in the wood.

Pruning Cuts

Proper pruning technique is critical to long-term tree health. The most important principle to remember is that each cut has the potential to change the architecture, dynamics, and health of a tree. Make each cut carefully, at the correct location, leaving a smooth surface with no jagged edges or torn bark. The ability of the tree to compartmentalize wounds is a function of the type and size of the cut, the age of the cut stem or branch, tree health, species, and the time of year.

Pruning cuts are typically classified by where they are made on the branch or stem.

Branch Removal Cut

A **branch removal cut** is a pruning cut that removes the smaller of two stems at their union, such as the removal of a branch from the trunk or a lateral branch from its parent branch. The cut should be made just outside the branch collar, or the branch bark ridge if the branch collar cannot be identified. A branch removal cut does not damage the branch collar, and it allows the branch protection zone to work effectively. Making cuts flush to the parent stem or branch is not an acceptable practice.

Reduction Cut

A **reduction cut** removes the larger of two or more branches, stems, or codominant stems to a live lateral branch or stem, preferably at least one-third the diameter of the branch being removed. The cut should be made just beyond the lateral branch, which will then assume an important role for support and survival. Reduction pruning is often used for improving branching structure, directing growth, removing branch defects, or decreasing plant size.

The smaller the cut and the healthier the tree, the better the closure rate and compartmentalization of the wound is likely to be. When possible, avoid large (greater than 10 cm [4 in] diameter) reduction cuts and cuts that expose heartwood, especially on species that are poor compartmentalizers. Trees do not always compartmentalize reduction-cut wounds as well as they compartmentalize wounds from removal cuts.

Heading Cuts

A **heading cut** removes a branch or stem between nodes, to a bud, or to a live branch less than one-third the diameter of the branch or stem being removed. Heading cuts are often used on small plants and shrubs to stimulate branching. They are not often used on mature, established trees, but they can be an acceptable and appropriate choice at times. One example is with the restoration of a severely storm-damaged tree when there are few branches left along a stem.

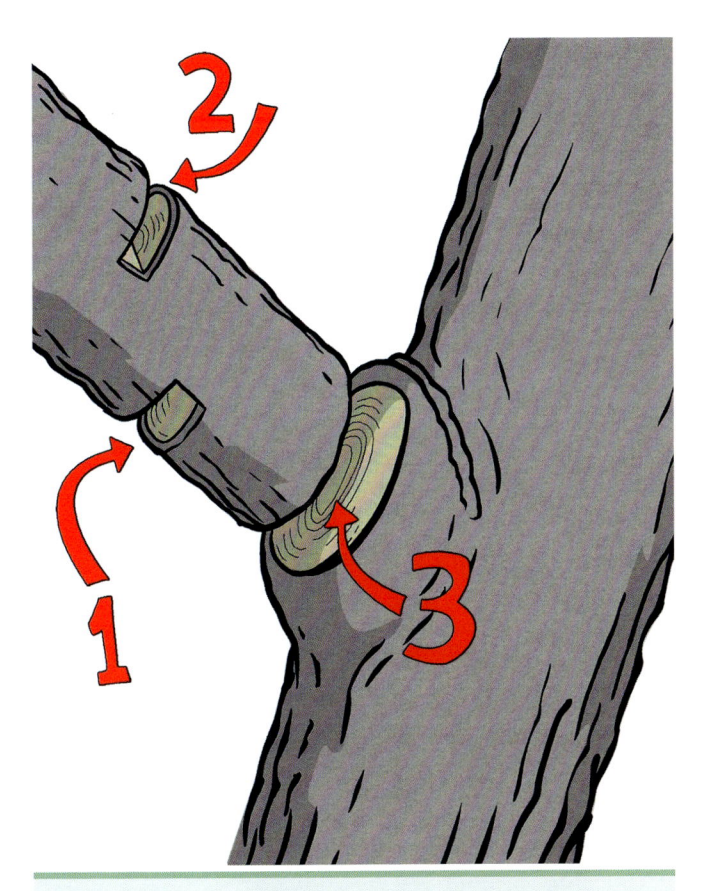

Figure 8.16 A branch removal cut using the three-cut method.

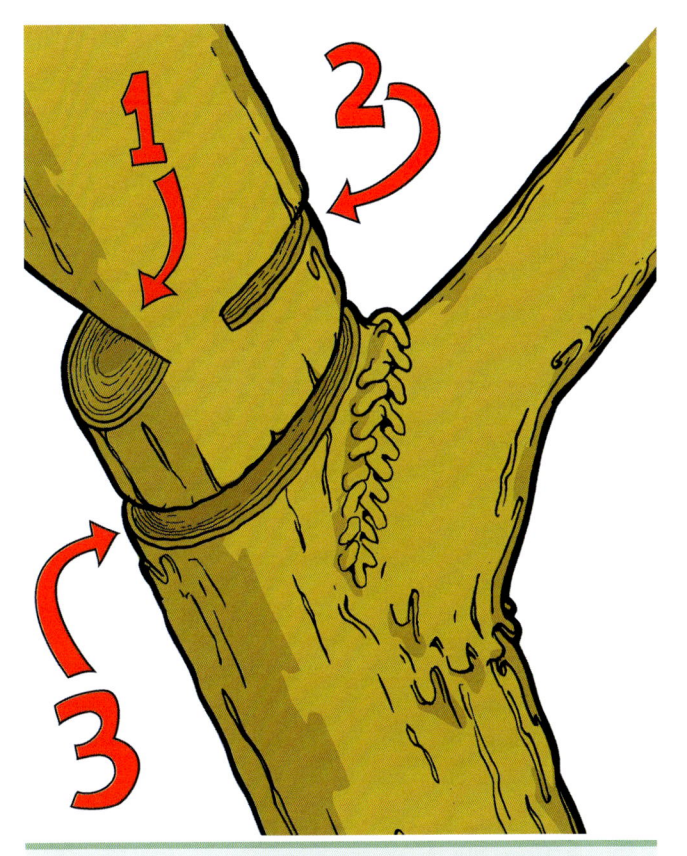

Figure 8.17 A reduction cut prunes a branch back to a lateral.

Shearing

Shearing is cutting of leaves, sprouts, and branches to a desired plane, shape, or form, as within the topiary system. Shearing is typically used to create a dense crown by stimulating growth from dormant and adventitious buds. It is a common practice on shrubs and also may be used on young trees and in nurseries to create multiple sprouts and branches for aesthetics. Shrub pruning with shearing cuts does not give attention to structural form and stability. Size management is the most common application.

Shearing cuts can be problematic to trees and shrubs. Most foliage is produced on the outside edge of the crown. Leaves and branches to the interior may die due to lack of sunlight. Long-term negative impacts on trees include poor branch taper, lack of a dominant leader, and poorly spaced scaffold branches. Another disadvantage is the requirement for frequent pruning to maintain the form.

Making Pruning Cuts

The practices used for pruning depend on the size of the branch to be cut; whether the branch is safely and easily supported by one hand while cutting;

Figure 8.18 Make each cut carefully, at the correct location, leaving a smooth surface with no jagged edges or torn bark.

Figure 8.19 When removing a dead branch, make the final cut just outside the collar of living tissue. If the collar has grown along a branch stub, remove only the dead stub that extends beyond the collar.

and if a simple, single cut can be made with hand pruners, loppers, or a handsaw.

Branches that are too large to be supported by hand should be removed using the **three-cut method** to avoid tearing or splitting the bark and damaging the branch protection zone. The first cut, called the undercut, begins on the bottom of the branch, usually about 15 to 30 cm (6 to 12 in) away from the branch union. The second cut, called the top cut, is made above or just outside of the undercut, from the top of the branch moving downward. After both cuts have been made, the branch should easily fall and be removed. The purpose of the first two cuts is to remove most of the weight of the branch before making the final cut. Finally, the third and final cut to remove the stub is made just outside the branch bark ridge and the outer portion of the branch collar on the bottom side of the attachment. If removing a dead branch, cut just outside any living tissue, taking care not to damage it.

Poor pruning cuts, leaving rips, stubs, or flush cuts, create many issues detrimental to the health of the tree. Pruning without damaging the branch collar and branch bark ridge encourages the formation of a woundwood that closes the wound and protects the tree. Avoid leaving a stub beyond

Figure 8.20 Flush cuts damage tissue beyond the branch that is being removed and can lead to more discoloration and decay.

the branch collar, except when the objective is to encourage new growth from the stub (such as following storm damage).

Wound Dressings

Wound dressings are compounds applied to tree wounds or pruning cuts. Research indicates that tree wound dressings are not needed on pruning cuts and provide no benefit to the tree, in most cases. Wound dressings are used primarily for cosmetic purposes. Some dressings are injurious and can inhibit closure of wounds. Some studies have shown beneficial effects in specific cases in reducing borer attack, oak wilt infection, or control of sprout production or mistletoe. If a dressing must be applied, use a light coating of a nonphytotoxic material.

Pruning Tools

Select pruning tools that are appropriate for the size of cuts being made in the process. Tools should be sharp so as to make clean cuts without jagged edges or tearing the bark. When using loppers or hand pruners, avoid anvil-type pruning tools, with a blade that cuts to a flat surface; tools with bypass blades are preferred. Do not use equipment and work practices that damage living tissue and bark beyond the scope of the work. This includes the use of chain saws for small, delicate cuts or in tight spaces where a handsaw is a better choice. Incidental damage to the trunk or other branches causes unnecessary wounds and is unprofessional.

Do not use climbing spurs to climb trees, including palms, for pruning operations. Although palms lack a cambium layer, they can still sustain significant damage from climbing spurs.

Sterilization of pruning tools between plants, or even between cuts, can minimize the chance of spreading some diseases. The likelihood of spreading diseases in this way varies with the disease organism, the plant, the pruning tools used, the environmental conditions, and the timing of pruning. If tools are sterilized, it is important to use a material that will not damage plant tissues.

Amount of Pruning

Pruning should remove only what is necessary to achieve the intended objectives. There are cases where a few small-diameter cuts are appropriate and cases where many or larger cuts are needed to accomplish objectives. The amount removed to meet objectives should be balanced against plant processes impacted by overpruning.

A tree's developmental stage (age), crown architecture, vitality (health), recent weather (e.g., drought), and site history (e.g., soil compaction or root severing) determine how tolerant the tree will be to the removal of branches. Young trees are more capable of quick and effective compartmentalization of wounds because they have relatively large quantities of stored resources and because pruning cuts are small. As trees age, this process may become more stressful for the tree due to larger cuts, inadequate stored energy, or environmental stress such as recent soil disturbance under the tree. Younger trees and healthy trees typically tolerate the removal of more living tissue more easily than mature trees do.

Large, mature trees often require little routine pruning, especially if they received the benefit of structural pruning when young. A tree's energy-producing capacity relative to its size decreases as it matures. Removing even a single, large-diameter branch can create a wound that a mature tree may not be able to close.

A healthy tree can tolerate more pruning than a weak or declining tree. During times of stress, such as drought conditions, severe pruning can have significant negative consequences on tree health.

Timing

Most routine pruning and removal of weak, diseased, undesirable, or dead branches can be accomplished at any time with little negative effect on the tree. Adjusting the timing of pruning based on where a tree is in its annual growth cycle can help achieve desired results. As a general guideline, plant growth is maximized if pruning is done just before the buds swell, in late winter or early spring; growth is minimized by pruning in late spring or summer. Research has provided some guidance for the optimal times to prune to reduce the impact on the tree and facilitate wound closure. Trees in temperate climates close wounds most quickly if pruned when they are actively growing.

Pruning when trees are dormant can minimize the risk of pest problems associated with wound entry and allow trees to take advantage of the full growing season to begin closing and compartmentalizing wounds. Plant growth can be reduced if pruning takes place during or soon after the initial growth flush, so pruning at this time is sometimes avoided. This is when trees have just expended stored carbohydrates to produce foliage and early shoot growth.

A few tree diseases, such as oak wilt, can be spread when pruning wounds allow insects and fungal spores access into the tree. Susceptible trees should not be pruned during active transmission periods.

Flowering can be prevented or enhanced by pruning at the appropriate time of year. Trees and shrubs that bloom on current-season growth, such as crapemyrtle (*Lagerstroemia* spp.) and linden (*Tilia* spp.), are best pruned in winter, prior to leaf emergence, or in the summer after bloom has occurred. Plants that bloom on last-season wood, such as many fruit trees, lilac (*Syringa* spp.), and azalea (*Rhododendron* spp.), should be pruned just after bloom to preserve next season's flowers. Often, fruit trees are pruned during the dormant season to enhance structure to direct fruiting wood. They can be pruned after bloom to reduce the amount of fruit production.

Some references recommend that certain species of trees, such as maples (*Acer* spp.) and birches (*Betula* spp.), should not be pruned in the early spring when sap flow is heavy. These trees tend to "bleed,"

Figure 8.21 Flowering can be maximized by pruning at the appropriate time of year, depending on when the species blooms.

or drain sap from the pruning cuts. Research has shown that sap released through pruning wounds, although unattractive, has little negative effect on tree growth.

Pruning during drought conditions should be minimized, if possible. Drought diminishes the capacity to compartmentalize wounds from pruning and could lead to an increase in the amount of decay. Heavy pruning of live branches should be postponed in severe drought conditions to improve wound recovery and closure of pruning wounds.

Pruning Palms, Conifers, and Shrubs

Palms

Palms should be pruned when **fronds**, flowers, fruit, or loose petioles might create a dangerous condition. Although palms shed fronds and reproductive parts naturally, this may pose a hazard in certain settings. Some palms (e.g., *Washingtonia* spp.) hold several years of dead fronds and may suddenly shed hundreds of pounds of them, which can be an extreme

Figure 8.22 Although palms do not have a cambium layer, they can still be damaged by using climbing spikes, and the wounds left behind are unsightly.

Figure 8.23 Palm pruning primarily removes dead or dying fronds. When live fronds are removed, it should only be those that are below horizontal.

hazard for arborists and people in the vicinity. For this reason, arborists performing this type of work should remove accumulations of dead fronds from the top down. Other species have such large flowers or fruit that pruning is required to preempt natural shedding. Some palms regenerate readily from seed, so developing flowers or fruit are removed to limit the spread of unwanted plants.

It is preferable not to remove live, healthy green fronds. Older fronds contribute to carbohydrate production through photosynthesis and may provide essential elements to younger fronds. Removing too many green fronds (overpruning) may contribute to mineral deficiencies such as magnesium or potassium. Overpruning can also result in a temporary reduction in stem diameter until the tree grows new fronds, which leads to hourglassing, a structural defect. If older fronds must be removed for clearance, avoid removing fronds that initiate above horizontal (above 90° from the stem). When removing fronds, sever them close to the petiole base without damaging living trunk tissue. "Skinning" palm trunks by cutting petioles flush with the stem can leave unsightly scars and has the potential to wound the stem and spread disease. Wounds on palms do not close over and

Figure 8.24 Size control and aesthetic shaping of shrubs are often achieved by shearing, although removing and shortening selected branches can produce a more natural shape. Shearing should be used only with species that are tolerant of this type of pruning, and only on stems and branches capable of generating new sprouts.

are present for the life of the tree. Practice precautionary measures when pruning palms susceptible to certain diseases in the region so disease does not spread to healthy palms.

Conifers

Pruning of conifers often includes the removal of dead, dying, and diseased branches. Conifers may be pruned to reduce the risk of failure, minimize disease spread, and improve appearance. Conifers can also benefit from structural pruning if several upright stems developed from a damaged dominant leader. At times, they may need clearance pruning. Pine (*Pinus* spp.) stem elongation may be managed by shortening new emerging growth shoots (candles) rather than through branch removal.

Some pruning strategies are not appropriate for conifers. For example, pruning for branch spacing and scaffold branch development is usually not necessary. Few conifer tree species respond well to pollarding. Reducing branch length beyond the zone of live foliage usually results in branch death, unless the species is capable of sprouting. Few species of fir (*Abies* spp.) or spruce (*Picea* spp.) produce sprouts; pine species vary widely in sprout production; and new growth on old branches is common in coast redwood (*Sequoia sempervirens*) and species of cypress (*Cupressus* spp.).

Shrubs

Most of the principles and practices used for pruning trees also apply to shrubs. Several of the pruning systems described, such as natural, topiary, and espalier, are applied when pruning shrubs. The choice of the system depends on the pruning objectives. The most common objectives for pruning shrubs include height reduction, rejuvenation, and improvement of aesthetics.

Objectives for shrub size control and aesthetic shaping are often achieved by shearing, although they can be achieved by removing and shortening selected branches to achieve a more natural shape. Shearing should be employed only with species tolerant of this type of pruning and only on stems and branches capable of generating new sprouts. Shearing tends to produce a dense outer crown that causes loss of foliage in the inner crown.

Shrub rejuvenation often combines two or more objectives including increasing the amount of foliage closer to the ground, reducing size, and improving health and appearance. It is applied to large-growing mature shrubs that are overly dense at the top or are declining in health due to previous shearing or the size or age of some of the stems. Rejuvenation is accomplished by removing or reducing large or declining stems to near ground level. This allows sunlight to reach interior branches of adjacent stems and new sprouts that emerge from the base of the shrubs. Removal, reduction, and/or heading cuts are made as low to the ground as possible.

Plant Growth Regulators

Plant growth regulators are substances, usually effective in small quantities, that enhance or alter the growth and development of a plant. In most cases, these chemicals either increase or decrease normal growth, flowering, or fruiting of plants and have been used to reduce the frequency of pruning necessary to maintain clearance. Utility arborists sometimes use growth regulators to reduce the growth rate of trees and other vegetation near or beneath utility lines. Growth inhibitors can be sprayed on the foliage, banded on the bark, applied into the soil, or injected into the tree. Certain growth regulators (**antigibberellins**) inhibit the synthesis of gibberellin, the naturally occurring cell-elongation hormone. These chemicals can significantly reduce growth rates—and therefore reduce the need for pruning trees and shrubs used as hedges.

Another use of growth regulators is to reduce watersprout production following pruning. Studies have shown that watersprout and sucker growth can be minimized, in certain instances, with the use of growth regulators. Interest in the use of plant growth regulators is strongest among utility arborists, but they have gained acceptance by some commercial and municipal arborists as well.

Pruning Specifications

Written specifications or pruning prescriptions are key to good pruning because they communicate what is to be done. Writing prescriptions is also a reliable means of communicating to the arborists that will be performing the work. They are written using language similar to that found in a country's national standards. The United States, United Kingdom, Germany, Sweden, Australia, and other countries have national standards for tree pruning.

Specifications protect both client and arborist by ensuring that everyone clearly understands the objectives and scope of work. Municipalities, utilities, commercial arborists, and property owners all benefit from using written specifications.

Specifications should include the pruning system(s), objectives, pruning cut type(s), size range of branches to remove, amount to remove, and location of branches. The amount to remove can be expressed as a percentage of crown volume, pruning cut diameter, length of branch to be removed, number of branches to be removed, or some combination of these (e.g., eight 8 cm/3 in reduction cuts). Specifications should also include the time frame for completion, the plan for disposal or repurposing of debris, and a recommendation for reinspection or repruning (pruning interval).

CHAPTER 8 WORKBOOK

1. True/False—Poor pruning practices can shorten the life of a tree and can increase the likelihood of tree or structural failure.

2. Name five objectives for pruning trees.

    a.

    b.

    c.

    d.

    e.

3. The majority of pruning by most arborists follows a _____ pruning system.

4. If two branches (stems) develop at the tip of the same stem, they may form _____ _____ .

5. _____ _____ is bark that becomes enclosed inside the attachment as the two branches grow and develop.

6. When training young trees, a dominant leader should be selected and competing leaders should be removed or _____ .

7. When practical, temporary lower branches should be left on a young tree to help develop trunk _____ .

8. Removing too much foliage can have adverse effects on the tree, resulting in production of _____ on interior branches, which is often a tree's reaction to overpruning.

9. Label the branch bark ridge and the branch collar on this drawing. Show where the undercut, top cut, and final cut should be made in removing the branch.

10. The swollen area at the base of a branch where it arises from the trunk is called the _____ _____ .

11. True/False—In the absence of included bark, the size of a branch in relation to the trunk is more important for branch attachment strength than is the attachment angle.

12. A _____ _____ cut removes the smaller of two branches at a union with the parent stem.

13. A _____ cut removes the larger of two or more branches, stems, or codominant stems to a live lateral branch or stem.

14. True/False—Tree wound dressings accelerate wound closure and prevent decay.

15. When pruning palms, if older, live fronds must be removed for clearance, avoid removing fronds that initiate above _____ .

16. Name three negative effects of topping a tree.

    a.

    b.

    c.

17. True/False—Arborists should be aware that removing or disturbing active bird nests, especially of protected species, is illegal in many locations.

18. True/False—Trees that tend to "bleed" should never be pruned in the early spring, because doing so is likely to cause a major decline in vigor.

Chapter 8: Pruning

19. True/False—As a general rule, mature trees are less tolerant of heavy pruning than juvenile trees.

20. _____ _____ _____ are substances, usually effective in small quantities, that enhance or alter the growth and development of a plant.

Challenge Questions

1. What are the effects of pruning on the growth and development of shoots of a tree?

2. Explain the response of a tree to topping. What steps should be taken to restore the crown? What information might be given to a client who wants a tree topped?

3. Why is good trunk taper important in the development of a tree?

4. How do antigibberellins reduce the growth of trees? What are the limitations to their use?

Sample Test Questions

1. When pruning young trees, it is important to train for a dominant leader and well-spaced scaffold branches to

    a. minimize the need for future pruning
    b. develop a structurally sound crown
    c. minimize codominant branching on the trunk
    d. all of the above

2. To prune trees that flower on the previous year's growth and to maximize flowering, you should prune

    a. anytime during the dormant season
    b. shortly after flowering
    c. in late summer after seed formation
    d. in the fall, just after leaf drop

3. When pruning a branch from a tree, the final cut should be

    a. flush with the parent stem
    b. at a 45° angle to the parent stem
    c. parallel to the branch bark ridge
    d. just outside the branch collar

4. When it comes to pruning, as a rule, mature trees are

   a. more tolerant of extreme pruning than young trees
   b. capable of tolerating heading better than young trees
   c. not as tolerant of severe pruning as young trees
   d. less likely to produce watersprouts than young trees

5. If the height of a tree must be reduced,

   a. branches should be cut to a lateral one-third or more the diameter of the branch removed
   b. all cuts should be made at internodes to avoid cutting through buds
   c. the tree should be root pruned to compensate for foliage loss
   d. pruning should only take place during the dormant season

Recommended Resources

(See Recommended Resources in back of book for detailed information.)

Best Management Practices: *Pruning* (Lilly et al. 2019)

An Illustrated Guide to Pruning (Gilman 2012)

ANSI A300 (Part 1) *Pruning* (American National Standards Institute 2017)

Tree Work. Recommendations. (BS 3998) (British Standards Institution 2010)

The CODIT Principle (Dujesiefken and Liese 2015)

Applied Tree Biology (Hirons and Thomas 2018)

CHAPTER 9

TREE SUPPORT AND LIGHTNING PROTECTION

▼ OBJECTIVES

- Identify tree conditions that might justify the installation of cables, guys, bracing rods, or props.

- Recognize the limitations of tree support systems.

- Explain the different ways trees can be supported with cables, braces, props, or guying hardware.

- Describe the techniques and materials used in cabling and bracing.

- Compare and contrast steel and rope support systems.

- Identify the circumstances under which a lightning protection system may be recommended for trees.

KEY TERMS

7-strand, common-grade cable
air terminal
amon-eye nut
anchor hardware
bend radius
cable stop
conductor
dead-end grips
extra-high-strength (EHS)
  cable

eye bolt
eye splice
ground rod
ground terminal
guying
lag eye
lag hook (J-hook)
lag-threaded
machine-threaded
propping

rope support system
ship auger
steel cable system
swage
thimble
threaded rod
wire rope

TREE SUPPORT SYSTEMS

Introduction

Tree cabling, bracing, guying, and propping all involve the installation of hardware in trees. This hardware provides a tree with supplemental support by limiting the movement of branches and helping to support loads. If properly designed and installed, tree support systems can be used to selectively reinforce weak tree parts, but they need to be inspected and maintained regularly. When used appropriately, tree support systems may extend the life of the tree by reducing the risk of structural failure.

Support may be needed due to the presence of split, defective, or decayed branch unions or those with included bark, which may pose an elevated likelihood of failure. Trees with multiple stems originating from the same point can be susceptible to breakage under wind loads or the weight of accumulated ice or snow. Branches that pose a potential risk to people or property may be candidates for cabling.

Tree support systems cannot eliminate all risk associated with tree failures, but they can, in some cases, reduce the likelihood of failure. A tree must be assessed carefully to decide when support is warranted and likely to be effective and whether to install supplemental support systems. If the root system is compromised, or if a significant amount of decay is present, removal of the tree may be preferable to installation of support systems. Keep in mind that anytime support hardware is installed into a tree, the tree's response to loads may be altered, and if wounds are created, the spread of decay may be accelerated.

Before installing a support system, pruning should be considered as an alternative or in conjunction with the system. In the case of small- and medium-sized trees, weak codominant stems and overextended branches can often be reduced or removed to lessen the need for a support system. For larger trees, reducing or removing some outer branches can reduce load on the support.

Installation of tree support systems should follow the appropriate standards for each country. ISA's Best Management Practices: *Tree Support Systems: Cabling, Bracing, Guying, and Propping* provides detailed information on the installation of tree support systems.

Figure 9.1 Tree support systems cannot eliminate risk of failure, but they can, in some cases, reduce it to acceptable levels. The large crack in this codominant stem was supported with bracing and cabling above.

Steel Cable Systems

An advantage of **steel cable systems** (sometimes incorrectly referred to as static systems) is the strength and longevity of the materials. Steel cables are known to last for many decades in trees without significant degradation in strength. A limitation is

Figure 9.2 An advantage of steel cable systems is the strength and longevity of the materials. A limitation is the requirement of drilling through branches for installation of anchor hardware.

the requirement of drilling through branches for installation of anchor hardware. In addition, steel cable systems can take longer to install than rope support systems.

Cabling Hardware

It is important to select the appropriate hardware for cabling a tree. Cables, eye bolts, and other hardware come in different sizes and types. Consider the size of the branches, the weight to be supported, and the presence of decay when choosing materials. If the hardware is too small or otherwise inadequate, the cable may fail. In the United States, the ANSI A300 Standards for *Supplemental Support Systems* specify minimum hardware sizes for various sizes of branches.

Two types of steel cable are commonly used in North America for cabling trees: **7-strand, common-grade cable** and **extra-high-strength (EHS) cable**. Common-grade cable is easy to bend and work with; EHS cable is much stronger but less flexible than common-grade cable. Both cables are available in a wide range of sizes.

Wire rope (aircraft cable), made of woven strands of galvanized wire, is widely used in Australia and the United Kingdom for tree support system applications. One advantage of this type of cable is that it is both strong and flexible. A potential drawback is the limited selection of termination hardware.

Anchor Hardware

Anchor hardware that passes through the branch and is fastened on the opposite side is generally preferred for attaching cables. **Eye bolts** or **threaded rods** with **amon-eye nuts** are commonly used. Drop-forged eye bolts are slightly stronger than threaded rods used with amon-eye nuts. However, an advantage of the amon-eye system is that the length of the rod can be adjusted for various diameters of branches.

To install the hardware for steel or hybrid cable systems (rope support systems with steel termination hardware), holes must be drilled. The drill bit should be a **ship auger** or a similar bit, which works more efficiently in green wood and will pull out the shavings from the hole. All bits should be sharp.

To install eye bolts or threaded rods, drill a hole equal to or 1.6 to 3.2 mm (1/16 to 1/8 in) larger than the diameter of the hardware, completely through the branch to be cabled. Install the eye bolt or threaded rod with a round washer and nut on the outside end, seating the washer directly against the bark. On trees with very thick bark, chisel away the bark or drill (using a Forstner bit) to countersink the washer against (but not into) the sapwood. To prevent the nut from unscrewing, flatten or otherwise secure the

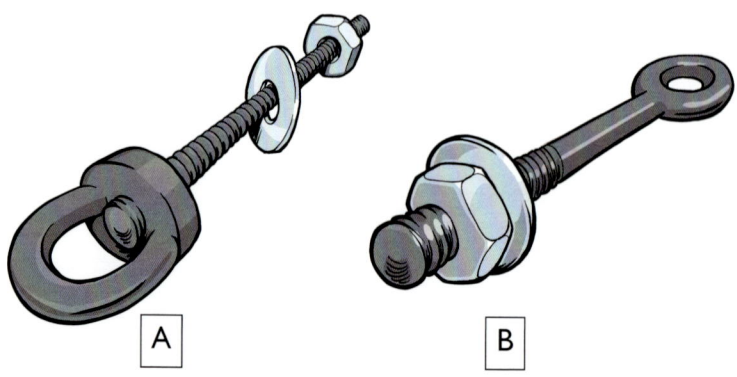

Figure 9.3 A: threaded rod with amon-eye nut. B: eye bolt.

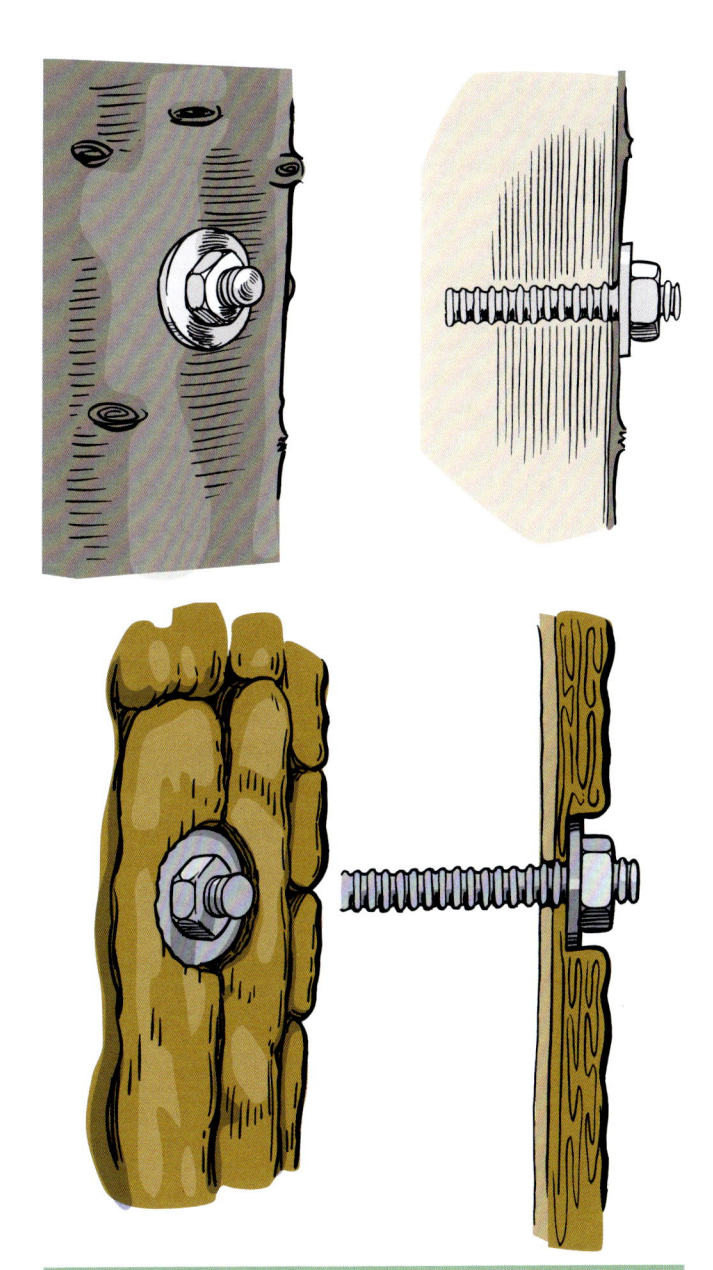

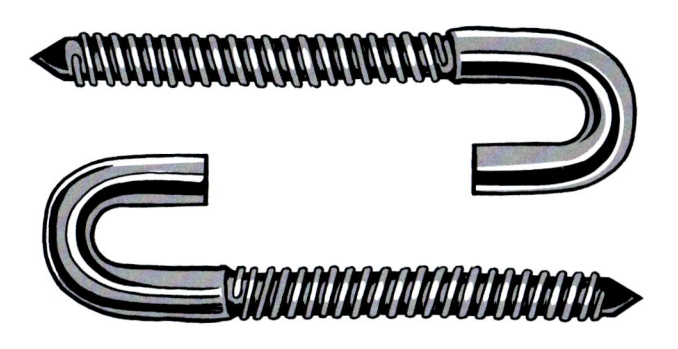

Figure 9.5 Left- and right-lay lag hooks.

Figure 9.4 On most trees, seat the washer directly on the bark. On trees with very thick bark, countersink the washer to the sapwood.

Figure 9.6 Lag hooks should **not** be installed in decayed wood.

exposed threads on the end of the eye bolts or on both ends of the threaded rod when amon-eye hardware is used.

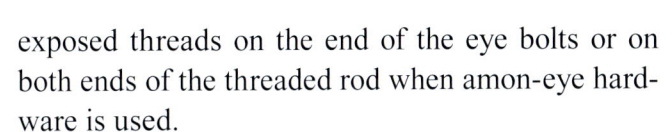

An alternative to through hardware is dead-end hardware, which does not go entirely through a branch or stem. A **lag eye** is a lag-threaded, drop-forged anchor with a closed eye. A **lag hook**, or **J-hook**, is a lag-threaded steel device in the shape of a "J." Lag hooks come with right- and left-handed threads so that when each end is twisted into the branch to tension the cable, the cable will not come unlaid or unwound. Install lag hooks by screwing into a predrilled hole that is 1.6 to 3.2 mm (1/16 to 1/8 in) smaller in diameter than the lag.

Lags work adequately on small branches (< 25 cm [10 in] diameter) and on trees with strong, decay-resistant wood, but they should never be used on large branches (> 25 cm [10 in] diameter) or on those with or prone to decay. Wood decay will limit the holding capacity of the lags and may spread into healthy wood. Additionally, do not use lags if the full length of the threads cannot be seated into the branch. If not properly installed, lags are more likely to pull out of the tree as it moves. Some jurisdictions and companies do not recommend the use of dead-end hardware.

Attachments and Terminations

When using eye bolts or threaded rods to install cables, anchor them with heavy-duty or heavy-duty, heat-treated round washers and nuts. The use of these washers reduces the risk of the nut pulling through the tree. When attaching the cable to its anchoring hardware, incorporate galvanized or stainless-steel, heavy-duty **thimbles** into the termination. The purpose of the thimble is to protect the cable from abrasive wear and to increase the cable's **bend radius**. If soft, common-grade cable is installed directly on the hardware, the steel-to-steel contact and abrasion may eventually cause wear and cable breakage.

Common-grade, 7-strand cable can be attached to the hardware using an **eye splice**. Always use a heavy-duty thimble when forming an eye in the end of the cable. Make the eye splice by wrapping the end of the cable around the thimble and then separating the cable strands. Individually wrap each strand around the cable with at least two complete turns per strand in the same direction. When complete, the finished splice will have a neat appearance and provide optimal holding power.

Extra-high-strength cables are difficult to bend and cannot be readily made into an eye splice. **Dead-end grips** are manufactured spiral-wraps designed to form a termination at the end of 7-strand EHS cable. Install dead-end grips over a heavy-duty thimble to reduce wear on the grip hardware. Place the cable in the thimble, wrapping the short side of the grip on the cable. Then wrap the longer side, fitting it between the wraps of the first side. Both ends should be completely wrapped and in contact with the cable when finished.

Manufactured **cable-stop** or **swage** termination devices are fasteners that clamp directly to the ends of a cable. When cable stops or swages are used, the cable itself is passed through the branch, and the hardware is installed on the back of the branch

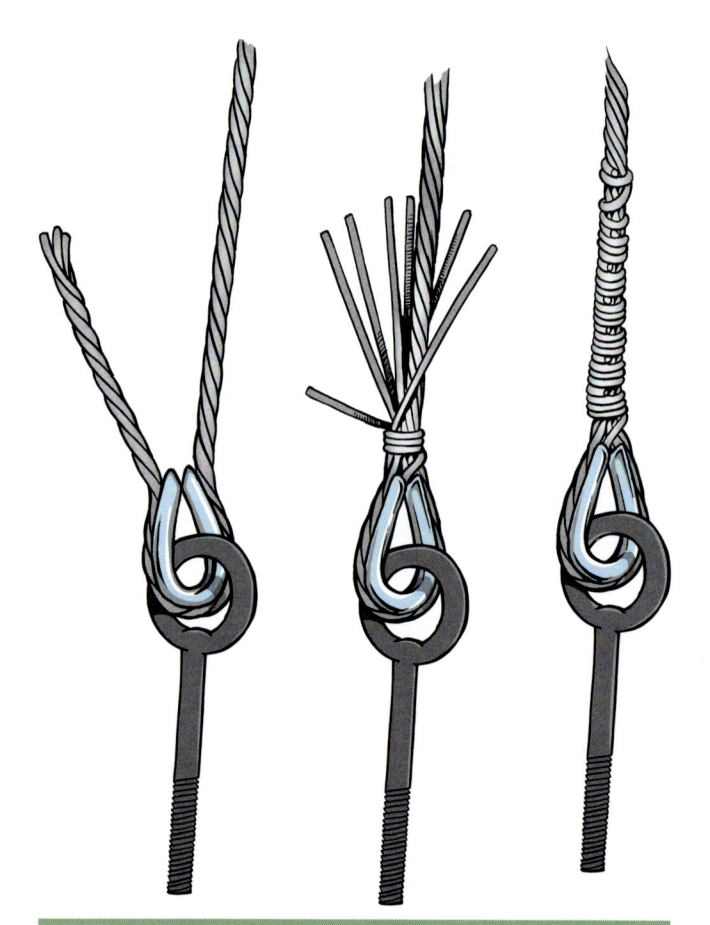

Figure 9.7 Forming an eye splice with common-grade, 7-strand cable. Each strand is wrapped two to three times around all of the remaining strands.

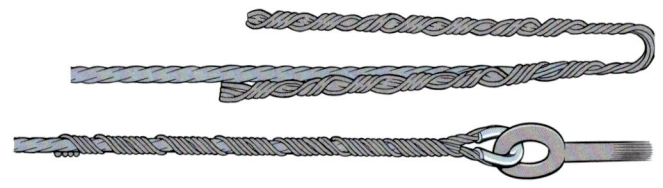

Figure 9.8 Wrap the dead-end grip onto the cable. A thimble must be installed before wrapping the second side.

to secure the cable. Wire rope is commonly terminated with a swage termination, which is crimped onto the cable with a swaging tool. The wire crimp is critical to attaining a strong attachment and must be done correctly and according to manufacturer's instructions. Wire rope clamps can also be used and must be installed according to manufacturer's instructions.

Rope Support Systems

An alternative to steel cable systems is **rope support systems** (sometimes called dynamic support systems). Rope support systems allow for more tree sway than steel cable support systems. In theory, more tree movement allows a tree to produce supporting wood where needed. These systems use ropes and belts, sheaths, or straps instead of cables and anchors. Several proprietary brands are available and are widely used in Europe. Rope and steel cable support systems share many properties, but their materials and installation techniques differ.

Rope support systems use some form of rope or strap that is wrapped around the stems and branches. Therefore, these cables do not require the arborist to drill into the tree. Several systems use a hollow-braid, polypropylene rope, which enables the use of simple splices for attachment.

Rope support systems may reduce the potential for shock loading the system, which can occur if two stems move in opposite directions with great force. Some systems include an additional shock-absorbing rubber component. Others incorporate an overload indicator, which can signal that the system was overstressed and prompt the arborist to perform a closer inspection or replace the system.

Because most rope systems are installed by wrapping the support cable around the stem, the potential for girdling the tree or damaging the rope is always present. With one commonly used system, special flat inserts are placed within the rope at the points of contact to create a wide band that distributes the load over a larger surface area. In addition,

Figure 9.9 Because most dynamic systems are installed by wrapping the support cable around the stem, the potential for girdling the tree or damaging the rope is always present. Flat inserts placed within the rope at the points of contact can distribute the load over a larger surface area. A protective, nonabrasive sheath covers the rope to reduce friction between the rope and the bark.

a protective, nonabrasive sheath covers the rope to reduce friction between the rope and the bark.

Some systems call for the formation of tension/compensation loops of cable, which are intended to accommodate tree growth and further minimize the potential for girdling. These loops should be inspected regularly to identify when growth exceeds the capacity of the expansion loop.

One hybrid system uses the traditional through-hardware installation with connecting hardware and rope between the anchors. The goal of this system is to provide the flexibility and shock-absorbing properties of rope without the potential for girdling.

Because rope support systems are relatively easy to install, they are sometimes used as temporary tree support systems. For example, they can be quickly installed to support trees or branches that have been damaged in storms, giving arborists time to take more permanent action.

The shock-absorbing and nondrilling advantages of rope support systems must be weighed against

Table 9.1
Comparison of traditional cabling systems.

Cabling system	
Eye-bolt-anchored steel cable	
Cable-stop-terminated steel cable	
J-lag-anchored steel cable	

Advantages	Limitations
• Greatest strength/diameter ratio • Not visually intrusive • High longevity	• No stretch • Requires drilling through branch • EHS cable can be difficult to work with
• Faster to install than an eye bolt • Especially effective for small branches • Just as effective on large branches • Not visually intrusive	• Branch movement may enlarge drilled hole if there is slack or great range of motion • Requires drilling through branch
• Faster to install than an eye bolt • Works well for small branches • Not visually intrusive	• Cannot be installed into decayed wood • Lag can open or pull out when overloaded • Installation limited to branches no larger than 25 cm (10 in) in diameter • Requires drilling into branch

(Table 9.1 continued on p. 174)

Table 9.1 (continued)

system	
Sling-terminated synthetic rope	
J-lag-anchored synthetic rope	

their potential limitations. The major concern is the potential for UV degradation. Because of this risk, control intervals or life expectancies are sometimes defined in regional tree care standards or by municipalities. Some systems incorporate color-coded indicators so that the year of installation can be determined easily. One European standard requires that manufacturers ensure adequate strength for a minimum of eight years. A close inspection on a one-to-three-year interval is appropriate in most locations.

Another potential limitation of rope support systems that has been noted by researchers and practitioners is that small animals, such as squirrels and birds, may chew on the rope support, which can significantly reduce its strength.

Some manufacturers also limit applications of rope support systems. For example, rope support systems may not be recommended for branch unions that are already split, because the branch movement can allow the split to enlarge. Also, rope support systems may not be suitable for static loads such as those associated with overextended branches. Cables spanning long distances require greater stiffness, so steel cables are sometimes preferred in those cases.

Advantages	Limitations
• Fast installation • Minimal tools required • No drilling • Good for temporary support • Absorbs some shock through stretching • Polypropylene is easy to splice during installation	• May require a branch union for installation • May girdle the branch or leader • May be damaged by animals or UV degradation • Large-diameter rope has high visibility from the ground
• Absorbs some shock through stretching without risk of girdling	• Requires drilling through branch • May be damaged by animals or UV degradation • Large-diameter rope has high visibility from the ground • Difficult to splice during installation

Cabling Installation Techniques

The angle of the cable and its distance from the branch union determine its effectiveness. A general guideline for supporting codominant stems is to install the cable at least two-thirds the distance from the union being supported to the ends of the stems. Farther out, the branches might be too small (in diameter) for cables to be installed. If installing a cable in an overextended branch, the anchor should be installed closer to the trunk, a distance of one-third to half the length of the branch from the trunk.

This is the center of gravity for most branches. If possible, within the limits of the branch configuration, a near-vertical orientation of the cable is preferred. For both codominant stems and overextended branches, the wood at the anchor point must be large enough and solid enough to provide adequate support of the anchor hardware.

On codominant stems, support can be maximized by installing the cable directly across the branch union being supported. "Directly across" can be determined by setting the cable perpendicular to (at a 90° angle to) an imaginary line through the center of the branch union. Attachment hardware

Figure 9.10 Install the cable at least two-thirds the distance from the defective (weak) branch union to the ends of the branches.

should be installed with the cable's pull in direct line with the attachment hardware. In most cases, the cable will not be installed perpendicular to either branch, and the cable will be near horizontal. Cables installed on overextended branches should be oriented as close to vertical as practical.

Once installed, the cable should be just taut—tightened to the point where there is no visible slack. A cable that is too tight may put excessive stress on the wood fibers, resulting in more damage at the defect or causing the hardware to pull out. To make installation easier, branches may be brought closer together with ropes or with slings and a come-along. After the cable is installed, the branches are released slowly until the cable tension is appropriate.

The most common cable installation is the simple, or direct, cable (one cable between two branches). Sometimes a tree will require more than one cable. If more than two stems or branches require cabling, the use of multiple cables is required. Extra stability can be added to the system by cabling the branches together in triangular combinations. If more crown movement is desired, a box or rotary system can be installed. A hub-and-spoke system can connect multiple leaders while allowing some independent movement, although it is the most complex configuration to install.

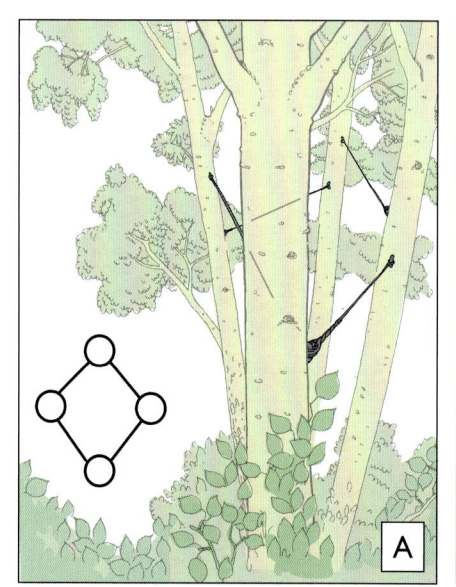

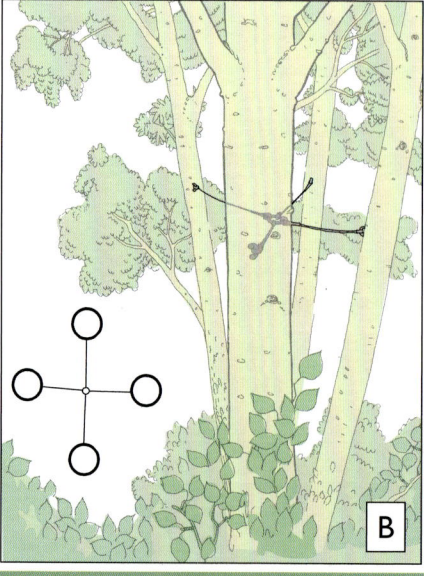

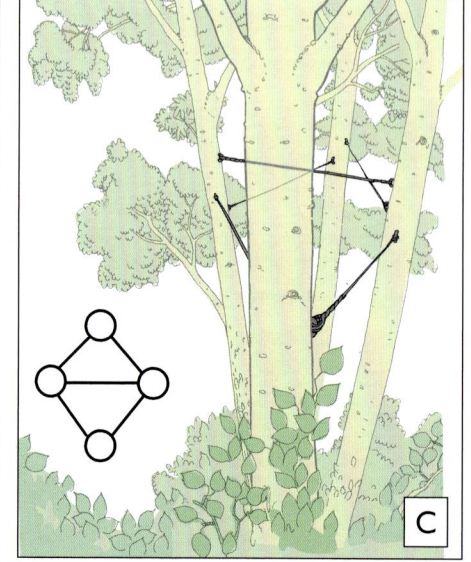

Figure 9.11 Cabling configurations. A: box; B: hub and spoke; C: triangular.

Figure 9.12 The hardware should be installed in a direct line with the pull of the cable.

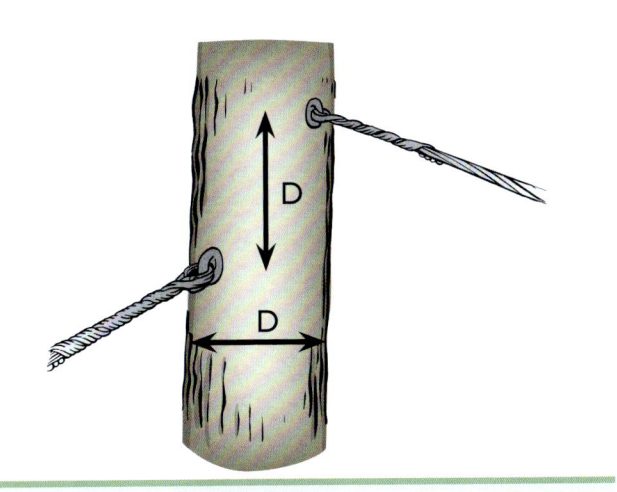

Figure 9.13 If more than one cable must be installed, keep the hardware at least one branch diameter's distance apart, if practical. Do not install the hardware in vertical alignment.

Figure 9.14 Do not install more than one cable on each eye bolt or lag.

When more than one cable is being installed on the same branch, space the cable attachments no closer than a distance equal to the diameter of the branch, or 30 cm (12 in), whichever is less. When installing hardware, avoid longitudinal alignment with other hardware. Internal decay from the hardware installation could merge, further weakening the support. Cables must not rub against wood or each other. Only one cable should be attached to any one anchor.

Do not install anchors into decayed areas where the amount of sound wood is less than 30 percent of the trunk or branch diameter. If the decayed portion is too large, consider pruning options, rope support systems, or removing the branch or the tree. If the branch is already decayed, the decay inside the branch may spread along the anchor after hardware installation, thus reducing holding capacity.

Bracing

Bracing is the use of steel rod(s) in branches, leaders, or trunks to provide support for a union or the branch itself. Bracing is used to reinforce weak branch unions, to strengthen decayed areas, or to reinforce longitudinal cracks. In most cases, bracing is used in combination with cabling, not as a substitute.

Two types of steel rods are used in tree bracing: **lag-threaded** rods and **machine-threaded** rods. Lag-threaded rods have fewer and deeper threads per inch (2.5 cm) than machine-threaded rods. Lag-threaded rods can be installed as dead-end hardware, with the hole drilled smaller than the rod. Drill the hole through the entire smaller stem and at least halfway into the larger stem. If a nut and washer are not used to secure the exposed end of a dead-end installation, break off the rod inside the bark so the rod end is not exposed: Turn the rod in most of the way, saw through the rod approximately two-thirds to three-quarters of the diameter of the rod, then carefully turn the rod in so that the precut portion is inside the bark. Bend the rod until it breaks at the cut.

In large trees, or trees with soft or decayed wood, use only machine-threaded rods. In this case, drill a hole equal to or 1.6 to 3.2 mm (1/16 to 1/8 in) larger than the diameter of the rod. Use a drill bit of sufficient length to penetrate all the way through the section(s) being braced. Never attempt to drill from opposite ends and hope to have the holes align.

Feed the rod through the tree and bolt on each end using washers and nuts. Seat the washers directly against the bark. If the bark is thick, countersink the washer against the wood. Cut off any excess rod beyond the nuts, and flatten the threads or otherwise secure the end of the rod to prevent the nuts from backing off. If rust is a concern, use galvanized or stainless-steel rod, or treat any portion of the steel rod that is exposed with a rust-resistant coating (paint, spray galvanization).

If a single rod is being used to support a branch union that is not split, install it at a distance of about one to two times the branch diameter above the branch union. If necessary, install additional braces below the union. In many cases, it may be advisable to install more than one bracing rod. Multiple rods provide added strength and reinforcement to a split or weak union. If more than one bracing rod is to be installed, stagger them if possible—longitudinal alignment of the rods may add to the formation of decay columns in the tree. Whenever practical, space the rods apart at a distance greater than the diameter of the wood being braced, or 30 cm (12 in), whichever is less.

Installation of bracing rods can provide significant support and can sometimes even close a split branch union. They do, however, require significant drilling for installation, so consideration must be given to the potential for the spread of decay into new wood as a result of this wounding.

Figure 9.16 Drill all the way through both stems to install the brace rod. Here the rod is set against the tree to align the drill.

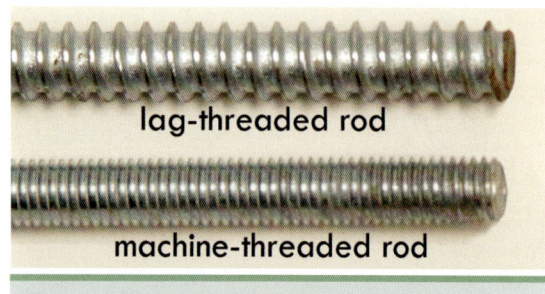

Figure 9.15 Lag- and machine-threaded rods.

Chapter 9: Tree Support and Lightning Protection

tree movement. Large transplanted trees often require guying during the establishment period, but the guy wires should be removed after allowing adequate time for the root system to develop. Trees that have serious root defects but cannot be removed due to historic importance or other reasons may require guying to keep them upright or to prevent failure in the direction of a target. Guying of mature trees tends to be permanent because the tree may not produce the roots or wood necessary to support itself once it has been guyed. The tree may become dependent on this mechanical support.

When considering the installation of guy wires, arborists should carefully assess the risk associated with the tree. If the objectives cannot be met by the additional support, other means of mitigating the risk should be considered.

Guy wires can be attached to trees using the same methods as for installing cables. As with cable installation, hardware should be installed in direct line with the guy wire. When installing tree-to-ground guys, make sure that the ground anchor (earth anchor) has sufficient strength to support the tree, even under wet conditions. Attach guy wires to the tree at a height at or above its midpoint. Place the ground anchor(s) no closer to the trunk than two-thirds the distance from the ground to the lowest attachment in the tree.

If the tree is to be guyed to another tree, carefully inspect the anchor tree to ensure it has enough structural integrity and strength to support the other tree. The guys installed on the anchor tree should be attached on the lower half of the trunk. Those installed on the tree to be supported should be attached above the midpoint, preferably at about two-thirds the height of the tree.

Figure 9.17 If a single rod is being used to support a branch union that is not split, install it at a distance of about one to two times the branch diameter above the branch union.

Guying

Guying is the installation of a wire cable (or, in limited cases, a rope) between a tree and an external anchor to provide supplemental support and reduce

A significant consideration in the installation of guying systems is public safety. This involves the additional support of the trees to be guyed, as well

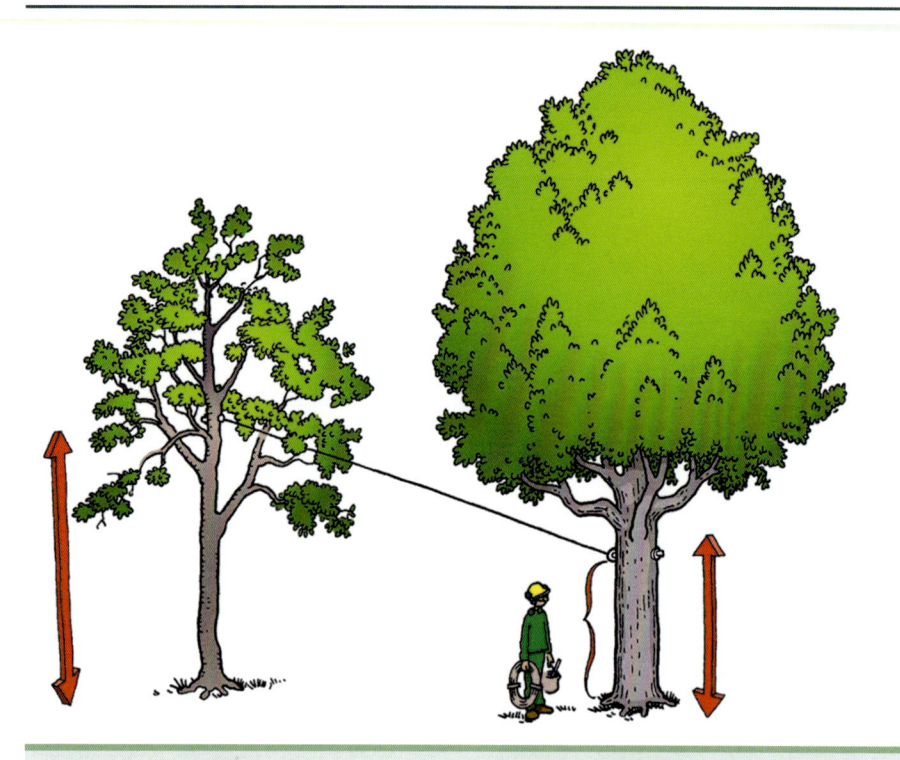

Figure 9.18 If a tree is to be guyed to another tree, the guy installed on the anchor tree should be attached on the lower half of the trunk. The guy on the tree to be supported should be attached above the midpoint, preferably at about two-thirds the height of the tree. The anchor tree should be larger than the supported tree.

Figure 9.19 Tree-to-ground guys should be clearly marked and protected to minimize contact. It is also advisable to place a mulch ring around the ground attachment point to avoid lawn mower damage.

as safety concerns related to the guy wires themselves. Tree-to-tree guy wires should be installed high enough so that people or vehicles will not encounter the wire. Clients should be informed of the risks associated with guy wire failure, pedestrian or vehicular contact, and the potential for tripping hazards.

It is preferable to place the guy wires on any support tree at a height greater than 2 m (7 ft) if there is pedestrian traffic under the guy wires, or greater than 4 m (14 ft) if there is vehicular traffic. Tree-to-ground guy wires should be clearly marked and protected from inadvertent contact. It is also advisable to place a mulch ring around the ground attachment point to avoid mower damage.

Propping

Propping is the installation of rigid structures between the ground and a branch or trunk to provide support. Props are used under branches or leaning trunks to keep the tree part off of the ground or a structure, to share the load of the branch, or to provide clearance. Typically, props are used under branches that are nearly horizontal or downward growing.

Little information about the use or design of props in arboriculture is available, and there are wide variations in design. A key facet of any prop design is that it should not damage or restrict future growth of the branch. Props can be made from wood, galvanized steel, concrete, or other materials. They must have sufficient strength to support the expected load. They also should be protected from

Figure 9.20 Props are rigid structures erected to support branches from below. It is important that the design does not restrict the growth of the branch.

deterioration. Over time, wood may decay or steel may rust and weaken a prop. This weakening may not be visible and could lead to an increased likelihood of tree failure.

The prop must have a provision to keep the branch secured to it. The prop must also be anchored in the ground to keep it from moving excessively. If a hole is dug for this footing, root damage beyond the scope of the work must be avoided.

Inspection and Maintenance

The installation of tree support systems in a tree creates an ongoing responsibility. The arborist should explain to the client that cables, braces, guys, and props must be inspected routinely. Periodic inspections should be made to check the structural integrity of the tree and branches, the condition of the hardware and support system, the cable tension, and the position of cable(s) in the tree. Some standards and equipment manufacturers specify the frequency of inspections.

As the tree grows taller, new cables may eventually need to be installed higher in the tree. Old systems should not be removed until the new system has been properly installed, and often it is better to leave them in place. Trees that have been cabled may also need to have the tension of the cable adjusted or may need to be pruned periodically to remove excess weight.

LIGHTNING PROTECTION SYSTEMS

Introduction

Lightning is an extremely powerful and pervasive force of nature. Thousands of forest trees are struck every year, occasionally starting forest fires. Lightning can damage or destroy trees valued for their shade, beauty, and other benefits. Lightning protection systems installed in trees are very effective at reducing damage to trees caused by lightning strikes. These systems protect trees by providing an alternate path for lightning to go from the top of the tree to the ground using a conductor.

Lightning and Trees

A lightning strike can instantly destroy the entire tree or may cause significant structural damage requiring removal of the tree. Most trees struck by lightning are not killed immediately, and many show no visible injuries. Nonetheless, trees that have no external indicators of lightning damage might have a damaged vascular system, making them prone to increased stress. The disruption of nutrient and water flow from a lightning strike may take a year or longer before visible symptoms of the injury appear. Factors that affect the extent of damage include bark thickness and wetness, wood porosity, and moisture content.

Because the extent of internal damage cannot be assessed immediately, treatments to consider include restoration pruning, water management, pest prevention, and removal of stripped bark, if applicable. The affected tree should be monitored for potential risks, but extensive arboricultural treatments should be delayed for a year or more, unless safety is an overriding factor. The injury may not immediately cause structural damage, although it might serve as an "open door" for borers, disease, and future structural degradation.

Figure 9.21 Lightning damage.

Candidates for Protection

Height and location increase the probability that a tree will be struck by lightning. Lightning is more likely to strike the tallest tree in a group, trees growing in the open, tall trees that border woods or line a street, trees near large bodies of water, trees on hilltops, or trees in geographic regions where lightning is common.

Trees that are close to houses or other buildings and that are much taller than the building should be considered for protection. Trees of historic interest and high economic value are also good candidates for lightning protection. Likewise, trees that are more prone to strikes and are located within recreational areas, parks, golf courses, or other locations where people congregate should be considered for lightning protection.

Lightning Protection Systems

A lightning protection system consists of a copper conductor that extends from the top of the tree, down the main branches and trunk, and out into the soil where it is grounded. Lightning protection systems create an alternate path of lower resistance for the electrical charge of a lightning strike to follow. A heavy copper conductor (cable) is connected from the highest part(s) of the tree to a grounding system.

A lightning protection system consists of an **air terminal** (blunt tip), **conductors**, and a **ground terminal**. The air terminal is the top end of the tree lightning protection system. Because any branches or trunk sections above the air terminals may be damaged by a strike, it is important to install air terminals as high on the tree as possible.

The ground terminal (grounding electrode) is the portion of a tree lightning protection system that is installed for the purpose of providing electrical contact with the ground. Ground terminals consist of ground conductor(s), **ground rod**(s) or plate(s), and all other associated connectors. It is best to select the ground terminal location before determining the path of the conductor in the tree.

Even though lightning protection systems are very effective in preventing or reducing tree damage, these systems are not intended to directly protect people from lightning strikes. Protected trees cannot be considered safe havens from lightning strikes during storms.

Figure 9.22 Lightning protection systems create an alternate path (of lower resistance) for the electrical charge of a lightning strike to follow. A heavy copper conductor (cable) is connected from the highest part(s) of the tree to a grounding system.

For more information on the installation of lightning protection systems, see ISA's Best Management Practices: *Tree Lightning Protection Systems*.

Inspection and Maintenance

As with tree support systems, lightning protection systems should be inspected periodically. When the tree has grown significantly past the air terminals, the conductor cables and terminals should be extended higher in the tree. Eventually, the tree may grow enough in girth to envelop the fasteners and conductor, if the conductor is not periodically refastened. An overgrown system will still be functional as long as there is no break in the conductor. However, because it is harder to inspect an overgrown system, refastening may be preferred. Periodic inspections (for example, annually on fast-growing trees, every two or three years on slow-growing trees) should include a check of all splices and connections to ensure continuity.

CHAPTER 9 WORKBOOK

1. Common-grade cable is relatively malleable (bendable) and easy to work with. _____ - _____ - _____ cable is much stronger but less flexible than common-grade cables.

2. One advantage of _____ _____ is that it's both strong and flexible. A potential drawback is the limitation of choices for attachment.

3. As a general guideline, cables to support codominant stems should be installed _____ - _____ the distance from the weak branch union to the top of the tree, as long as the wood is solid and large enough to install the hardware.

4. Branches may be brought closer together while installing the cable so that when released, the cable will be just _____ .

5. True/False—Attachment hardware should be installed with the cable's pull in direct line with the attachment.

6. True/False—When more than one cable is installed on the same branch, the hardware should be spaced at least as far apart as the diameter of the branch.

7. True/False—When installing multiple cables, use common anchors whenever possible.

8. Extra-high-strength cable should be attached to hardware using _____ - _____ _____ .

9. True/False—The installation of steel cables, if done properly, will not wound the tree.

10. True/False—Rope support systems may reduce the potential for shock loading the system, which can occur if two stems move in opposite directions with great force.

11. True/False—An advantage of most rope support systems is that they do not require drilling the tree to install.

12. True/False—If a single rod is being used to support a branch union that is not split, it should be installed just below the branch union.

13. _____ is the installation of a cable between a tree and an external anchor to provide supplemental support and reduce tree movement.

14. Rigid structures mounted or built on the ground to support a branch or trunk are called _____ .

15. Name three circumstances in which lightning protection for trees might be recommended.

   a.

   b.

   c.

Challenge Questions

1. What are the differences between 7-strand, common-grade cable and extra-high-strength (EHS) cable? What are the advantages and disadvantages of each?

2. Ideally, a cable should be installed in a tree perpendicular to the line that bisects the angle of the branch union. Diagram this installation and discuss how it affects hardware installation.

3. In what ways are steel cable systems different from rope systems, and what are the advantages and limitations of each?

4. Discuss lightning protection for a grove of several large and small trees.

Sample Test Questions

1. An advantage of the amon-eye system over the use of an eye bolt is

   a. no washers are needed on the terminations
   b. the length of the rod can be adjusted
   c. they are considered stronger than eye bolts
   d. the eye bolt is not drop forged

2. The purpose of a lightning protection system is to

   a. reduce the voltage of the strike
   b. prevent the tree from being struck
   c. conduct the electrical charge into the soil away from the tree
   d. all of the above

3. If two bracing rods are installed to support a weak union, they should be placed

   a. no more than 10 cm (4 in) apart
   b. in vertical alignment, one above the other
   c. staggered and no closer together than the diameter of the trunk
   d. below the union and perpendicular to one another

4. To install a cable directly across a branch union, install it perpendicular to

   a. an imaginary line that bisects the union
   b. the main stem or larger branch
   c. the branch being supported or the smaller branch
   d. both stems arising from the union

5. An advantage of rope support systems is

   a. the cables are more durable than steel cables
   b. they don't require drilling for installation
   c. they last a very long time
   d. they don't require follow-up inspections

Recommended Resources

(See Recommended Resources in back of book for detailed information.)

Best Management Practices: *Tree Support Systems: Cabling, Bracing, Guying, and Propping* (Smiley and Lilly 2014)

Best Management Practices: *Tree Lightning Protection Systems* (Smiley et al. 2015)

Introduction to Arboriculture: Tree Support Systems online course (ISA)

Introduction to Arboriculture: Lightning Protection Systems online course (ISA)

CHAPTER 10
DIAGNOSIS AND PLANT DISORDERS

▼ OBJECTIVES

- Distinguish between plant problems caused by biotic and abiotic agents.
- Describe diagnostic principles, and develop a systematic approach to plant diagnostics.
- Explain how stress can predispose a tree to additional problems.
- Identify the symptoms and signs of tree disorders.
- Describe the various physiological disorders and injuries that can affect trees, and explain what treatments are appropriate.
- Discuss common types of insect and disease problems, and describe their impact on trees.

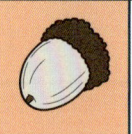

KEY TERMS

abiotic agents
acute
aerobic
allelopathy
alternate host
anaerobic
biotic agent
blight
blotch
canker
chlorosis
chronic
decay
defoliation
dieback
disorder
eriophyid mites

frass
gall
girdling
girdling root
gummosis
honeydew
infectious
necrosis
nematode
noninfectious
oozing
opportunistic
pathogen
pest
phytotoxic
plant disease triangle
scorch

sign
skeletonized
sooty mold
spot
stippling
stunting
symptom
systemic
tree stress
trunk flare
vascular discoloration
vector
wilt
witch's broom

Introduction

Diagnosing plant disorders requires a combination of knowledge, experience, skilled observation, and deductive reasoning. Arborists must use their knowledge and the clues at hand to arrive at a diagnosis. A **disorder** may be defined as an abnormal condition that impairs the performance of one or more vital plant functions. Deterioration of health and, sometimes, tree death can result from the effects of a combination of factors. The correct diagnosis often is not a matter of simply identifying an insect or pathogen. Insects and pathogens are often **opportunistic**, attacking trees that are already weakened. It takes some detective work to piece together all the clues. Arriving at a diagnosis is only the first step. The second, and equally important, step is to consider treatment options and make appropriate recommendations.

Figure 10.1 Diagnosing plant problems requires a combination of knowledge, experience, keen observation, and deductive reasoning.

General Diagnosis Principles

A big challenge facing an arborist in diagnosing tree disorders can be the lack of available information. Obviously, the tree cannot describe its symptoms; it is up to the diagnostician to discover the factors that have affected the tree, and involving the client is important. The history of the tree and its environment are of utmost importance, and an arborist must learn to ask key questions about the history of the site and the tree: How long has the problem been going on? What were the early symptoms? Has there been any construction, excavation, or chemical treatment in the area? Asking open-ended questions and asking the same question in different ways may help reveal more information. Keep in mind that the information received may not be entirely accurate. It is not unusual to be told by a client that a tree died suddenly when in reality its health had been deteriorating for years. Often, the arborist is not brought in to treat the tree until it is already dying or dead.

It is helpful to categorize factors affecting plant health into two major groups: living and nonliving. **Biotic** (living) **agents** include plant **pathogens** such as fungi, bacteria, viruses, phytoplasmas, parasitic plants, and nematodes, as well as insect pests, mites, and other animals. **Pest** is a broadly defined term than can refer to an organism (including, but not limited to, weeds, insects, bacteria, or fungi) that is damaging, noxious, or a nuisance. Pathogens are considered to be **infectious** because they can spread from one plant to the next. **Abiotic** (nonliving) **agents** are **noninfectious** and include causes of mechanical injury and environmental conditions such as temperature and moisture extremes, soil compaction, salt accumulation, mineral deficiencies, and many others.

Symptoms and Signs

When diagnosing tree and shrub disorders, look for symptoms and signs to determine the cause. **Symptoms** are the effects of the causal agents or factors on the plant. Examples of symptoms include chlorosis, wilting, and leaf scorch. Rarely can the cause of a problem be identified by presence of a single symptom, and many different disorders can have symptoms in common. The symptom of wilting, for example, can be the result of drought, root damage, water-logged soil, or pathogens that colonize xylem. **Signs** are direct indications of primary or secondary causal agents, or something "left behind" by the causal agent. Signs might include fungal fruiting bodies such as conks; insect **frass**; emergence holes; or discarded insect parts. Some of the most common tree disorder symptoms and signs are defined here:

Blight—symptom, regardless of the causal agent, characterized by rapid death of flowers, leaves, or young stems that typically remain on the plant

Blotch—irregularly shaped dead area on leaf, stem, or fruit

Canker—localized, usually dead area on stems, roots, or branches; often sunken and discolored

Chlorosis—whitish or yellowish leaf discoloration caused by lack of chlorophyll

Decay—altered wood (white rot or brown rot, for example) resulting from the process of wood decomposition by fungi

Dieback—progressive death of twigs and branches from the tip down or back toward the main stem

Gall—abnormal, enlarged plant structure that develops from proliferation of the cells in leaves, stems, or roots colonized by certain parasitic organisms such as bacteria, fungi, nematodes, mites, or insects

Gummosis—exudation of sap or gum, often in response to disease or insect damage

Necrosis—localized or general death of cells or parts of a living organism

Oozing—seeping or exudation from a tree cavity or other opening

Scorch—browning and shriveling of foliage, especially along leaf margins and/or between veins

Spot—discrete, localized, and usually small dead area of a leaf or needle, stem, flower, or fruit

Stippling—speckled or dotted areas on leaves where tissue has been damaged, such as by piercing-sucking insects

Stunting—abnormally reduced plant growth

Vascular discoloration—darkening of the xylem or phloem of woody plants in response to disease, insect boring, or injury

Wilt—drooping of leaves or shoots, often due to lack of water in the tree

Witch's broom—plant structure resulting from shortening of internodes and proliferation of buds to form a dense, brushlike mass of shoots

Many symptoms of tree disorders are nonspecific. They may not suggest a particular diagnosis, because

Figure 10.2 Symptoms are the effects of the causal agents or factors on the plant. Chlorosis, or yellowing of the leaves, is an example of a symptom. In this case, interveinal (between the veins) chlorosis is a symptom of iron deficiency.

Figure 10.3 Signs are direct indications of primary or secondary causal agents, or something "left behind" by the causal agent. These frass "toothpicks" are an example of a sign—of ambrosia beetle, in this case.

some symptoms can be induced by many different abiotic agents, insects, and pathogens. In developing a diagnosis, analyze nonspecific symptoms in combination with additional information about the tree and the site. Keep in mind that some of the information collected may not be related to the disorder.

Diagnosis is a systematic process that involves information gathering, careful observation, and logical analysis. Unfortunately, there is a tendency to shortcut the process, resulting in an incomplete diagnosis or a misdiagnosis. Correct diagnosis of tree disorders requires careful examination of the situation and systematic elimination of possibilities by following a few important steps.

1. **Accurately identify the plant.** Many insects and diseases are specific to certain hosts. In addition, certain species of plants are prone to particular abiotic disorders. Knowing the identity of the plant may limit the number of suspected causes. To master diagnosis, learn to recognize the species, hybrids, and cultivars of trees and shrubs.

2. **Look for a pattern of abnormality.** Inherent in identification is the knowledge and recognition of what is normal. For example, a common client complaint in the autumn is that the needles of their pine trees are yellowing and dropping. Because it is normal for pines to shed their two- or three-year-old needles, it would be important to first determine whether it is the inner, older needles or the new-growth needles that are dropping.

Figure 10.4 A: Leaf blotch; B: Leaf scorch; C: Leaf spot; D: Leaf galls; E: Witch's broom; F: Bronzing.

It may be helpful to compare the affected tree with other plants on the site, especially those of the same species. Differences in shape, color, and distribution of symptoms on leaves and within the crown may present clues to the source of the problem. Nonuniform damage patterns among individuals of the same species or tissues within a plant often indicate biotic factors, such as insects or pathogens—problems caused by living agents rarely show uniform patterns. In contrast, if there is a uniform pattern throughout the plant, the cause is more likely to be abiotic. Uniform damage over a large area (perhaps involving several plant species) usually indicates nonliving factors, such as physical injury, water availability, or weather.

3. **Carefully examine the site.** Consider light levels, soil characteristics, water availability, prevailing winds, and any other factors that might affect tree health. Check the contour of the land, and note the structures present. If affected plants are restricted to a walkway, road, or fence, the disorder could be a result of deicing salts, wood treatments, or other toxic products. Examine the area or community where the tree is located. How do plants of the same species or other species look? Is the area newly built with stresses associated with compacted soil, construction damage, and recently transplanted trees? Is it an older community where age and overcrowding can be limiting factors? Or is it a new housing division in an old woodlot? The environmental conditions of a general area can often set the stage for certain types of problems, as can microclimatic conditions associated with the location in which the plant is growing.

4. **Note the color, size, and thickness of the foliage.** The foliage is the first part of the plant that most people notice. Changes and irregularities in the leaves tend to be easy to see, and symptoms are often obvious. Many disorders first apparent in the foliage begin elsewhere, without obvious symptoms. For example, chlorosis or death of leaves is often the result of mechanical damage to the roots or vascular system. Twisted or curled foliage may indicate viral infection, insect feeding, or exposure to herbicides. Early autumn color can result from girdling roots or other root-related problems. The list can be very long. The confusing part is that many different abiotic agents, insects, and pathogens can produce similar symptoms on the leaves.

5. **Check the trunk and branches.** Additional clues, symptoms, and signs can be found on the trunk, branches, and twigs. Examine the buds, new growth, and bark for any discoloration, drying, and oozing. Look for loose bark and wounds on the branches and trunk. Wounds provide entrances for insects and pathogens that rot wood, produce cankers, and disrupt the transport of materials between roots and leaves. Small holes may indicate the presence of borers or bark beetles. Profuse development of watersprouts can result from a variety of stresses. A good arborist is thorough and systematic in this investigative process, being careful not to miss any clues.

Assessing tree growth over time may provide an indication of potential sources of

Figure 10.5 The root system is the most frequently overlooked portion of the tree in the diagnostic process. Stress caused by poor site conditions, site changes, or poor management practices account for many root-related plant health problems in urban sites.

stress. By measuring the distance between terminal bud scale scars from year to year, it may be possible to determine whether growth is steady, increasing, or decreasing, and when a reduction in growth began. Slow growth over an extended period may be due to factors related to the site or climate. Sudden growth reduction can result from disturbances such as construction damage, addition of soil over roots, or defoliation.

6. **Examine the roots and root collar.** The root system is the most frequently overlooked portion of the tree in the diagnostic process. This is unfortunate because the belowground part of the plant is often the "root" of the problem. Stress caused by poor site conditions, human-caused site changes, poor selection of plants, or poor planting and management practices accounts for many root-related plant health problems in urban sites.

Take the time to examine the site and soil, the root collar (**trunk flare**), and the rest of the root zone. Examine the soil texture and moisture conditions, and note any soil odor. Check for symptoms and signs of decay in the major roots and collar. Note the color and feel of the fine roots: healthy roots are generally light-colored inside and are flexible and firm. Healthy roots indicate the **aerobic** soil conditions with sufficient oxygen. Brown roots may indicate dry soil conditions or the presence of toxic chemicals. Black roots, a sour smell, and bluish-gray soil may reflect wet and **anaerobic** (low oxygen) soil conditions. Death and decay of fine roots can also result from disease-causing organisms.

When diagnosing a tree disorder, try to evaluate the relative role of the many factors that might affect a tree's health and systematically rule out certain possibilities. Eliminating factors and conditions can help the arborist to reduce the possibilities and to think deductively through diagnosis. The majority of tree health problems are caused by factors other than insects, mites, or pathogens. Approximately 70 to 90 percent of all plant problems result from adverse cultural and abiotic environmental conditions, such as soil compaction, drought, moisture fluctuations, temperature extremes, mechanical injuries, or poor species selection. Opportunistic insects and pathogens often attack and may kill trees stressed by adverse conditions. While these agents might be identified in the diagnostic process, it is important to recognize the underlying conditions that have allowed these opportunists to act.

The tree "decline disease" concept provides an explanation for the gradual, progressive deterioration in tree health that often ends in death. Decline diseases are viewed as the result of multiple, interacting and interchangeable factors. These predisposing, inciting, and contributing factors sequentially alter trees and result in recognizable symptoms that intensify over many years. Diagnosis of a tree decline disease involves identifying these

Figure 10.6 Early autumn color and dieback from the ends of branches are indications of stress, which may be caused by a belowground problem.

factors and understanding their relative roles in the prolonged deterioration of tree health, as opposed to merely blaming an obvious insect, pathogen, or other agent for the disorder. For example, an oak tree predisposed by old age might be severely damaged by an inciting factor such as defoliation by gypsy moth. The opportunistic two-lined chestnut borer or *Armillaria* spp. root disease fungi may exploit the compromised tree and contribute to a "spiral" of deteriorating health and eventual tree death. However, effects of factors involved in decline diseases are not necessarily irreversible. If the predisposing, inciting, and contributing factors can be identified, it is sometimes possible to prevent or reduce their effects to improve the health and extend the longevity of the tree.

Diagnostic ability is a skill developed through practice and experience. The best diagnosticians continue to learn every time they look at a tree. Learn to keep an open mind and gather as much information as possible before committing to a diagnosis. Concentrating only on the most obvious symptoms and signs may not reveal the underlying reason for their presence. For example, consider the arborist who finds wood decay, cankers, or borers on a tree and thinks only about fungi or insects. Many insects and fungi are secondary and only become a problem on trees that are already stressed. If the source of the problem is still unknown or unclear after going through the diagnostic process, it may be time to seek help. Guessing at the diagnosis and recommending treatment based on that guess are serious mistakes. Part of the process is to know when additional expertise is needed and where to get it.

Tree Stress

Factors that promote plant health include sufficient water, air movement, drainage, optimal temperature and light, and nutrient availability. Too much or too little of any of these factors may cause stress in the plant. **Tree stress** refers to the condition of the plant when it is not able to fully utilize its genetic potential to maintain normal functions such as growth and defense. Tree stress is one of the most important causes of increased susceptibility to biotic agents that contribute to further deterioration in tree health.

Stress may be classified into two broad categories: **acute** and **chronic**. Acute stress caused by lightning, improper pesticide applications, or untimely frosts or freezes, for example, occurs suddenly and causes almost immediate damage. Chronic stress results from factors that affect plant health over a longer time, such as mineral nutrient deficiency, improper soil pH, poor drainage, soil compaction, long-term weather changes, suboptimal light exposure, or pollution.

Early symptoms of stress might include reduced growth, abnormal foliage color, vigorous watersprouting, or premature leaf drop. The most common causes of tree stress are related to site conditions. If a tree is not well suited for the site in which it has been planted, it is more likely to become stressed. Common causes of stress related to other site or abiotic environment conditions include excess or inadequate water, extreme cold or heat, mechanical injury from construction activities, and soil conditions such as too high or too low pH or soil compaction. Trees stressed by such conditions often change their metabolic processes to compensate for deficiencies, resulting in

Figure 10.7 Irrigation directly on the trunk of this tree can be a chronic stress that may lead to both abiotic and biotic problems.

Diagnostic Guide for Landscape Plants Exhibiting Nonspecific Symptoms

1. Possible causes of brown or scorched leaves or progressive dieback of branches:
   a. poor root health from inadequate drainage, excessive soil dryness, excessive fertilizer, girdling roots, excessive soil salinity, compaction, or poor water penetration into soils
   b. specific nutrient toxicities or imbalances
   c. excessive heat or light reflected onto leaves from pavement or buildings
   d. pesticide, herbicide, or mechanical injury
   e. air pollution
   f. winter drying
   g. vascular disease

2. Possible causes of leaf spots, blotches, blisters, or scabby spots:
   a. excessive soil dryness coupled with high temperatures
   b. frost injury
   c. chemical spray injury
   d. fungal or bacterial infections
   e. herbicide injury
   f. insect damage
   g. sunscald

3. Possible causes of yellow-green or off-color foliage:
   a. nutrient deficiencies
   b. poor root health due to compacted soil, poor drainage, or girdling roots
   c. winter drying
   d. root or root collar injury
   e. air pollution
   f. soil pH that is not appropriate for the plant species
   g. herbicide injury
   h. mites or sucking insects such as scales, leafhoppers, and aphids
   i. excess or inadequate light
   j. pathogen infection

4. Possible causes of the foliage of one branch or section of a tree dying:
   a. fungal or bacterial canker
   b. mechanical injury
   c. insect and animal damage
   d. winter damage
   e. chemical spray injury
   f. lightning
   g. localized herbicide exposure
   h. root damage or disease
   i. girdling root

5. Possible causes of leaf drop:
   a. poor root health from inadequate drainage, excessive dryness, excessive fertilizer, compacted soil, or girdling roots
   b. heat and drought stress
   c. insect or mite infestation
   d. herbicide injury
   e. mechanical injury
   f. vascular pathogens
   g. foliar disease

6. Possible causes of wilting or drooping foliage:
   a. poor root health from inadequate drainage, excessive dryness, excessive fertilizer or other soluble salts in the soil, compacted soil, soilborne pathogens, or overwatering
   b. toxic chemical poured into soil
   c. fungal or bacterial infection of the vascular system
   d. fungal cankers

> e. insect infestation
> f. mechanical injury
> 7. Possible causes of leaves with tiny yellow speckling or yellow banding of needles:
>   a. insect or mite infestation
>   b. air pollution
>   c. fungal or bacterial infection
>   d. chemical injury
> 8. Possible causes of deformed or misshapen leaves:
>   a. herbicide injury
>   b. late frost or freeze
>   c. insect or mite infestation
>   d. anthracnose disease
>   e. leaf galls
>   f. fungal or viral diseases

Figure 10.8 Severe soil compaction and restricted root space can cause tip dieback and may lead to other problems on an already stressed tree.

the delay of visible symptoms until the disorder is in an advanced stage and it may be too late to be effectively treated.

Disorders Caused by Abiotic Agents

A large variety of abiotic agents affect the normal growth and development of a tree. They may disrupt water and mineral uptake or other vital functions, which sometimes creates a chain of events of physiological dysfunction within the tree. Many abiotic agents of tree disorders are predisposing factors in the development of a decline disease. Each successive factor progressively alters the tree in the decline disease spiral, often further reducing the possibility of reversing the deteriorating tree health.

Symptoms of abiotic disorders are not always obvious and may be difficult to recognize. Learning the history of the tree and site often provides the best clues. Unfortunately for the diagnostician, however, it can take a long time for the tree to react to these abiotic initial factors inducing stress. Because of this, the link between cause and effect is often obscured or lost.

Soil and Site Problems

A common mistake in diagnosis is to carefully examine the trunk and crown of the tree while overlooking the soil conditions and root health. Root-related problems are often difficult to diagnose due to the limited accessibility of tree roots. Symptoms observed in the upper portions of a tree often result from poor root health. If a transplanted tree exhibits dieback within the first year, the most likely cause is either a lack of or an excessive amount of water—which may be related to planting depth, soil compaction, tree condition, or poor species selection. If the tree's root ball is too small or if a saturated soil restricts new root formation or water absorption, the tree will suffer from the lack of water and reduced nutrient uptake.

Another common site problem is soil compaction. Without sufficient pore space, water and oxygen will be limited. Root growth and water absorption will decrease, causing deterioration of

Figure 10.9 These roots have been damaged repeatedly by lawn mowing equipment. Mulching the area can reduce or eliminate this problem.

Figure 10.10 Injury at the base of the trunk is a common result of mower or string-trimmer damage. Mulching around trees can reduce the incidence of injuries such as these.

tree health or tree death. Beneficial biological activity in the soil can also be inhibited by insufficient or excessive water. Too much water may also restrict oxygen availability to the tree's roots.

Some species of trees and palms are particularly prone to mineral deficiencies in the urban landscape. For example, some palms are especially sensitive to deficiencies of boron, manganese, or magnesium. Laboratory soil and foliar analyses can help determine whether a mineral deficiency is present. A soil analysis will give the pH as well as the levels of some essential mineral elements in the soil. Some mineral deficiencies, such as iron and manganese, are related to high pH, causing elements to be "tied up" in forms that trees cannot absorb. Certain mineral toxicities, such as aluminum, are associated with low-pH soils. Changing the soil pH, however, is often not practical.

Physical and Mechanical Injuries

Physical or mechanical injuries, such as lightning, lawn mower damage, and vandalism, occur suddenly. Often the full extent of damage due to physical injuries cannot be immediately assessed. In some cases, initial treatment should be limited until the extent of tree damage can be determined.

Lawn mowers and string trimmers can damage trees by wounding or girdling the lower trunk and sometimes damaging surface roots, which may cause decline or even death. Trees suffering from mechanical injury may have less than ideal structural support. Decay near the place where the root system and trunk join (the trunk flare) or in the structural roots can be a serious condition that can lead to tree failure. Some indicators of decay in the roots or trunk flare include dieback, abnormal leaf color, fungal fruiting bodies, and uneven form. Excavating soil near the base of the trunk to expose buttress roots can help locate decay.

Mechanical injuries may take years (if ever) to become apparent in the crown. Trees in good health may compartmentalize mechanical wounds successfully without complications. Trees with thick bark are more tolerant to trunk and branch damage than trees with thin bark.

Sometimes the causes of physical damage can be eliminated to prevent injury to the tree. For example, where lawn mower injury is a problem, remove the turfgrass from the area around the tree and replace it with mulch. It is always preferable to prevent physical damage than to treat a tree after the damage has occurred, because treatment options are typically very limited. Broken or damaged branches can sometimes be removed using proper pruning, which may contribute to successful wound closure and compartmentalization of subsequent decay.

Weather-Related Problems

Extremes in temperature, either high or low, and other weather-related problems can be damaging or even lethal to trees. Weather-related problems include moisture extremes; burn from deicing salts; and damage caused by snow, ice, wind, lightning, and hail. Low-temperature injury can affect all plant parts, although injury to the roots is not common unless the tree is in a raised planter with insufficient soil volume. Species and time of year have a large effect on the extent of damage caused by low temperatures. Some species are able to withstand temperatures far below freezing, while others may be killed at much warmer temperatures. Also, plants that have not been given a chance to become acclimated or have not been hardened off are more likely to experience freezing and chilling injury.

Freezing injury can occur when temperatures drop below the freezing point. When a temperature is below the freezing point, ice crystals can form inside plant tissues, which may potentially rupture cell membranes. This often leads to dehydration and death of the tissue. Injury that occurs from low temperatures that are above the freezing point is referred to as chilling injury. Chilling injury often affects the integrity of cell membranes, causing cells to leak. Trunks can develop vertical cracks (sometimes erroneously called frost cracks) on the sun-facing side of a tree as a result of wood tissues under the insulating bark expanding due to freezing and subsequently thawing. Most of these types of cracks are associated with preexisting trunk wounds or pruning cuts. Although not considered serious, once a tree has been damaged in this way, the crack often reopens in the same place in subsequent years, and the injury may provide an entry point for pests and pathogens.

Another weather-related bark injury is sunscald, which typically occurs on the south or southwest side of trunks or branches (north side in the Southern Hemisphere) and is caused by rapid freezing of tissues warmed on sunny winter days. Thin-barked trees, trees with bark that has been suddenly exposed to sunlight (due to removal of shade), or previously injured trees are more susceptible to sunscald. Young trees are also more susceptible to sunscald.

High temperatures can also stress trees, especially when combined with drought. In urban settings, reflected heat from pavement and buildings can add to the heat load on the plant and lead to wilting, leaf scorch, sunburn, and tissue death. Plants with limited soil volume or in raised planters are especially susceptible to stress due to elevated temperatures and insufficient soil moisture.

Figure 10.11 Freezing injury (left) can occur when temperatures drop below the freezing point. Hail damage (right) includes torn or tattered leaves and broken, dented, or scarred branches.

Storm damage is a common problem and can be classified in several ways. Windthrow is when trees are physically pushed over by high winds, typically when the soil is saturated. Stem failure is breakage of the trunk, sometimes at a point of decay or other weakness. Root failure results from roots being pulled or snapped, causing trees to fall or lean. Branch failure is common during strong wind events. Trees are often broken from the heavy load of snow or ice. Ice can add thousands of pounds of weight to the branches of a tree. Ice damage can be particularly devastating because it often affects almost all of the trees in an area. Tree failures can lead to loss of life, power outages, and millions of dollars in economic losses.

Lightning strikes can cause serious damage or even kill a tree. An indication of a lightning strike is a spiral scar running down the length of the trunk. The scar follows the grain of the wood where the bark or sapwood has exploded off the trunk. The full extent of internal damage may not be obvious. There can be structural weaknesses in the wood, and, if the cambium is killed, the tree may not leaf out the following spring. Some trees show very little, if any, adverse effects from the strike. A long-term effect of a lightning strike can be infection and invasion of the wound by wood-decay organisms.

Hail causes immediate damage to the leaves, bark, and branches of a tree, predominantly on the windward side. Symptoms include torn or tattered leaves and broken, dented, or scarred branches. Thicker leaves may show pitting on the upper surface. Severe hail may completely kill the bark on one side of a stem and lead to **defoliation** of the tree.

Competition and Allelopathy

Tree disorders can result from competition with other trees and plants, including competition for sunlight. Shade-intolerant trees growing under the canopy of larger trees may exhibit death of lower branches, stem curvature, and reduced growth. Trees may also compete with turf and other plantings for available soil nutrients, water, and growing space. Tall fescue and other grasses can stunt the growth of young trees when grown over a tree's root system.

Allelopathy is the chemical inhibition of growth and development of one plant by another. Many trees and other plants produce chemical substances that affect the growth of other plants. Allelopathic chemicals may exude directly from the plant or be released indirectly through decomposition. These chemicals may inhibit growth, seed germination, flowering, or fruiting of nearby plants. A few trees are considered highly allelopathic, including tree-of-heaven (*Ailanthus*), walnuts (*Juglans* spp.), sugar maple (*Acer saccharum*), and black locust

Figure 10.12 Lightning is an acute, weather-related injury. Internal damage is sometimes difficult to assess. The scar from a lightning strike will often spiral down the entire length of the trunk.

(*Robinia pseudoacacia*). In most cases, the inhibition is minor and not easily diagnosed. Young trees and stressed plants are more susceptible to allelopathic effects than non-stressed plants or plants with large, established root systems.

Pollution Damage

Pollution (air, water, and soil) damage to trees may be divided into two categories: acute and chronic. Acute toxicity results from exposure to high concentrations of toxic compounds over a relatively short period. Chronic injury is caused by long exposures, usually in much lower pollutant concentrations. Other factors that influence development of pollution damage are type of pollutant, species tolerance, wind, temperature, humidity, soil grade and type, precipitation, and the general condition of the tree.

Pollution damage is often difficult to diagnose unless the source of the pollutant is known. The symptoms may mimic other problems such as insect injury and mineral deficiencies. Most of the symptoms of pollution damage will be evident on the foliage. Necrotic areas may appear on the leaf or needle tips, along the leaf margins, or between the veins. The leaves may appear whitish or silvery or may exhibit stippling or spots.

Chemical Injury

Although any number of chemicals can kill or injure a tree, misapplied herbicides are the most frequent culprits. Herbicides are formulated to kill plants. The growth regulator herbicides, such as 2,4-D and dicamba (used to kill broad-leaved weeds in the lawn), may harm nontarget broad-leaved plants such as trees and shrubs. These materials are **systemic**, which means they move throughout the plant. Exposure may be a result of accidental application, drift, volatilization, or movement of the pesticide through grafted roots or the soil.

If a tree has been exposed to systemic herbicides, the leaves will often begin to curl and cup, and the shoot tips may become twisted. New foliage may show varying degrees of parallel venation.

Figure 10.13 Curling of the leaf margins is one symptom of some systemic herbicides.

The foliage may appear wilted and become chlorotic before dying and falling off. Other symptoms include veinal or interveinal chlorosis, marginal chlorosis, stunting of buds and new growth, and leaf drop. Some herbicides have been shown to cause canker-like bark cracks when applied to tree trunks. Herbicides that are nonselective or that function as soil sterilants have greater potential to cause serious damage than most other growth regulator herbicides.

Water Availability

All plants can suffer from water deficit, which results in slow growth or even plant death. Chronic water deficit refers to a prolonged water shortage lasting for weeks or months. This leads to a decrease in photosynthesis and other physiological processes, slowing the tree's growth and reducing carbohydrate storage. Symptoms that result may include smaller leaves, less intense leaf color, early autumn coloration, and defoliation. Prolonged periods of below-normal rainfall can lead to extremely low soil moisture and drought conditions. When soil moisture decreases below the permanent wilting point, water is bound so tightly to the soil particles that the roots can no longer absorb water, causing death of foliage, branches, or the entire tree. Prolonged water stress can leave the plant highly susceptible to attack by opportunistic insects and pathogens.

Some species, known as xerophytes, are more adapted to regions with drier climate or sites where soil has low water-holding capacity, such as sand. Xerophytes have special adaptations to dry environments that allow for better water retention and/or water uptake. Adaptations include photosynthetic adaptations, extensive root systems, and thicker, leathery leaves, often with a waxy cuticle or hairy coating.

Just as water deficit can be a serious threat to the health of trees, so, too, can excessive irrigation and flooding. In fact, symptoms of excessive water can sometimes be similar to those of water deficit. When soil becomes saturated with water, most of the oxygen is excluded from the pore spaces of the soil. Lack of oxygen stops aerobic respiration in the roots, causing a switch to anaerobic fermentation. During anaerobic fermentation, a great deal of storage reserves are used to release a small amount of energy, thereby "starving" the tree. In addition, byproducts of anaerobic fermentation can be toxic. Flooded conditions also promote the growth of bacteria and some fungi that can be harmful to trees and kill beneficial soil microorganisms. Trees will generally tolerate flooding for several days. If the flooding persists for weeks, then roots will be damaged.

Girdling

Girdling is the restriction, compression, or destruction of the vascular system within a root, stem, or branch that inhibits movement of materials in the phloem and xylem. Girdling is sometimes caused by poor planting techniques such as not removing

Figure 10.14 Stem-girdling roots can restrict vascular transport, especially in the phloem. Early fall color may be a symptom.

twine from the base of balled-and-burlapped trees following planting and staking/guying attachments that are left in place too long. Another common cause of girdling is damage caused by lawn mowers and string trimmers repeatedly striking the lower trunks of trees.

Girdling roots grow around or across the stem or buttress roots. As the trunk and roots increase in diameter, girdling roots may begin to constrict (or girdle) the tree. Phloem and, sometimes, xylem tissues in the trunk become compressed, which may cause swelling directly above the point of girdling. In severe cases, compression can kill the cambium, weaken the stem, and make the tree more prone to failure. Symptoms of girdling roots may be difficult to detect because they often resemble disorders associated with other stressors. Affected trees may show crown symptoms similar to those of water deficiency, and early autumn leaf color and abscission are very common. Eventually the tree may also show significant dieback.

Girdling roots often develop on container-grown plants or result from improper planting. Trees planted too deeply do not develop a normal trunk-to-root flare and are more likely to develop girdling roots. In addition, some species are more likely to develop girdling roots than others. Often girdling roots are not detected until symptoms begin to appear in the aboveground portions of the tree, when the girdling roots are too large to be removed without severely damaging the structural support and uptake capabilities of the root system.

The best way to deal with girdling roots is to prevent them from developing. Carefully inspect nursery stock for girdling or circling roots, and reject stock with girdling roots or sever the girdling roots before installation. Trees should be planted so that roots radiate outward from the stem. To minimize the development of adventitious roots that may encircle the stem, take care not to plant the tree too deeply, and make sure that soil or mulch does not collect around the base of the trunk.

Girdling roots are not always visible above the soil line. If girdling roots are suspected but not obvious, shallow excavation with a hand tool may be sufficient to expose them. If girdling roots are present, it may be possible to remove the portion in contact with the stem. Roots that have become grafted to the trunk or roots that are too large (larger than one-third of the trunk diameter) should be left in place because severing them may remove a large portion of the tree's root system.

Disorders Caused by Biotic Agents

Biotic agents are living organisms including pathogens, nematodes, insects, mites, parasitic plants, and vertebrate animals. Damage from biotic agents tends to be nonuniform with irregular borders and lacking any regular or even pattern on the tree or plants in the landscape. Usually a single plant, a few plants, or one type of plant are affected, while other species in the same location are not. Most tree species are susceptible to at least one insect, mite, or pathogen; some species, such as those in the rose family, are prone to many.

Not all disorders caused by biotic agents seriously threaten tree health. Correct identification is important, and many biotic agents leave signs such as eggs, excrement, silk, wax, or fruiting bodies

Figure 10.15 The Japanese beetle is an example of a leaf skeletonizer.

that can be clues to their identification. But remember, diagnosis is usually more involved than simply identifying a single pest, and there are often abiotic factors affecting plant health as well.

Insects

At any given time, there may be several species of insects present on a tree, although only a few, if any, may be harmful. Many insects are beneficial predators or parasites of harmful insects. However, almost every tree species has at least one insect pest that can cause some problems. Some insects feed on a variety of host plants (Japanese beetles, aphids, scales of some varieties, gypsy moths, and cankerworms), while others (emerald ash borers, hemlock woolly adelgids, and holly leafminers) are specific to certain hosts.

Insects have complex life cycles; one developmental stage may cause problems, while another does not. Knowledge of the insect life cycles is important in identification and treatment. An arborist must be able to identify both harmful and beneficial insects and know the extent of damage the plant can tolerate before any control is necessary. Reference books and diagnostic labs can be helpful in this regard, or it may be necessary to consult an entomologist (insect expert).

Most insect damage to trees is the result of either feeding or egg-laying activity. Cicadas can destroy small twigs by ovipositing (laying eggs) in slits under the bark. Because periodical cicadas may be present in large numbers during years when major broods emerge, oviposition injury can be a major problem, especially to young and newly planted trees.

Insect feeding damage is characterized by the type of mouthparts the insect has. Insects such as caterpillars, webworms, sawfly larvae, beetles, and weevils have chewing mouthparts. Chewing insects eat plant flowers, leaves, buds, and wood. Some insects, such as gypsy moth caterpillars, eastern tent caterpillars, and cankerworms, eat the entire leaf. In contrast, black vine weevils feed on the leaf margins, and their damage appears as uneven or broken margins or as notches on leaves. Other insects, such

Figure 10.16 The holly leafminer, like other leafminers, feeds between the upper and lower leaf surfaces, making tunnels.

Figure 10.17 Eastern tent caterpillar larvae spin silken webs in the branch unions of trees.

as Japanese beetles and elm leaf beetles, eat only the tissue between the veins, creating a **skeletonized** leaf. Leafminers feed between the upper and lower leaf surfaces, excavating tunnels inside the leaves.

Bark beetles and borers are larvae of beetles, moths, or wasps that have chewing mouthparts. The adults tunnel under the bark to lay eggs, or they may lay eggs on the bark and the larvae will bore into the host. The larvae eat the inner bark, phloem and cambium, and/or xylem, thereby destroying the tree's ability to transport water and nutrients between the roots and crown of the tree. Some borers, like the Asian longhorned beetle, tunnel into the wood of the plant to cause structural damage and may allow entry of wood-decay fungi. Trees infested with bark beetles or borers typically exhibit crown symptoms. These can include thinning, wilting, chlorosis of foliage, and death of twigs and branches. Because bark beetles and borers differ in feeding style, gallery pattern, and exit hole shapes, these insects often are identified by their work even after they have left the scene. Evidence includes small emergence holes in the trunk or branches with frass (semi-digested wood). Bark beetles and borers often infest trees stressed by other factors, including drought and disease.

Other insects feed by piercing plant tissues and sucking out the liquid contents. These include aphids, adelgids, scales, leafhoppers, psyllids, and true bugs. Symptoms of this type of feeding include chlorosis, stippling, drooping, and distortion.

Figure 10.18 Treehoppers are one example of insects that feed with piercing-sucking mouthparts.

Some symptoms are caused by **phytotoxic** chemicals secreted into the plant during feeding. Some scales can cause serious disorders of trees, partially because they may not be detected for years. Generally, however, most sap- or leaf-feeding insects do not kill trees outright, although they can be an additional stress factor.

Some sucking insects such as aphids, soft scales (not armored scales), and mealybugs use their piercing-sucking mouthparts to feed in phloem, excreting excess sugar in the form of a liquid called **honeydew**. Honeydew serves as a substrate for the growth of a dark, unsightly but entirely superficial fungi that cause **sooty mold**. The black sooty mold coating of leaves can impact plant health by interfering with photosynthesis, in serious cases.

Some insects are **vectors** of plant pathogens. This means that they carry a disease-causing organism from one tree and introduce it into another tree. Dutch elm disease is an example of a disease in which a fungal pathogen is often spread by an insect vector—bark beetles. Fire blight bacteria can be spread by bees to other trees they visit while collecting nectar from flowers. Aphids and leafhoppers are among the insects that are known to transmit viruses.

When considering damage from insect pests, it is important to consider the harm to the overall health of the tree. For example, late-season defoliation of broad-leaved trees can be visually striking. But defoliation at this time has much less of an impact on overall tree health compared to early-season defoliation, which requires trees to utilize carbohydrate reserves to produce new leaves. Feeding damage to evergreen conifers can be more harmful than to deciduous trees because many lack the ability to produce new foliage until the following year.

Mites

Mites are actually arachnids, close relatives of spiders and ticks and distant relatives of insects. The mites that cause significant damage to woody plants are very small, and hand-lens magnification is

Figure 10.19 Spruce spider mite feeding causes discolored needles and can lead to branch dieback.

usually required for identification. There are many species of mites, but they all have piercing-sucking mouthparts. Some species, such as the spider mite, discolor leaves by extracting leaf fluids, resulting in symptoms similar to those caused by sucking insects. Another type, called **eriophyid mites**, may cause galls to form on foliage and twigs because of feeding or egg laying. These galls are abnormal plant tissues that develop in response to chemicals produced by the insect or mite. Eriophyid mites are so small that they cannot be seen with the average hand lens, and they are typically diagnosed based on the galls they produce. Not all mites are harmful to woody plants. In fact, some mites prey on other mites in the landscape.

Figure 10.20 Maple bladder gall damage is caused by a tiny eriophyid mite and is typically just an aesthetic problem.

Nematodes

Another group of common plant pests is **nematodes**. Nematodes are animals, but because problem-causing nematodes require entering a plant to complete their life cycle, they are also considered pathogens. Nematodes are small, unsegmented, often microscopic, wormlike creatures. Thousands of species have been identified, but not all are harmful to plants. Most nematodes are found in the soil, although some can live in other environments. Some beneficial nematodes are important in the breakdown of plant debris. Plant parasitic nematodes commonly attack the root system of the plant, although others can damage any part of the plant. Symptoms of nematode diseases include swelling, deformation, galls, stunting, chlorosis, and wilting. Nematodes can also be vectors of other pathogens, but some nematodes are beneficial and can be applied to soil, foliage, or bark to kill insect larvae.

Other Animals

When animals feed on phloem, cambium, and underlying xylem, stems and branches can be girdled and killed. When the phloem is damaged, translocation of photosynthates is disrupted, which can lead to the death of localized foliage or the entire canopy if the trunk has been completely girdled. The absence of bark removes the protective covering, allowing drying of the vascular tissues and, potentially, entry of pathogens and insects.

Woodpecker and sapsucker damage is fairly easy to identify, although it can be confused with borer damage. Woodpecker holes are typically much larger than borer exit holes and are caused when the bird pecks into trees in search of borers or other insects. Sapsuckers bore a series of smaller holes, usually in horizontal rows, in order to feed off sap and insects. Generally, this feeding does not significantly injure the tree. However, trees that are repeatedly damaged may begin to show stress symptoms.

Mammals can also cause problems to trees. Deer browse on buds, twigs, and leaves of woody plants, destroying new growth and permanently deforming shrubs and trees. Deer feeding looks jagged or torn compared to the clean-cut feeding of rabbits and rodents; it also occurs higher on the plant. During the winter when food runs short, squirrels may chew the bark of some species and clip short twigs from conifers. In the spring, squirrels prefer swelling tree buds. Rabbits, mice, voles, and other small mammals can damage and kill valuable trees and plants by gnawing on bark and clipping off leaves and stems. Damage is especially severe in the winter during long periods with snow cover. Young plants and trees can be completely clipped off at snow height, while older trees can be girdled and killed.

Beavers feed on a wide variety of plants, especially leaves, young twigs, and the inner bark of older trees. They can also harm trees indirectly when they colonize areas and dam waterways, flooding large areas of land. If the existing vegetation remains submerged, it usually dies.

Trees and associated plants provide shelter and food for a variety of birds and other animals. Nowadays, clients are more aware of the need for wildlife habitat preservation. Therefore, arborists should target only nuisance animals when prescribing management solutions for avoiding tree damage. For example, if rodents are a problem during the winter, do not leave coarse layers of mulch close to the base of trees where animals can hide and chew. In addition, or alternatively, place a protective plastic sleeve around the lower trunk.

Diseases

Three requirements are necessary for a tree to become diseased: the tree must be susceptible to the pathogen, the pathogen must be present, and the environment must be suitable for disease development. These three factors are the elements of the **plant disease triangle**. In contrast to injury, which occurs very quickly or only intermittently, time is a fourth factor in disease development, resulting in referral to the plant disease pyramid.

Some pathogens are host specific, meaning they attack a specific plant species, genus, or family. Other pathogens attack a broad range of species. The part of the tree affected often greatly influences the severity of the disease. For example, diseases that affect only the foliage may not have a major effect on deciduous tree health unless defoliation occurs in several consecutive years. However, pathogens

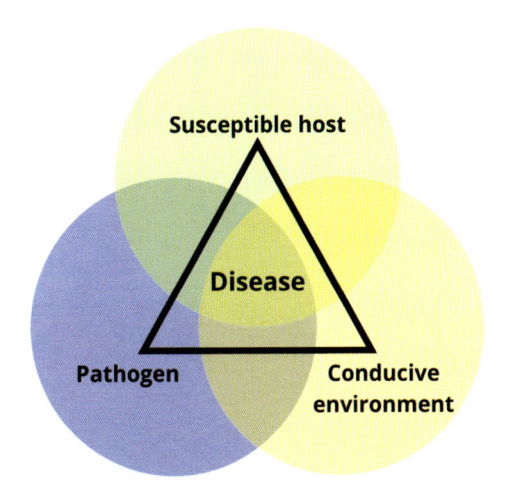

Figure 10.21 Plant disease triangle. Three requirements are necessary for a tree to become diseased: the tree must be susceptible to the pathogen, the pathogen must be present, and the environment must be suitable for disease development. Some plant pathologists consider "time" as the fourth requirement, making the triangle a pyramid.

Figure 10.22 Apple scab is an example of a fungal leaf spot disease that can also infect fruit. Severe cases can lead to complete defoliation.

Figure 10.23 Verticillium wilt is an example of a vascular wilt disease. Note the discoloration in the xylem.

colonizing vascular tissues, such as the fungi that cause oak wilt and Dutch elm disease, tend to cause mortality more rapidly.

Because foliar diseases often produce visible damage to foliage (and sometimes fruit and young twigs), they can pose significant aesthetic problems. Often, however, by the time a client reports the problem, it could be too late to treat for that growing season. Signs of the pathogens that cause these diseases are usually tiny but may be found in dead portions of killed leaves. Diagnosis might require microscopic examination and identification of fungal spores from samples. The first lines of defense against leaf diseases are to plant resistant varieties, and avoiding prolonged moisture on leaf surfaces inhibits infection. Other cultural measures used to maintain good growing conditions for trees affected by leaf diseases may help reduce their impacts.

Symptoms of shoot and twig blights include wilting and necrosis of leaves and stems and cankers on blighted shoots and twigs. The tissues of the bark of a canker are dead, along with the leaves and stems around the canker. These diseases are most often caused by fungi and sometimes bacteria. Some require microscopic identification of spores. These fungi are often opportunistic pathogens that attack trees stressed by abiotic environmental or site factors.

Vascular wilt diseases are usually the most difficult to treat. Vascular wilting affects tissues that conduct water and nutrients. It occurs when pathogens enter the plant—often through wounds (including pruning wounds or wounds produced by vector feeding) or through grafted roots—and colonize the xylem or phloem. Disruption of the flow of water and nutrients between the roots and foliage leads to eventual death. Vascular diseases are, for the most part, impossible to eliminate from an individual after infection occurs. Dutch elm disease and oak wilt are examples of vascular diseases that may result in rapid death of large trees, although there are some vascular diseases that progress relatively slowly. Cutting into affected stems will sometimes reveal vascular discoloration, although diagnosis of vascular disease often requires laboratory analysis of samples containing the pathogen.

Some pathogens attack fine roots only, whereas others can cause decay of large, woody roots, resulting in increased risk. Trees with root disease may or may not develop crown symptoms even after significant portions of the root system have been affected. Reduced growth, wilting, and dieback are typical symptoms for root diseases and other disorders affecting roots.

A few plant pathogens require **alternate hosts** to complete their life cycle and produce different

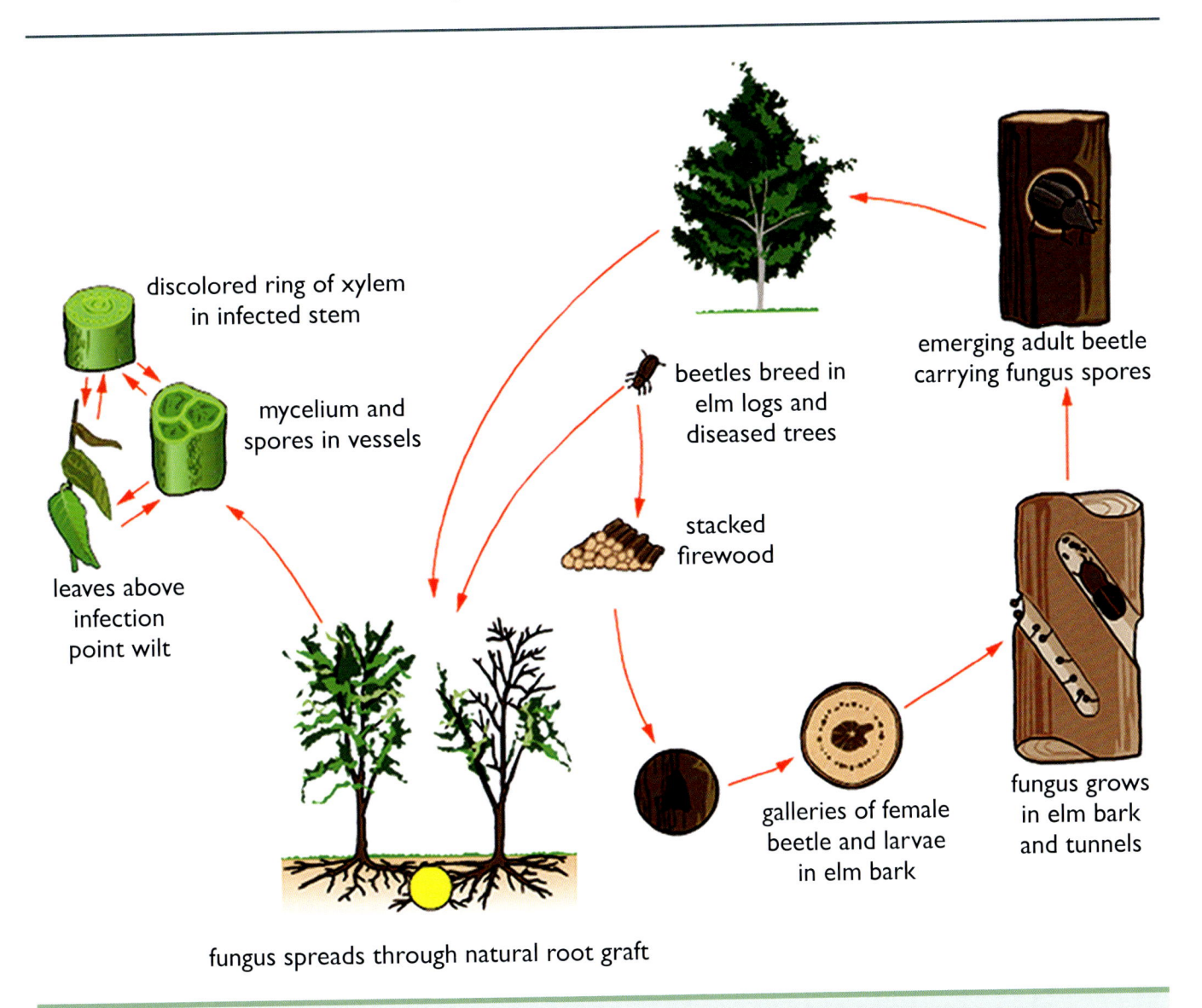

Figure 10.24 Cycle of Dutch elm disease. Understanding the life cycle of a disease or pest can be important in developing an effective management plan.

types of spores on each host. Many of the rust fungi are included in this category. Some rust fungi attack leaves and fruits and other stems, and some attack all of these plant parts. In most cases, management strategies are targeted at planting resistant varieties, improving tree health, and not planting the alternate hosts together in a landscape.

Fungi are responsible for the majority of plant diseases, and most trees are susceptible to at least one type of fungus. However, there are many beneficial fungi that play an important role in the natural cycling of nutrients. Mycorrhizae are symbiotic relationships between certain species of fungi and trees, aiding in the uptake of water and nutrients and providing other benefits. For most fungi to exist, they must receive plenty of moisture and require specific temperatures.

Besides fungi, there are many other disease-causing organisms, or pathogens. Three very common diseases are caused by bacteria: fire blight, bacterial leaf scorch, and crown gall. Fire blight and bacterial leaf scorch pathogens are vectored by insects, and crown gall is caused by a soilborne bacterium. Temperature and adequate moisture are important factors in the growth and spread of bacteria. Sometimes plant tissues colonized by bacteria

Figure 10.25 Cedar-apple rust is an example of an alternate-host disease of plants in the Rosaceae family, shown here on the alternate host *Juniperus virginiana*.

appear water soaked and may have an unpleasant odor, although many of the most common tree-pathogenic bacteria do not.

Some plant diseases are caused by phytoplasmas. Phytoplasmas are tiny bacteria-like organisms that can live only in phloem tissue. They are responsible for many of the "yellows" diseases, such as elm yellows, named for the symptom of extreme chlorosis.

Viruses are frequently spread by the feeding activity of insect vectors. Viruses seldom completely kill their tree hosts, although tree health might deteriorate over time. Viral disease symptoms are most often seen in the leaves as ring spots, mosaic, and mottling. Fruit trees, like plums and apples, have more viral diseases than shade and ornamental trees, though it is unknown whether fruit trees are more prone to viral disease or whether there is just more research on fruit trees due to their economic importance.

Management of plant diseases must be considered an exercise in prevention because very few curative treatments exist for tree diseases. Planting species or cultivars that are resistant to prevalent diseases is the best start. Most fungicides and antibiotics are best used to prevent the establishment of a pathogen rather than to eliminate existing disease problems. Steps taken to maintain tree health and improve cultural conditions are often the most effective disease prevention measures.

The Threat of "Exotic" Pests

As global commerce has increased, the importation of nonnative species of insects, pathogens, and even invasive plants is becoming a bigger issue. When an invasive pest is introduced to a new area, native trees may have little or no resistance, and there may be no native predators or parasites to control the pest. The same pest might be of minor importance or may significantly damage only stressed plants in their native land, but in the new locations, they can be very aggressive and kill otherwise healthy plants. The pest may then spread rapidly, causing a devastating loss of affected tree species. Invasive alien plants may displace native ones and change the composition and function of natural ecosystems.

The problem is worldwide, and experts are divided on how best to respond. Currently, most countries attempt to limit the accidental import of pests through shipping regulations, inspection protocols, and other means. Once a pest is detected, the area may be quarantined to limit its spread. Too often, though, the pest has already spread by the time it is detected. Localized infestations may be treated, or the host species may even be intentionally destroyed in attempt to eradicate the infestation while it is limited. These strategies have had limited success with a small number of species, but other pests have spread extensively. Much current research is focusing on the introduction of predators, parasites, and pathogens of exotic pests, often imported from the pest's native range. Researchers and policy makers are understandably cautious, however, about introducing new species and the potential to further affect, and possibly disrupt, the local ecosystem.

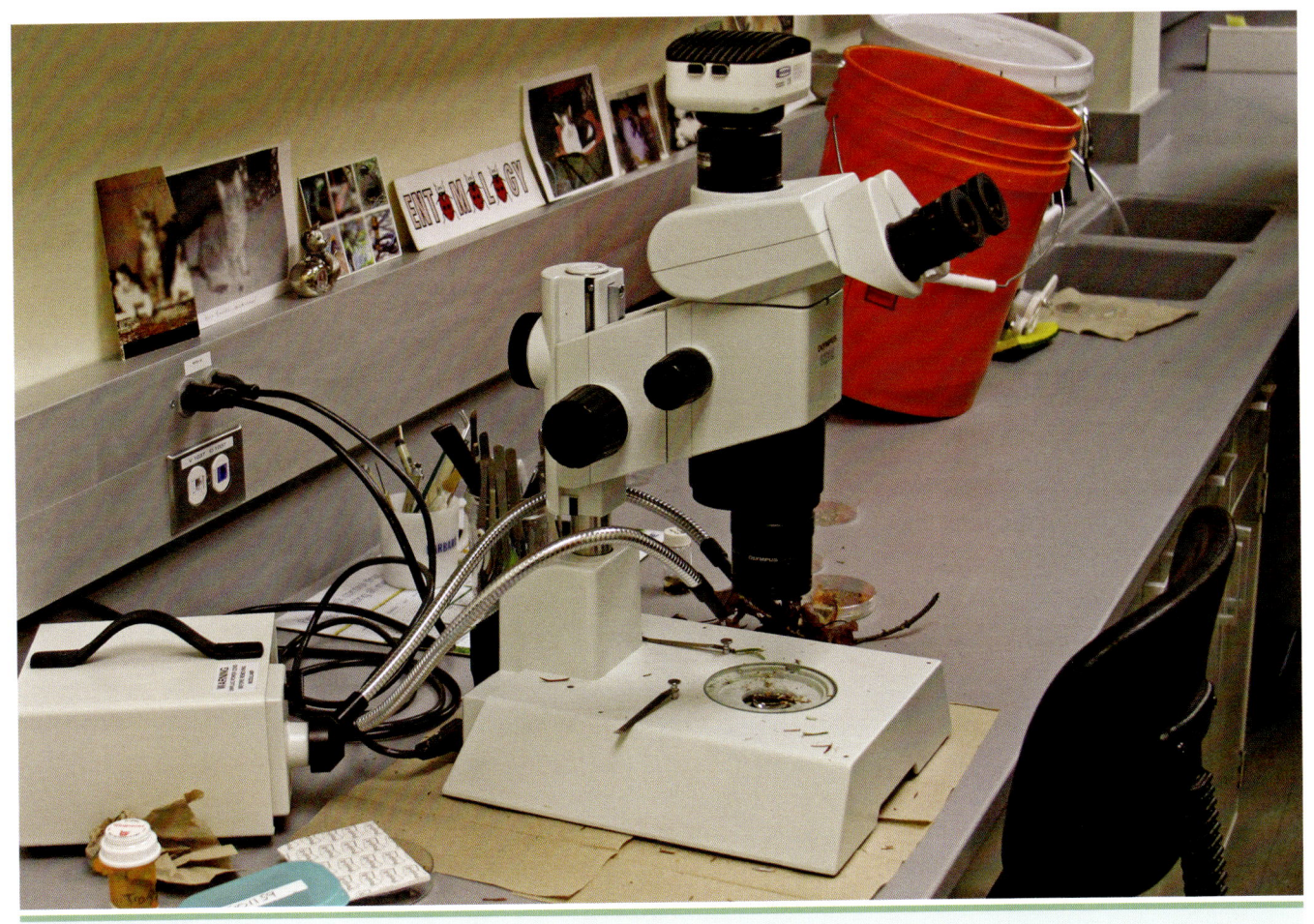

Figure 10.26 It may be necessary, at times, to send samples to a plant lab to confirm a diagnosis.

Getting Laboratory Assistance

In diagnosing tree health problems, it may be helpful to request laboratory assistance. A new or unusual pest or plant may be difficult for anyone to identify, and in some cases, testing may be the only way to confirm a diagnosis. Personnel at some laboratories can also recommend management strategies, in addition to culturing fungi, identifying other potential pests, and keeping arborists informed of potential problems of the season. Many countries and most states in the United States have diagnostic facilities available to the public through universities, the local extension service, or the USDA.

Plant diagnostic facilities use a variety of laboratory tests and tools including microscopes, culturing of pathogens, and some of the same analyses used for human pathogen diagnoses. RNA/DNA analysis, for example, has become common, relatively fast, highly accurate, and inexpensive. A limitation is that the techniques test for specific organisms and so are not useful as a first step to narrowing down the possibilities.

Samples for laboratory testing should be taken from portions of the plant exhibiting representative symptoms and signs, if present. If laboratory culture of a pathogen is to be attempted, the specimen should contain the interface (transition zone) between diseased and healthy tissue. Insects and mites should be placed in sealed vials, jars, or plastic bags. Specimens such as leaves or pieces of branches should be freshly collected, wrapped in dry paper towels to absorb moisture, and sealed in a plastic bag. Soft specimens such as mushrooms and fruit should be placed into dry paper bags or boxes. Wet or moist plant samples placed in plastic bags may arrive at the laboratory in poor condition. If practical, it is best to bring plant samples to the lab and describe the problem to the laboratory technician. If the sample must be mailed, minimize damage by properly packing the sample and by shipping early in the week and using the fastest shipping method available. Include a detailed description of the problem, the host plant, any known history, and the surrounding site conditions (including color photographs). Remember, rarely is the correct diagnosis simply a matter of isolating a fungus or identifying an insect or mite.

CHAPTER 10 WORKBOOK

1. True/False—Information about a tree's history and symptoms gained from a client can always be considered accurate.

2. If a tree is not well suited for the site in which it has been planted, it may become _____ , predisposing it to other problems.

3. A common mistake in diagnosis is to carefully examine the aboveground portion of the tree while ignoring the _____ .

4. True/False—If a tree declines or dies within the first year following installation, a likely cause is excess or insufficient water.

5. Pollution damage, girdling roots, and mineral deficiencies are examples of _____ disorders.

6. Name five causes of physical or mechanical injuries to trees.

    a.

    b.

    c.

    d.

    e.

7. Insect damage to trees is usually the result of feeding or _____ .

8. Name five insect pests of trees with chewing mouthparts. Name five with piercing-sucking mouthparts.

    Chewing   Piercing-Sucking

    a.   a.

    b.   b.

    c.   c.

    d.   d.

    e.   e.

9. Insects that carry plant pathogens and introduce them into hosts to result in disease are known as _____ .

10. True/False—Mites are not insects.

11. Microscopic wormlike organisms that sometimes feed on or in trees, causing disease, are called _____ .

12. Name the four factors required for a tree disease.

    a.

    b.

    c.

    d.

13. True/False—Vascular diseases of trees are rarely fatal.

14. True/False—Diseases that affect only the foliage of a deciduous tree may not be a serious problem unless defoliation occurs in several consecutive years.

15. True/False—Most fungi cause plant disease.

16. True/False—The pathogens that cause plant diseases are primarily fungi.

17. Fire blight is an example of a disease caused by a _____ .

18. _____ is the chemical inhibition of growth and development of one plant by another.

19. True/False—Pollution damage is often difficult to diagnose because the symptoms may mimic other problems such as insect injury and mineral deficiencies.

20. Curling and cupping of the foliage, and parallel venation, are common symptoms of _____ damage.

Matching

____ witch's broom   A. abnormal, enlarged plant structure, often insect or mite induced

____ vector   B. carrier of pathogens

____ canker   C. localized dead tissue, often sunken and discolored

____ gall   D. abnormal growth of multiple shoots

____ stunting   E. may predispose a plant to other problems

____ stress   F. causal agent of disease

____ pathogen   G. natural chemical inhibition of growth

____ dieback   H. abnormally reduced growth

____ allelopathy   I. progressive death of twigs and branches from the tips back

Challenge Questions

1. Outline the steps to be taken in the process of diagnosing a tree problem.

2. List the most common disease and insect problems in your region, and categorize them by severity.

3. Explain the procedure used in collecting plant samples to be sent to a laboratory for diagnosis.

4. Why is pollution damage often difficult to control? What other disorders have similar symptoms?

5. Describe the ways that humans are contributing to the reduced life span of trees in the urban environment.

Sample Test Questions

1. A condition characterized by a cluster of dwarfed shoots on affected twigs is called
   a. witch's broom
   b. anthracnose
   c. chlorosis
   d. Verticillium wilt

2. Twig dieback from periodical cicadas is primarily a result of

    a. ovipositing (egg laying)
    b. adults feeding on the foliage
    c. larvae feeding on the roots
    d. feeding-induced galls on the twigs and foliage

3. Plant damage associated with a sap-feeding insect pest might appear as

    a. leaves that have been skeletonized
    b. distorted leaves or shoots
    c. leaf mines or blotches
    d. webs or tents in the tree

4. Scale damage to plants is the result of

    a. fungal spore growth depleting xylem reserves
    b. a type of sucking insect causing a loss of vigor
    c. vascular damage from fungal invasion
    d. a physiological disorder due neither to insects nor to disease

5. Damage caused by nonliving factors tends to be

    a. uniform and may affect more than one species
    b. uniform but generally not affecting the new growth
    c. random and concentrated on the new growth
    d. random with irregular borders

Recommended Resources

(See Recommended Resources in back of book for detailed information.)

Abiotic Disorders of Landscape Plants (Costello et al. 2003)

Introduction to Arboriculture: General Diagnosis online course (ISA)

Introduction to Arboriculture: Abiotic Disorders online course (ISA)

Introduction to Arboriculture: Biotic Disorders online course (ISA)

Plant Health Care for Woody Ornamentals (Lloyd 1997)

Pest Management in the Landscape (Luley and Ali 2009)

Insects That Feed on Trees and Shrubs (Johnson and Lyon 1991)

Diseases of Trees and Shrubs (Sinclair et al. 2005)

CHAPTER 11
PLANT HEALTH CARE

▼ OBJECTIVES

- Explain the philosophy of Plant Health Care (PHC), and describe its relationship with Integrated Pest Management (IPM).

- Explain the concept of the appropriate response process (ARP) and how it is used in the diagnosis and treatment of plant health problems.

- Compare and contrast ways in which plants allocate resources depending on seasons and environmental stresses.

- Describe the components of an effective monitoring program, and explain why monitoring is an integral part of plant health management strategies.

- Identify the range of pest management options and the advantages and limitations of each.

- Explain why preventive strategies are often essential in Plant Health Care.

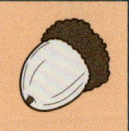

KEY TERMS

action threshold
allelochemicals
appropriate response process (ARP)
augmentation
bactericides
biological control
biorational control product
botanical pesticide
cellulose
chemical control
compost
contact pesticides
cultural control
degree day
eradication

fungicides
herbicides
herbivore
horticultural oils
insect growth regulators
insecticidal soaps
insecticides
Integrated Pest Management (IPM)
lignin
microbial pesticides
miticide
monitoring
monoculture
mortality spiral
parasite

pathogen
pesticide
pesticide resistance
pest resurgence
phenology
photosynthate
phytotoxic
Plant Health Care (PHC)
predator
prevention
resource allocation
root:shoot ratio
secondary pest outbreak
suppression
systemic pesticide
threshold

Figure 11.1 Plant Health Care is a holistic, comprehensive program to manage the health, structure, and appearance of plants in the landscape. This proactive approach to managing plant health involves proper planning, plant selection, and a wide range of cultural practices aimed at improving site and soil conditions.

Introduction

In the past, broad-spectrum **pesticides** and quick-release fertilizers were commonly used components of tree care. These practices attempted to control pest populations and increase plant growth. Today, sound management practices have shifted focus to maintaining plant health and managing pest populations to tolerable levels. The **Plant Health Care (PHC)** approach to managing trees recognizes that, in most cases, tree health problems are the result of many factors, not just a single agent. Plant Health Care takes a holistic approach when making management decisions that focuses on plants and their interactions with the living and nonliving elements of the landscape.

Plant Health Care practitioners often have a limited ability to treat disorders once they become noticeable to the client. Whereas past management practices were typically reactive in nature, PHC attempts to prevent problems before they start. This proactive and more cost-effective approach to managing plant health involves proper planning, plant selection, and a wide range of cultural practices aimed at improving site and soil conditions. When combined with careful monitoring to identify pests in the initial stages, these practices greatly reduce dependence on pesticides and other therapeutic treatments.

Definition and Philosophy

Plant Health Care is a holistic, comprehensive program to manage the health, structure, and appearance of plants in the landscape. Trees and other woody plants under the care of arborists are a part of a larger landscape ecosystem. Often, they are elements of a landscape that includes soil, shrubs, herbaceous plants, turfgrass, and hardscape. Trees interact, and sometimes compete, with other plants in the landscape, for example, with other trees and vines for light above ground. Below ground, trees compete with turf and other plants for water, minerals, and space. Additionally, cultural and maintenance practices that are employed to help one plant can sometimes harm the health and development of another plant.

Arborists have come to realize that they cannot view trees in isolation; they must consider the entire ecosystem. Consider the example of a golf course, where trees are growing in a "landscape" of highly maintained grass. Improper irrigation of turfgrass might create an environment favorable for root rot pathogens; excessive nitrogen application rates could lead to elevated populations of foliage-feeding insects as well as tree borers; and mowing activities may damage tree trunks, root collars, and roots. At the same time, the golf course superintendent will probably be frustrated by tree shade that limits light penetration to the turf and by tree branches that interfere with play. Maintenance options for either trees or turf can impact the health of nearby plants. Thus, the arborist and golf course superintendent should work together to prescribe maintenance practices that will maximize plant health while minimizing harmful side effects for the trees and turf sharing the same environment.

In a suburban landscape, it is common for property owners to contract with a tree service to maintain their trees, a lawn care company to manage their lawn, and a landscape maintenance firm to care for the other elements of the landscape. Without a coordinated effort, the maintenance practices performed by each company can affect the success of the others. For example, if each company is applying fertilizer, the potential exists for excessive application, which can lead to increased insect pest populations, disease incidence, unwanted growth, excess fertilizer salts in the soil, waste due to leaching, and groundwater pollution. This approach is inefficient and costly for the homeowner, and it may be harmful to the plants and the environment.

Plant Health Care is a proactive approach that, ideally, begins with the design of the landscape and selection of plants. Many plant health problems can be prevented if plants are matched appropriately to their sites. In addition, if trees are planted properly and well cared for in their early years, less maintenance may be needed as they mature. Unfortunately, arborists are usually called in long after these early stages, often when trees are in irreversible decline. Tree owners or managers then have a difficult time understanding why the arborist cannot just "fix" the problem. One of the main challenges for Plant Health Care professionals is getting involved in the early stages of the design process.

What Is a Healthy Plant?

Plant Health Care is the management of plant appearance, structure, and vitality. The appearance of landscape plants is a legitimate objective because landscapes are planted and valued as much for their aesthetic qualities as their functional utility. Structural integrity is extremely important when it comes to trees, not only for their preservation but also for safety issues. The subject of tree risk is given more attention in Chapter 12 (Tree Risk Assessment and Management). But what constitutes a healthy plant? A plant is considered healthy when it is free of significant disorders and pests and has sufficient ability to respond to stress, pests, and disorders.

Some plant care professionals differentiate between vitality and vigor, although the terms are frequently used interchangeably. Vitality is sometimes defined as a plant's ability to deal effectively with stress and thrive in a given environment. Vigor is considered by some to deal with the plant's

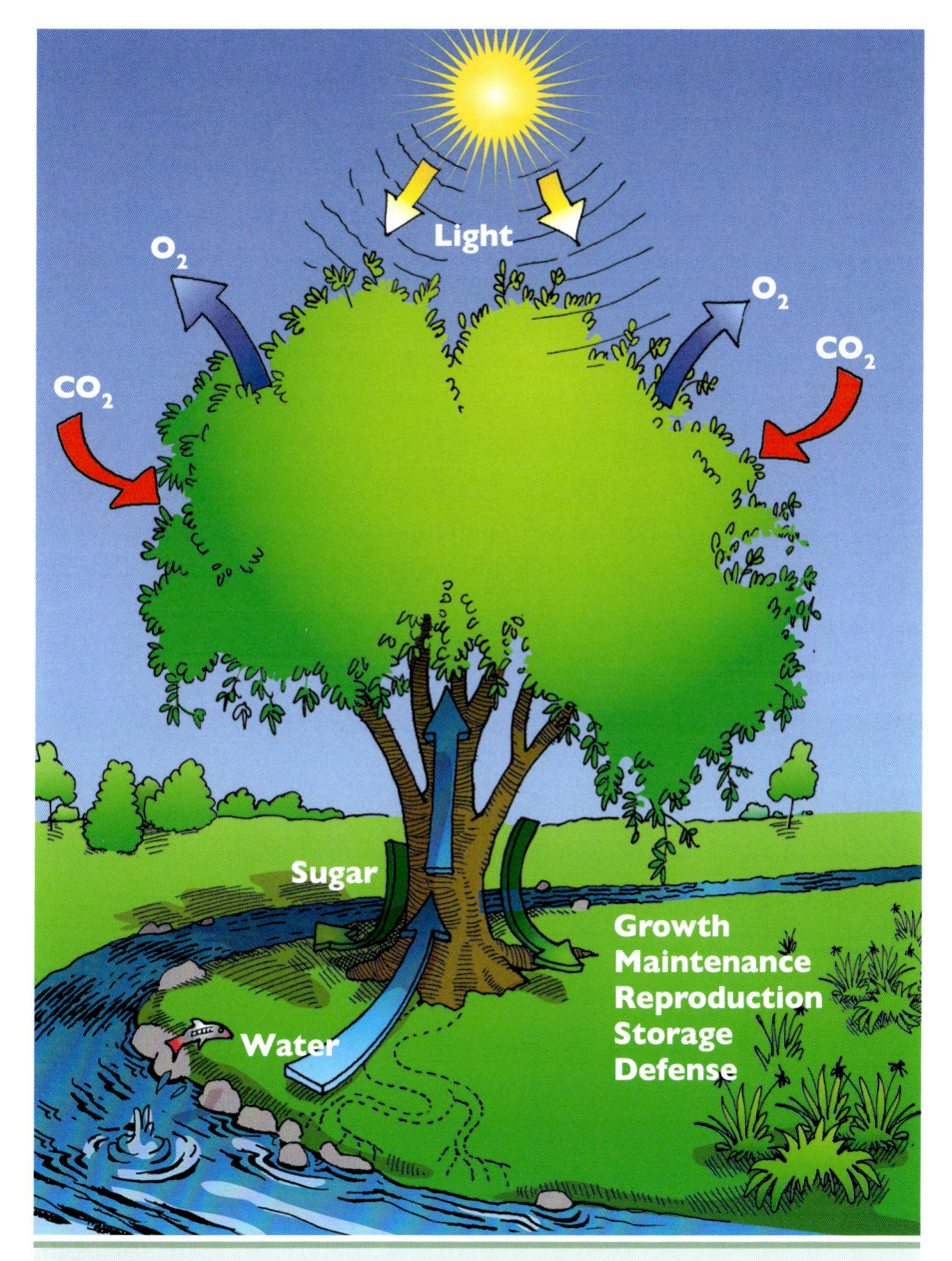

Figure 11.2 Resource allocation. Trees "budget" their limited resources among the processes that must be supported, including growth, maintenance, reproduction, storage, and defense. Not all processes can be fully supported at the same time.

long after selection and planting have taken place.

Some arborists, landscape professionals, researchers, and clients equate rapid growth with a healthy plant. The success of various treatments and practices has typically been measured by increased height growth, twig growth, trunk diameter, root mass, or total leaf area. As researchers have gained a better understanding of plant physiology and **resource allocation**, it is apparent that other aspects of plant health are often more important.

Trees and other plants produce sugars, which serve as their basic source of energy and building blocks, through the process of photosynthesis. This energy resource is allocated among five primary functions: maintenance, growth, storage, reproduction, and defense. Energy allocation to each function varies by tree age and health, seasonal and environmental conditions, stress factors, cultural practices, and even ecological strategies. For example, some tree species grow rapidly, reserving fewer resources for defense. Others grow at a slower rate but make greater allocations to defense and, therefore, may have a higher capacity to tolerate stressful conditions.

Photosynthesis requires light, water, CO_2, chlorophyll, and a few essential minerals. If any of these are limited, the plant will be stressed. Any condition that limits a plant's ability to obtain or use these key resources (or that leads to excessive amounts)

inherent genetic capacity to resist stress. A plant's inherent tolerance of various soil types, moisture conditions, cold or heat, and other factors is what defines its vigor. The only way to manage plant vigor is to select plants well suited for the environment and resistant to pests. Instead, most plant maintenance professionals are engaged in promoting the vitality of plants because they are usually called in can be considered a stress factor. Stress factors are often directly related to soil quality or other environmental conditions (for example, pH, drought, compaction, or excess water), and many can be attributed to human activity such as land development and construction. Some stress factors reduce a tree's ability to photosynthesize, decreasing the amount of sugar produced. This can cause the tree to enter a "survival mode," redirecting resources away from growth, storage, and defense and toward reproduction and maintenance. On the other hand, mild stress can redirect resource allocation with the potential effects of increasing drought or pest resistance.

Plant Defense Mechanisms

Trees may possess chemical and/or physical traits that deter or reduce insect or other animal feeding, mechanical damage, and pathogen infection and resultant diseases. Some trees have thorns, spines, tiny hairs, or leaves with thick, toughened cuticles that serve as physical defense mechanisms.

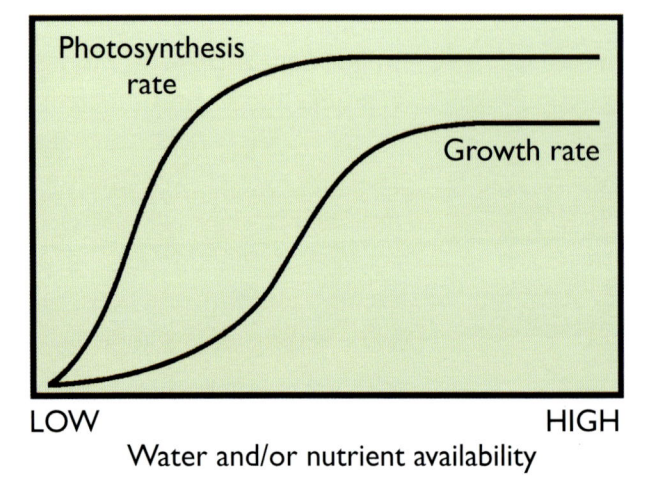

Figure 11.3 The effect of stress on growth and photosynthesis. The growth of trees is sensitive to the availability of moisture and nutrients and becomes limited even when these are mild deficiencies. Photosynthesis is much more resilient and does not become limited until nutrient and moisture deficits become more severe.

Cellulose and **lignin** in tree cells serve as structural defense because many **herbivores**, and even some pathogens, cannot digest these compounds. Compartmentalization processes are both anatomical and chemical defense mechanisms that resist insect damage and disease spread. Trees also produce **allelochemicals** such as tannins, other phenols, and other compounds that have toxic or deterring effects on certain herbivores and pathogens. In fact, extracts of some allelochemicals are used in certain pesticides. Nicotine, pyrethrin, neem, and rotenone are examples of plant-produced allelochemicals that function as natural insecticides. These allelochemicals are thought to be a major reason that most insect pests are host specific; the types and amounts of chemicals produced vary by species. Many herbivores have developed special enzymes or processes that allow them to detoxify or safely eliminate specific compounds.

Many research studies have investigated the relationship between environmental conditions and allelochemical production in trees. For example, moderate drought stress can increase levels of allelochemicals, which have been shown to boost a tree's defense system. Conversely, some sun-adapted tree species that are grown in shade have reduced levels of allelochemicals and lowered resistance to some pests. Much research has concentrated on the effects of nitrogen fertilization on host-pest relationships. Researchers have determined that rapidly growing trees are often less resistant to certain insects and diseases. This finding is contrary to the traditional belief that a rapidly growing tree is always a healthier tree. An explanation supported by this research is that **photosynthate** (plant sugars) is diverted from defense compound production to increased growth of succulent tissues. Photosynthesis is less sensitive to water and nutrient availability than is growth. Increased growth without a corresponding increase in photosynthetic rate of individual leaves results in trade-offs among different plant functions. Also, foliage with high-nitrogen content (such as succulent, new growth) generally benefits insect population growth and the development of certain pathogens because it is more nutritious.

Trees can adjust to adverse environmental and site conditions, and to moderate levels of stress, in a variety of ways. For example, when available water is limited, stomata may close, the levels of certain dissolved substances in cells may increase to maintain turgor pressure, and the **root:shoot ratio** may increase, meaning that roots will grow faster than shoots. Longer-term moisture stress triggers more dramatic responses such as leaf drop. Trees respond to low light levels by increasing shoot growth toward the light source and by adjusting leaf size and thickness to maximize photosynthesis of the whole plant. If nitrogen is deficient, trees may decrease shoot growth and increase root growth to maximize mineral uptake and availability for existing foliage. They may also produce smaller, thicker leaves, which has the effect of maintaining high rates of photosynthesis even as growth is decreased.

While mild levels of stress can sometimes increase resistance to certain pests, severe or multiple stressors can trigger a state of tree decline. Stressed trees can continue to decline over time. Chronically stressed trees are more likely to succumb to drought, defoliation, borers, bark beetles, or certain diseases. Intervention to manage tree health and improve vitality should occur as early as possible. Cumulative stresses can compound to the point that few management practices will be effective, at which point the tree is said to be in a decline or a **mortality spiral**. As already stated, the options an arborist has for reversing the effects of stress are often limited, so the emphasis must be placed on preventing or minimizing stress. As trees age, their patterns of resource allocation change as well. Young trees allocate relatively more energy to growth than to defense. As they mature, the situation changes so that defense increases in priority.

Plant Health Care Process

The key to the success of any PHC program is monitoring and examining the plants and their growing conditions. Monitoring is an important process of observing, identifying, recording, and analyzing what happens to plants in the landscape. Early detection and correction of situations that may cause plant stress can prevent more serious problems in the future. If a disorder is detected, its severity and potential impact must be assessed before deciding on an appropriate response. The process of gathering information, assessing the severity and implications of the problem, determining client expectations, formulating options, and deciding on a course of action is called the **appropriate response process (ARP)**.

Arborists have borrowed the concept of **thresholds** from Integrated Pest Management (IPM) applications with agronomic crops. When plants are grown as crops, the threshold for treatment of a pest population is determined by doing a cost-benefit analysis of the value of yield loss compared to treatment expense. The thresholds for intervention measures on landscape plants can vary according to client expectations and tolerance; they are often aesthetic thresholds. Although some pest or disease problems have little effect on the overall health of a plant, the aesthetic qualities of the plant may be temporarily or permanently compromised, or the pest may create a nuisance and warrant intervention.

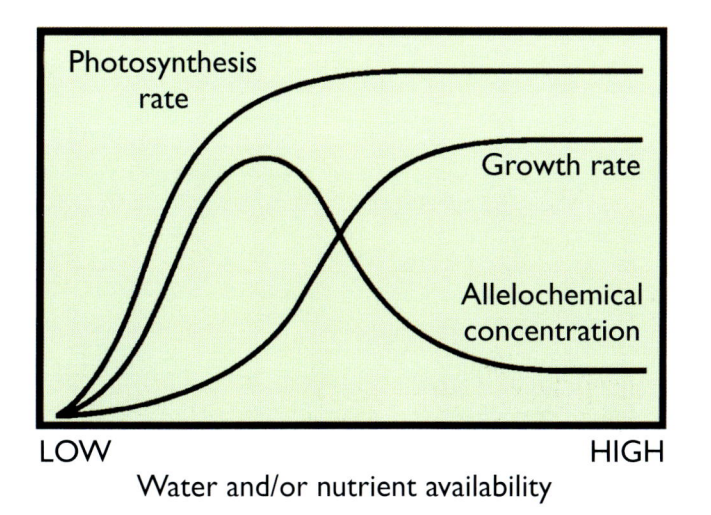

Figure 11.4 Moderate nutrient and water stress does not impact photosynthesis but does limit growth, making carbohydrates available to support other processes such as defense.

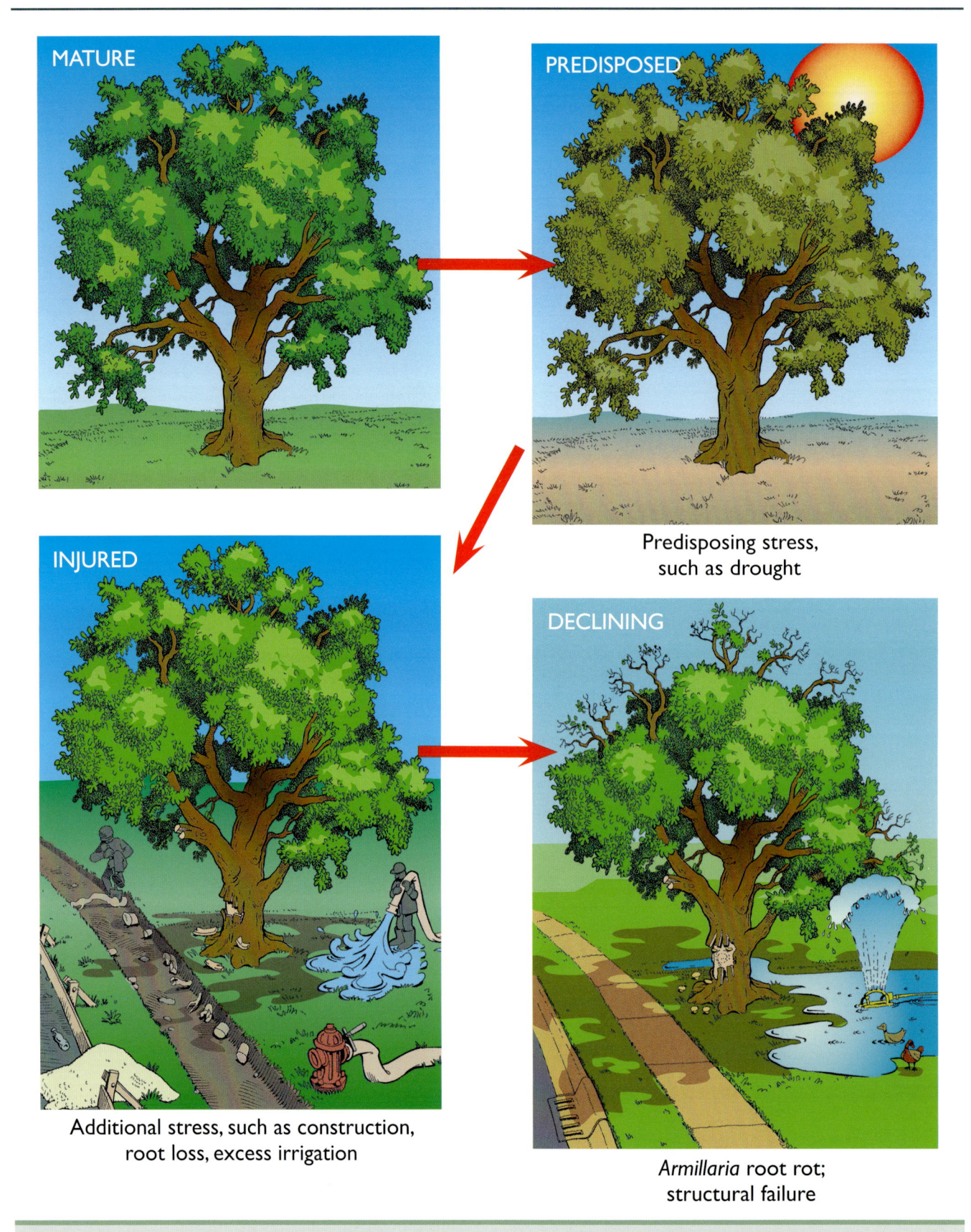

Figure 11.5 The mortality spiral illustrates how stress factors can compound, which can predispose a tree to additional problems and lead to decline.

Figure 11.6 An appropriate IPM strategy will avoid harmful effects on beneficial and nontarget organisms and cause minimal disturbance to the environment.

Another important aspect of PHC is arborist and client education. In some cases, it is preferable not to treat cosmetic problems but rather to help the client understand the nature of the pest and its minimal impact on tree health. First, however, the arborist must be knowledgeable. It is up to professionals to make educated decisions and avoid short-term treatments at the expense of long-term solutions. This is part of the theory and practice of appropriate response. Each situation can vary. What is most appropriate for one client or situation may be very different for another. Remember, Plant Health Care is a philosophy, an educational process, and a decision-making process.

Integrated Pest Management

Pest management is an important component of PHC. Pest management in the 20th century after World War II was dominated by pesticide use. More recently, there has been increased concern about the use of pesticides in the environment. **Integrated Pest Management (IPM)** is a method that manages plant pests by combining various control tactics. This method was developed as a way to address the ecological, social, and economic implications of overreliance on chemical pest control, and it provides a more effective means to manage pests and enhance plant health and appearance. Plant Health Care embraces IPM as the preferred approach to managing landscape pests. Think of IPM as one of the components of PHC.

The vast majority of organisms found in the landscape are not pests. Moreover, of those that are pests, most are not serious threats to plant health. In fact, many organisms are beneficial, making positive contributions to the landscape by pollinating plants, suppressing pest populations, or facilitating organic-matter cycling. In the context of managed landscapes, a pest is any organism that meets any of the following criteria:

- It competes with desirable plants for resources.
- It threatens the health, structural integrity, or appearance of desirable plants.
- It diminishes personal enjoyment, utility, or safety in the landscape.

An organism may be a pest in some circumstances while not in others. In addition, the mere presence of a pest may not necessarily warrant treatment. The task of the landscape manager is to determine whether an organism is a pest and whether the population is high enough to warrant intervention.

Integrated Pest Management is a method for managing pests that combines appropriate preventive and control tactics into a single management strategy. The goal of landscape IPM is to manage pests and their damage at tolerable levels. Pest eradication usually is not a necessary, achievable, or even desirable IPM goal except when dealing with highly infectious or damaging pests and certain exotic organisms. Typically, Plant Health Care focuses on pest prevention and suppression rather than eradication.

When devising an IPM strategy, the ecological, social, and economic implications of pest management should be considered. An appropriate IPM strategy will

- emphasize decision-making based on monitoring and action thresholds;
- complement other PHC practices to promote plant health, structural integrity, and appearance;
- avoid harmful effects on beneficial and nontarget organisms (people, animals, beneficial insects, and plants);
- cause minimal disturbance in the built and natural environments; and
- achieve the goals of the client in a cost-effective manner.

Figure 11.7 Monitoring should collect and record information about the site, the plants, and any disorders noted.

Monitoring

Monitoring is a program of regular landscape inspections to make observations and collect information to aid in making decisions about the management of pests and other disorders. Three types of information should be collected during a landscape inspection: site information, plant information, and disorder information.

Site information is valuable for the diagnosis and management of landscape pest problems. Valuable site information includes recent weather trends, landscape management practices, changes in drainage or land contour, addition or removal of nearby plant material, and hardscape construction or repair. These events can influence plant vitality and pest activity both positively and negatively. In many instances, plants are predisposed to pest infestation by adverse site conditions. Because woody plant response often lags behind environmental or cultural change by months to years, site history is extremely important. Identifying and correcting an adverse site condition, where feasible, can be an integral part of an effective PHC strategy.

Collecting information on plant condition is another important aspect of a landscape inspection. Leaf number, size, and color; twig growth; symptomatic reactions; and any signs of disorders are just a few indications of a plant's health. A plant's condition can influence its pest susceptibility and its injury tolerance. As a result, it may be appropriate to adjust the **action threshold** for a potential pest based on plant condition. For example, the action threshold for a pest may be relatively low for a plant

Figure 11.8 Inspect plants closely to observe signs and symptoms of disorders. Monitor for stages within the life cycles of pests and note population levels.

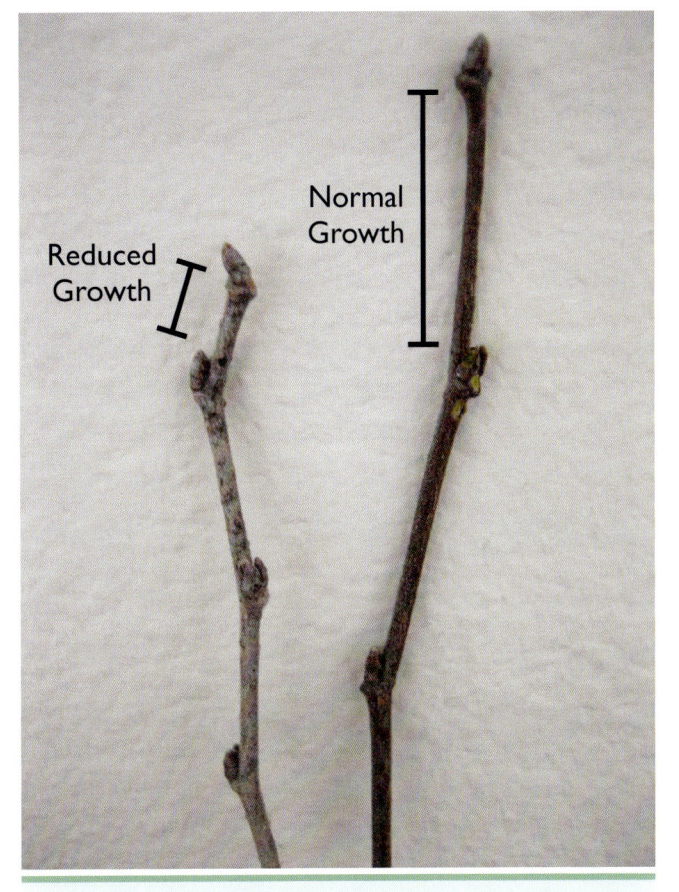

Figure 11.9 Measuring the extension growth of twigs is one way to observe stress. Look for decreased growth compared to previous years and compared to other plants of the same species.

in poor condition because the addition of even a relatively mild stressor may lead the tree to an irreversible state of decline.

Plant phenological stage can also influence injury tolerance and impact pest control decisions. **Phenology** is the study of recurring biological events (pest emergence, flowering, fruiting, leaf drop, etc.), which are often related to seasonal patterns of weather, temperature, and day length. For instance, a late-season, defoliating pest on a deciduous tree may not warrant control, because the leaves will soon fall. Monitoring pest phenology is critical to accurate timing of management tactics.

Plant pathogens (fungi, bacteria, etc.) may infect plant tissue but may not be noticed through monitoring until symptoms occur. Symptoms may include wilting, browning (necrosis), yellowing (chlorosis), or dieback of the plant or plant part. Sometimes, signs of the causal agent, such as fungal fruiting structures or bacterial ooze, are evident. It is important to identify highly susceptible plant species that are prone to specific diseases and monitor for signs and symptoms of known diseases or treat preventively if records indicate a previous history of disease.

Several types of pest information should be collected during a landscape inspection, including

- identification of the pest or other disorder;
- population level or severity;
- life stage(s) present, symptoms, and signs; and
- the potential for natural control of the pest.

Assessing the pest population level or the severity of a disorder is required for determining the action threshold. Many insect pests and diseases can be effectively managed only during a specific, vulnerable life stage. When pest control is being considered, it is imperative that the life stage(s) be accurately identified so that appropriate management tactics can be chosen. If there is high potential for natural control of a pest, then supplemental control may not be warranted. Instead, the appropriate response may be to continue monitoring the pest to gauge the efficacy of natural controls and then apply additional control tactics only as needed.

Numerous tools and techniques are available to aid in on-site landscape inspection and monitoring. A hand lens, binoculars, a field guide, a soil probe, and sample collection containers are valuable tools for on-site identification of plant species and disorders. If field identification is not possible, a sample should be collected for identification by a colleague or diagnostic clinic.

Often it is difficult to evaluate a pest problem based solely on visual inspection of the landscape. A pest may be active only at certain times (for example, black vine weevil adults feed at night), or current damage levels may provide insufficient information to guide management decisions. In addition, some pests are vulnerable to control only during a specific life stage that is difficult to detect (such as newly hatched larvae of clearwing moths). In these circumstances, trapping devices can be used to monitor pest abundance and life stage, to evaluate action thresholds, and to properly time management tactics. Trapping devices are commercially available for various (but not all) insect pests.

A simple method for tracking the seasonal development of pests is a phenology calendar. Annual natural events such as plant development and pest emergence are often better correlated with seasonal patterns of weather than with specific calendar dates. As a result, the development of insect and mite pests can be reasonably predicted by observing the budding, leaf expansion, flowering, or fruiting stages of common native or landscape plants. With a few years of careful observation and record keeping, a PHC practitioner can develop a reliable phenology calendar. A phenology calendar may also be available through local university or governmental agencies.

A **degree-day** model is a quantitative method for accounting for the effect of seasonal warming on pest development. There is a lower-threshold temperature below which insect and mite development slows dramatically or ceases. For insect and mite pests of woody plants, a commonly accepted lower threshold is 10 °C (50 °F). A simple degree-day model uses this lower threshold temperature (base temperature) and the daily average temperature (the average of the recorded high and low temperatures for a given day) to predict pest development stages.

To calculate degree days (also known as heat units), the base temperature (10 °C [50 °F]) is subtracted from the daily average temperature (sum of the maximum and minimum temperatures

Figure 11.10 Trapping devices can be used to monitor pest populations.

Figure 11.11 If unsure about the identification of a pest or other disorder, collect key information and samples and send them to a plant diagnostic lab.

divided by two). The difference equals the degree-day units for that day (negative values are recorded as zero). A running total of the cumulative degree days is maintained throughout the growing season. In temperate zones, degree days usually begin accumulating in late winter or early spring after chilling requirements for development have been met. Hence, cumulative degree-day calculations often begin on January 1. Time of year will vary with different regions and countries. When the cumulative degree-day number reaches a known target for a specific pest, intensive monitoring or preventive control tactics should be implemented if warranted. Degree-day targets have been established for the life cycles of many insect pests and are commonly available through university and government agencies. But it should be noted that degree days can vary somewhat within a small geographic area or even on the same property due to microclimate effects.

Key Stressors and Key Plants

To facilitate Plant Health Care, practitioners should become familiar with the key stressors and key plants in a particular region or on a specific property. Key stressors are pests or site conditions that are frequently encountered in landscapes and predictably cause injury to landscape plants. Key pests may also include particularly noxious pests in an area.

Key plants are defined in two ways. First, a key plant can be a species that has a high incidence of pest problems due to inherent susceptibility or common mismanagement. Second, a key plant can be a specimen that has significant value to the client and/or is particularly common in the landscape. A specimen plant may be highly valuable due to its location, function, size, appearance, historical significance, or cultural significance. Because value is often subjective, the PHC practitioner may not readily distinguish key plants in the landscape and should consult with the client. By identifying key stressors and key plants and understanding client expectations, the PHC practitioner can devise PHC programs more effectively.

Health Management Strategies

As part of an overall PHC program, pest prevention is an important strategy. Key aspects of pest prevention in the landscape include

- minimizing plant stress by encouraging favorable plant development conditions and
- minimizing pest activity by discouraging favorable pest development conditions.

Plant Health Care practitioners should identify short- and/or long-term stress factors and remediate them using appropriate management techniques. Plants that are in a severe state of decline or that are not appropriate for the site should be recommended for removal or transplanted to an acceptable location.

Landscape sanitation practices such as removing diseased, dead, and fallen twigs, leaves, and fruit can reduce some pest populations. In some cases, sterilization of pruning tools is recommended. The amount and timing of pruning, irrigation, fertilization, and pesticide application should be carefully considered. Improper prescription and application of

Figure 11.12 Excessive irrigation, especially on the trunks of trees, can increase susceptibility to *Phytophthora* and *Armillaria* root rot.

these and other management practices can increase plant susceptibility to certain pests. For example, improper timing of pruning can increase susceptibility to oak wilt, while excessive irrigation can increase susceptibility to *Phytophthora* and *Armillaria* root rot. However, timely irrigation to avoid moisture stress is critical to reducing the incidence of borers, bark beetles, and canker diseases. Pruning is a key cultural treatment for managing fire blight and other canker-causing diseases.

Practitioners should encourage clients to choose pest-resistant plant species and cultivars, high-quality nursery stock, and a diversity of plant species. Proper plant selection is very important in minimizing pest problems and can considerably reduce the need for control actions, particularly pesticide use. Certain plant species have a high incidence of pest problems; therefore, clients should be cautioned about the use of such plants, particularly if they have high expectations for plant performance or limited resources to invest in plant maintenance.

Matching species to the site also is important in avoiding pests and other problems. For example, planting shade-loving plants in the understory can avoid damage from high light and temperature, whereas planting them in full sun would cause drought and heat stress. Certain diseases can also result from planting sun-loving plants in shaded areas.

High-quality nursery stock is much more tolerant of transplant and establishment stress, which can predispose plants to secondary problems such as borers and canker fungi. The cultural needs of plants during establishment, particularly irrigation, should be carefully addressed to minimize predisposing stresses. Inspecting nursery stock for insects and diseases also minimizes the risk of introducing pests from the nursery to the landscape.

A landscape with a diversity of plant species is more stable than a landscape of few species. In a diverse landscape, one fatal infestation or infection will only kill a few trees and plants rather than a large percentage of them. **Monocultures** (extensive plantings of the same species) can have catastrophic consequences if a serious pest problem emerges. A high density of a vulnerable plant species may favor insect and pathogen spread. Additionally, overreliance on species susceptible to environmental conditions such as drought, ice storms, or high winds can be equally devastating. Plant diversity may promote natural pest control by providing alternate food sources and refuge for natural enemies of pests. Unfortunately, uniformity is often favored as a desirable characteristic in landscape design. Traditionally, aesthetics play a larger part in landscape design than ecology. Such potential conflicts underscore the importance of cooperative efforts among green-industry professionals.

Plant Health Care practitioners must choose from three pest management goals: prevention, eradication, and suppression. For some pests, prevention may be the only feasible management goal. For example, many fungal leaf diseases (such as leaf spots) and insect pests (such as borers) are most effectively managed with preventive cultural practices or, failing that, pesticide applications. Although preventive and cultural strategies are usually preferred, **prevention** strategies are sometimes insufficient for managing pest populations and plant injury at tolerable levels. When monitoring reveals that an action threshold has been exceeded, supplemental control tactics are often warranted.

Eradication (total elimination of a pest from an area) is an uncommon goal in landscape IPM because it is usually unwarranted and difficult to achieve. However, it may be considered when dealing with highly noxious, harmful, or introduced pests, such as the Asian longhorned beetle in North America. In extreme examples, trees that are infested by the pest, or even those that might become infested, are destroyed as part of an eradication program. Unfortunately, eradication programs often do not fully achieve their goals, but they have been successful for Asian gypsy moth and several Asian longhorned beetle infestations.

Suppression is a more common insect pest control goal in Plant Health Care. The intent is to reduce the pest population and associated plant injury to a tolerable level. Suppression is a preferred pest management approach because it usually can be achieved with limited applications of narrow-spectrum pesticides, other products such as horticultural oil and insecticidal soaps, or nothing at all and still meet expectations for landscape appearance and health. This approach can be both cost effective and environmentally responsible.

When devising a PHC management strategy, several considerations should be made:

- Communicate with the client to establish management goals and identify the management options for the pest, disease, or disorder, if any.
- Consider the advantages and disadvantages of each management plan.
- Whenever possible, choose the management option that has the least detrimental impact on the landscape ecosystem.
- Keep in mind that a management strategy that integrates complementary tactics is most effective.

Management Options

Cultural and Mechanical Control

Cultural control refers to landscape management practices that either promote plant health (thereby reducing pest susceptibility) or deter pest development. Cultural control tactics are particularly

Figure 11.13 Correcting improper mulch application minimizes damage caused by pathogens, stem-feeding rodents and insects, and stem-girdling roots.

desirable because they usually have limited negative impact on the environment and can help prevent future pest problems. A major component of cultural control is proper plant selection. It is important to choose a species or variety suitable for the existing landscape conditions to minimize chronic plant stress due to poor site adaptation. Always select high-quality nursery stock to minimize establishment-related stress and introduction of nursery pests to the landscape.

Cultural practices that promote soil health are integral to maintaining plant health. The physical, chemical, and biological properties of landscape soils should be evaluated, and deficiencies should be corrected according to best management practices. For example, use of **compost** can add organic matter to the soil, help retain soil moisture, and reduce the need for fertilizers. Excessive fertilization should be avoided because it can contribute to pest problems.

Proper mulching, with a 5 to 10 cm (2 to 4 in) layer of organic mulch, is perhaps the single best thing that can be done for a tree in many regions and climates. Mulch retains soil moisture during dry periods, moderates temperature extremes, reduces mowing injuries, and can even combat certain diseases on tree roots. Improper mulching and irrigation practices, however, can create favorable conditions for landscape pests. Correcting improper mulch application minimizes damage caused by pathogens, stem-feeding rodents and insects, and stem-girdling roots.

Figure 11.14 Wheel bugs are predatory insects. They commonly feed on soft-bodied insects such as caterpillars and sawfly larvae and can help keep certain pest populations in check.

Pruning can be an effective cultural management practice for suppressing localized pest populations. This is often a recommended tactic for certain canker diseases. Removal of dead and declining branches can remove reproduction and refuge sites for certain pests. Sanitation is another important cultural management tactic that involves the removal and proper disposal of dead or infested plant material that accumulates in the tree or on the ground. This debris provides ideal habitat for certain pests, particularly overwintering pathogens, insects, and mites.

The direct, physical removal of tree pests should not be overlooked as an effective cultural control tactic. Picking bagworms off of a single evergreen shrub or small tree, for example, might take only a few minutes. A few unsightly fall webworm nests can be removed with the swipe of a broom followed by crushing. Mechanical management might also consist of installing physical barriers to exclude pests from the tree.

Biological Control

Biological control is the suppression of pest populations by three main types of natural enemies: **predators**, **parasites**, and **pathogens**. Most native landscape pests have natural and native biological controls that regulate their populations. In many instances, natural biological control maintains pest populations at tolerable levels, and additional management strategies are not necessary. In urban settings, natural biological control may be insufficient to suppress pest populations, and applied biological control tactics could be a viable option. Moreover, traditional pesticide applications may be inappropriate in landscapes near bodies of water, food crops, playgrounds, or patios. In addition, some clients may oppose pesticide use in their landscape due to personal beliefs or environmental concerns.

There are three approaches to applied biological control: introduction, conservation, and augmentation. Introduction of nonnative predators, parasites, and pathogens attempts to create natural checks and balances for exotic pests where none previously existed. Introduction can be particularly useful against nonnative pests that threaten native ecosystems where there are few available management options. Introduction is, understandably, heavily regulated. Conservation involves strategies to conserve existing predators, parasites, and pathogens, such as providing supplemental habitat and limiting use of pesticides. **Augmentation** of natural biological control entails the rearing and release of beneficial organisms to supplement existing populations in the landscape. Many types of beneficial organisms are commercially available for use in IPM programs. These include lady beetles, lacewings, predatory mites, and parasitoid wasps. There have been few studies on the effectiveness of augmentation as a biological control in landscapes.

Figure 11.15 Parasitic wasps are important agents of biological control for many species of scale insects. Here a wasp prepares to lay a lethal egg inside a soft-scale pest.

Biological controls are generally slower and less consistent in managing pests than other controls. Pesticides can kill the targeted pest within minutes or hours of application, whereas biological control can take days or weeks to suppress a pest population. There is also greater uncertainty about the outcome of a biological control application.

Chemical Control

Chemical control of landscape pests entails the use of pesticides. Commonly used pesticides include

Figure 11.16 Chemical control may be necessary when other applied controls or natural controls fail to suppress a pest population adequately or when other pest control options are not available, practical, or economical.

insecticides, **miticides**, **fungicides**, **bactericides**, repellents, and **herbicides**. Chemical control may be appropriate when other applied controls or natural controls fail to suppress a pest population adequately or when other pest control options are not available, practical, or economical. The use of chemical pesticides is regulated in most jurisdictions and is restricted or illegal in some places. Arborists must be familiar with local laws and regulations and obey all pesticide laws.

Pesticides are generally differentiated as either contact or systemic pesticides. **Contact pesticides** suppress or kill pests through direct physical contact with the material or its residue, usually following treatment application. **Systemic pesticides** are absorbed and move within the plant by various means following application to the foliage, stem, or soil. Systemic pesticides often have contact properties as well. Systemic insecticides can generally target defoliating, sucking, leaf-mining, and wood-boring insects. Systemic fungicides are available for a number of foliar, vascular, and soilborne diseases. Both contact and systemic herbicides are used to manage weeds.

The trade-offs of contact and systemic pesticides should be considered carefully before selecting a product. Systemic pesticides generally take longer to work. Contact pesticides generally kill surface pests very rapidly, but their effectiveness depends on thorough spray coverage and proper timing. This can be a challenge when treating large plants, dense plants, or plants in sensitive landscapes or when rain or wind does not permit application. Contact and systemic insecticides can also impact nontarget and beneficial insects, including plant pollinators and insects that provide biological control of pests. In addition, some contact pesticides have limited residual effectiveness compared to systemic pesticides and may require repeated applications for sustained pest control. Repeated use of pesticides increases exposure of nontarget organisms, including wildlife and beneficial organisms such as those providing biological control. Other negative consequences of repeated use include development of pesticide resistance and replacing one pest with another.

When systemic pesticides are applied properly by soil or stem injection, there is less aboveground environmental exposure than with spraying. However, there is potential for groundwater contamination from soil application, depending on soil and site characteristics, and for plant injury from stem injections. It's important to note that decay may occur following repeated trunk flare injection treatments. Also, injection treatments to trees in drought conditions or with a compromised transport (vascular) system may not yield desirable results, as uptake and distribution of product within the tree may be slow and nonuniform. When selecting a product, all attributes and limitations should be considered. Choosing between systemic and contact pesticides is just one aspect of the decision.

For effective, responsible pest management, PHC practitioners should follow this process when using pesticides:

1. Identify the pest or disorder.
2. Evaluate the potential severity of the pest problem.

Figure 11.17 Systemic pesticides are absorbed and move within the plant following application to the foliage, stem, or soil. This is a trunk application.

3. Choose an appropriate pesticide that primarily targets the intended pest.
4. Use the correct dosage.
5. Employ the correct application technique.
6. Apply at the recommended time and frequency.
7. Always adhere to pesticide label instructions and all applicable pesticide regulations.

Pesticide selection is based on pest identification, host plant identification, sensitivity of the environment, client expectations, and government regulations. Correct pest and plant identification is of paramount importance in pesticide selection. Correct identification confirms that the organism is the cause of injury and permits the selection of a narrow-spectrum pesticide. Mixing several broad-spectrum pesticides together to target an unknown pest is expensive, environmentally irresponsible, likely illegal, and may worsen the problem. Pesticides must be employed using correct dosage, timing, frequency, and application technique, according to pesticide label directions. Remember that it is illegal to apply pesticides outside the restrictions of the label, and the applicator can be held responsible for consequences of misapplications.

A relatively new development in pesticide application is electrostatic spraying, which applies tiny droplets that are electrically charged to adhere to plant surfaces. This application technique has the potential to reduce the amount of pesticide used as well as runoff.

Indiscriminate use of broad-spectrum pesticides can kill both pests and their natural enemies, which can exacerbate pest problems. Pest resurgence and secondary pest outbreak are often the result of decreased populations of natural enemies. **Pest resurgence** can occur when the pest population rapidly rebounds in the absence of natural enemies that are slower to repopulate than the pest. Resurgence can occur because some pests do not have natural biological controls to suppress populations or because environmental conditions occur that favor continued development of the pest. The latter commonly occurs with diseases during growing seasons with cool temperatures and frequent rainfall that favor foliage disease development. With **secondary pest outbreaks**, both the primary pest and natural enemies remain suppressed, and a secondary pest takes advantage of the lack of competition and predators to become a prominent pest. A classic example of a secondary pest outbreak is the increase in mite population following insecticide applications. Pest resurgence and secondary pest outbreaks can be minimized by choosing narrow-spectrum pesticides and by making precise, targeted pesticide applications based on landscape monitoring.

Pesticide resistance occurs when a pesticide selectively kills members of a pest population that

Figure 11.18 Following some pesticide treatments, populations of secondary pests such as the two-spotted spider mite may increase.

are susceptible to the pesticide, leaving only resistant individuals to pass along their genes to future generations. Repeated, frequent applications of a single pesticide may lead to resistance. To reduce the risk of pesticide resistance, multiple pesticides with different active ingredients or modes of action can be used in a rotation system. Resistance is most likely to develop when a very large proportion of the pest population is treated repeatedly with the same or similar pesticide. This can occur where pest populations are enclosed, such as in a greenhouse, or where large crop areas are treated repeatedly with the same chemical.

Although pesticides are valuable for managing pests when used appropriately, their misuse can have negative consequences for the practitioner, the plant, and the ecosystem. Plant Health Care practitioners should not use pesticides without appropriate training. The United States and most other countries require training and licensing for commercial pesticide applicators.

Other Pesticide Options

Effort should be made to avoid injury to nontarget organisms and the environment. If using pesticides, PHC practitioners should choose the least toxic, most narrow-spectrum pesticide necessary to meet the pest management objective. Many of the newer synthetic pesticides meet this requirement. In addition, many biorational control products are available to provide effective suppression of pests.

Biorational control products include insecticidal soaps, horticultural oils, botanicals, insect growth regulators, microbial-based products, and microbial

Figure 11.19 Insecticidal soaps are highly refined soaps that disrupt the cell membranes of soft-bodied insects and mites. Horticultural oils have insecticidal properties; they suffocate certain insects or disrupt their membranes.

agents. These products tend to be less toxic because they have short residual activity, can be pest specific, or have unique modes of action. Of course, caution must still be exercised when handling, mixing, and applying these materials, because their toxicity may be similar to traditional chemical pesticides.

Insecticidal soaps are highly refined soaps that disrupt the cell membranes of soft-bodied insects and mites. They are effective on scale crawlers, aphids, mealybugs, and spider mites. These soaps have no residual effects and may require repeated applications for sustained pest control. Insecticidal soaps may injure understory flowering plants and may also cause injury to certain plants during drought periods.

Horticultural oils have insecticidal properties because they suffocate certain insects or disrupt their membranes. Horticultural oil applications are divided into two categories, summer oil applications and dormant oil applications, which vary based on timing and rates. Dormant oil treatments are applied before budbreak and are highly effective on some pests without damaging beneficial insects. Summer oils can be sprayed on the foliage of some plants, although other plants may suffer **phytotoxic** injury if horticultural oils are applied during dry periods and when temperatures exceed 29 °C (85 °F), especially if humidity is also high, delaying drying of the product. Some horticultural oils are labeled for both dormant and summer applications. In these cases, the summer applications are applied at a lower mix rate.

Botanical pesticides are plant extracts used for insecticidal purposes. Examples include neem, pyrethrin, and rotenone. Botanicals range from highly toxic (e.g., nicotine) to quite mild for both insects and people. Even mild botanicals, however, can be toxic to aquatic wildlife.

Insect growth regulators are synthetic compounds that act like insect hormones, disrupting the molting or growth processes. They have also been used to control populations by disrupting the mating process of certain insects or by making the adults sterile. Hormones can also be used as bait for traps, which are valuable cultural and control monitoring tools.

Microbial pesticides contain insect pathogens or lethal microbial byproducts that are derived from extracts of bacterial pathogens of insects. The most common example is proteins synthesized by *Bacillus thuringiensis* (Bt), a bacterium that is deadly to many insects. Different strains are effective against different groups of insects. One type is used for leaf-feeding beetles, another affects caterpillars, and a third is effective against mosquito larvae. These products have shown almost no toxicity to nontarget organisms or the environment.

CHAPTER 11 WORKBOOK

1. True/False—Plant Health Care and Integrated Pest Management are essentially the same thing.

2. Carbohydrates, produced through photosynthesis, are allocated to these primary functions:

    a.

    b.

    c.

    d.

    e.

3. _____ factors are often directly related to soil quality or other environmental conditions (for example, pH, drought, compaction, or excess water), and many can be attributed to human activity such as land development and construction.

4. The _____ and _____ in tree cells are indigestible to many insects and other animals and even to some pathogens.

5. Trees produce _____ such as tannins and other phenols that have toxic or deterrent effects on certain insects.

6. True/False—Rapidly growing trees may be less resistant to certain insects and diseases.

7. _____ is the process of observing, identifying, recording, and analyzing what happens with plants in the landscape.

8. The process of gathering information, assessing the severity and implications of the problem, determining client expectations, and deciding on a course of action is called the _____ _____ _____ .

9. _____ _____ _____ is a systematic approach to insect and disease management that incorporates a combination of techniques including resistant plants as well as cultural, biological, and chemical control tactics.

10. True/False—A simple degree-day model uses an established threshold temperature and the daily average temperatures to predict pest development stages.

11. When possible, arborists should select trees that are _____ to known insects or diseases.

12. _____ _____ are organisms that are frequently encountered in landscapes, predictably cause injury to landscape plants, and may include particularly noxious pests in the area.

13. Extensive plantings of the same species, known as _____ , can have catastrophic consequences if a fatal insect pest or disease is introduced.

14. Plant Health Care practitioners often choose from three pest management goals: _____ , _____ , and _____ .

15. True/False—Chemical pesticides often kill the targeted pest within minutes or hours of application, whereas biological control can take days or weeks to suppress a pest population.

16. _____ pesticides are taken up by the plant and translocated throughout the branches and into the leaves.

17. _____ _____ occurs when the pest population rapidly rebounds in the absence of natural enemies, which are slower to repopulate than the pest.

18. True/False—The use of multiple pesticides with different active ingredients or modes of action in a rotation system will increase the incidence of pesticide resistance.

19. True/False—Insecticidal soaps disrupt the cell membranes of soft-bodied insects and are effective on some scales, aphids, mealybugs, and spider mites.

20. True/False—Horticultural oil applications are always safe to use on trees in leaf because they have no phytotoxic properties.

21. _____ _____ _____ are synthetic compounds that act like insect hormones.

22. _____ _____ are derived from certain bacterial pathogens of insects.

23. Products that contain proteins of _____ _____ (Bt) are examples of microbial pesticides that utilize insect pathogens or lethal microbial byproducts derived from extracts of bacterial pathogens of insects.

24. The biological control strategy is based on the concept that many insect pests live in a natural, dynamic balance with _____ , _____ , and _____ that control pest populations.

25. True/False—Plant Health Care practitioners should identify short- and/or long-term stress factors and remediate them using appropriate management techniques.

Challenge Questions

1. Select two species of trees in your area that have different genetic strategies for resource allocation. Compare and contrast the advantages and limitations of each.

2. Describe some of the defense mechanisms found in trees, and discuss how they can be effective in limiting certain pests.

3. Discuss the various treatment and control strategies that can be implemented to control pest problems on plants, and explain the advantages and limitations of each.

Sample Test Questions

1. Plant Health Care is a comprehensive program to manage

    a. insects and diseases of plants
    b. tree health without the use of pesticides
    c. the appearance, structure, and health of plants
    d. pests, pathogens, and abiotic disorders of trees

2. The mortality spiral describes the

    a. process of infection and spread of disease in a tree
    b. cumulative effects of stress causing decline of a plant over time
    c. process in which pesticides eradicate both pests and beneficial insects
    d. allocation of resources among growth, storage, and defense

3. The process of gathering information, assessing the severity and implications of the problem, determining client expectations, and deciding on a course of action is called

    a. the appropriate response process
    b. Integrated Pest Management
    c. the cultural control mechanism
    d. Plant Health Care

4. A systemic pesticide is one that

    a. kills all living organisms
    b. kills insects on direct contact
    c. is translocated throughout the plant
    d. has no harmful effect on the environment

5. Releasing predators or parasites of an insect pest is an example of a

   a. cultural control
   b. mechanical control
   c. biological control
   d. chemical control

Recommended Resources

(See Recommended Resources in back of book for detailed information.)

Best Management Practices: *Integrated Pest Management* (Wiseman and Raupp 2016)

Introduction to Arboriculture: Plant Health Basics online course (ISA)

Introduction to Arboriculture: Plant Health and Pest Control Practices online course (ISA)

Introduction to Arboriculture: Pest Management Techniques online course (ISA)

Plant Health Care for Woody Ornamentals (Lloyd 1997)

CHAPTER 12
TREE RISK ASSESSMENT AND MANAGEMENT

▼ OBJECTIVES

- Define tree risk assessment, including its component factors.
- Compare and contrast the three levels of assessment.
- Explain the steps of the assessment process from data collection to analysis and categorization of risk.
- Describe the tree conditions that can increase or decrease the likelihood of failure.
- Describe some of the ways that tree risk can be mitigated.
- Contrast risk assessment and risk management, including roles and responsibilities.

KEY TERMS

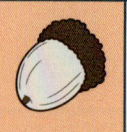

acceptable risk
advanced assessment
adventitious branch
aerial inspection
air-excavation device
basal rot
basic assessment
bracket
brown rot
cavity
cellulose
codominant stems
compression wood
conk
consequences
consequences of failure
crack
definite indicator
failure mode
failure potential

flexure wood
fruiting body
heartwood rot
included bark
lean
level of assessment
lignin
likelihood
likelihood of failure
likelihood of impact
limited visual assessment
load
mitigation
occupancy rate
potential indicator
protection factors
reaction wood
residual risk
resistance-recording drill
response growth

risk
root collar excavation (RCX)
sapwood rot
shear plane crack
soft rot
sonic tomography
species failure profile
structural defect
target
target zone
tension wood
time frame
tomogram
transverse crack
tree risk assessment
tree risk management
white rot
woundwood

Introduction

Trees provide many benefits to people living and working in the urban environment. These benefits increase as the tree's age and size increase. However, as a tree gets older and larger, it is also more likely to shed branches or develop decay or other conditions that can predispose it to failure.

Tree risk assessment is a systematic process used to identify, analyze, and evaluate tree risk. Risk is usually assessed by categorizing both the likelihood (probability) of occurrence and the severity of consequences.

Arborists perform tree risk assessments to maintain public safety, protect workers on the jobsite, and promote tree longevity. They assess trees and make recommendations to reduce risks. The primary risk related to trees is associated with structural failure. Injuries and fatalities that result from tree failures are relatively uncommon. Property damage and disruption can also occur. Tree owners or risk managers should base their tree management decisions, in part, on the tree risk assessments made by qualified arborists. Arborists who possess a strong understanding and skill set in tree risk assessment preserve trees more frequently through knowledgeable assessments and confidence in their decisions. This chapter provides introductory information of

Figure 12.1 Tree risk assessment is a systematic process used to identify, analyze, and evaluate tree risk. Risk is assessed by categorizing both the likelihood of occurrence and the severity of consequences.

Table 12.1
Likelihood matrix.

Likelihood of Failure	Likelihood of Impact			
	Very Low	Low	Medium	High
Imminent	Unlikely	Somewhat likely	Likely	Very likely
Probable	Unlikely	Unlikely	Somewhat likely	Likely
Possible	Unlikely	Unlikely	Unlikely	Somewhat likely
Improbable	Unlikely	Unlikely	Unlikely	Unlikely

Table 12.2

Risk-rating matrix.

Likelihood of Failure and Impact	Consequences of Failure			
	Negligible	Minor	Significant	Severe
Very likely	Low	Moderate	High	Extreme
Likely	Low	Moderate	High	High
Somewhat likely	Low	Low	Moderate	Moderate
Unlikely	Low	Low	Low	Low

the concepts and terminology of tree risk assessment and management. Arborists are encouraged to seek further training and qualifications to become proficient in the discipline.

What Is Tree Risk?

Risk is the combination of the likelihood of an event and the severity of the potential consequences. Trees can pose a variety of risks, most of which are categorized into two basic groups: conflicts (e.g., blocked views of traffic signage, pedestrian hazards, etc.) and structural failures. This chapter focuses on structural failures.

Assessing tree risk has two components: likelihood and consequences. **Likelihood** is the chance of a specified event occurring (within an identified time frame). In the context of tree failures, likelihood is the combination of (1) the chance of a tree failure occurring and (2) the chance of that failure impacting a specific target. **Consequences** are the effects or outcome of an event assuming a target is impacted. In tree risk assessment, consequences include personal injury, property damage, or disruption of activities due to the event.

Risk Assessment Basics

Tree risk assessment is a systematic process used to identify, analyze, and evaluate tree risk. Risk is assessed by categorizing both the likelihood of occurrence and the severity of consequences. Various systems for assessing tree risk exist. Most of the assessment methods that arborists use categorize risk into levels of severity. The International Society of Arboriculture offers a course and credential called TRAQ (Tree Risk Assessment Qualification), which trains and assesses candidates using a specific methodology.

Whichever method is chosen, users should recognize the limitations of the method as well as the

Figure 12.2 Injuries and fatalities that result from tree failures are relatively uncommon. Property damage and disruption can occur. It is important to balance risk with the social, aesthetic, environmental, and economic benefits that trees provide.

nature and degree of uncertainty in the data and information available. There is typically a considerable amount of uncertainty associated with tree risk assessment due to arborists' limited ability to predict natural processes (rate of progression of decay, response growth, etc.), weather events, traffic and occupancy rates, and potential consequences of tree failure. Sources of uncertainty should be understood and communicated to the risk manager/tree owner.

There are four guiding premises of risk assessment:

1. All trees pose some amount of risk.
2. All trees provide at least some benefits.
3. Often the benefits provided outweigh the risks posed.
4. The only way to completely eliminate all tree risk is to eliminate all trees.

It is important to balance risk with tree benefits. Trees provide important social, aesthetic, environmental, and economic benefits, and if people desire these benefits from trees, then some level of risk must be understood and accepted.

Levels of Assessment

Tree risk assessments can be conducted at different levels and employ various methods and tools. The level(s) should be appropriate for the assignment. There are three levels of tree risk assessment:

- Level 1: Limited Visual Assessment
- Level 2: Basic Assessment
- Level 3: Advanced Assessment

If conditions cannot be adequately assessed at the specified level, the assessor may recommend a higher level or a different assessment. However, the assessor is not obligated to provide the higher level if it is not within the scope of the original assignment, without additional compensation or without modifications to the agreement or contract. The level of assessment should also be proportionate to the assignment.

Level 1: Limited Visual Assessment

Level 1 risk assessment is a **limited visual assessment** from a specified perspective of an individual tree or a population of trees near specified targets. It is conducted to identify obvious defects or specified conditions.

A limited visual assessment assesses a large number of trees in a short period of time. Often the objective is to look for obvious defects such as dead trees, large **cavity** openings, large dead or broken branches, fungal fruiting structures, large cracks, and severe or uncorrected leans in areas with sufficient usage to warrant mitigation. It is common to consider the likelihood of impact and consequences to be the same for all trees in a specific location, such

Figure 12.3 A Level 1 risk assessment is a limited visual assessment from a specified perspective of an individual tree or a population of trees near specified targets. It is conducted to identify obvious defects or specified conditions.

as along an urban street. In some circumstances, such as transmission line corridors, the focus is on likelihood of impact; assessors look for any tree that could impact the utility wires.

Limited visual assessments are the fastest, but least thorough, means of assessment and are intended primarily for quick assessment of large populations of trees. When conducted by trained professionals, limited visual assessments can sometimes provide tree managers with a sufficient level of information for them to make their risk management decisions.

A limitation of limited visual assessments is that some conditions may not be visible from a one-sided inspection of a tree, nor are all conditions visible on a year-round basis. The assessor may use the Level 1 assessment to determine which trees require further inspection at the basic or advanced levels.

A Level 1 assessment may be performed in one or more ways:

- Walk-by: a limited visual inspection of one or more sides of the tree, performed as the inspector walks past a tree
- Drive-by: a limited visual inspection of one side of the tree, performed from a slow-moving vehicle, driven by another person
- Aerial patrol: observation made from an aircraft (or drones) overflying utility rights-of-way, other areas, or individual trees
- LiDAR (light detection and ranging): a remote sensing method that uses laser technology to measure tree size and location in relation to the target of concern

Level 2: Basic Assessment

A Level 2 assessment or **basic assessment** is a detailed visual inspection of a tree, its surrounding site, and its potential targets, as well as a synthesis of the information collected. It requires that a tree risk assessor inspect all sides of the tree—looking at the site and at visible buttress roots, trunk, and branches. This is the **level of assessment** that is commonly performed by arborists in response to clients' requests for individual tree risk assessments.

Figure 12.4 A Level 2 or basic assessment is a detailed visual inspection of a tree and its surrounding site, and a synthesis of the information collected. It requires that a tree risk assessor inspect completely around the tree—looking at the site, and at visible buttress roots, trunk, and branches. A mallet is commonly used to "sound" a tree.

A basic assessment may include the use of simple tools to gain additional information about the tree or its defects. Simple tools may be used for measuring the tree and acquiring more information about it or any potential defects. However, the use of these tools is not mandatory unless specified in the scope of work. Measuring tools may include diameter tape, a clinometer, or a tape measure. Other basic inspection tools include binoculars, a mallet (for sounding), a probe, and manual digging tools. Using these simple tools does not elevate the level of inspection. Use of more advanced tools would classify the assessment as a Level 3, advanced assessment.

The primary limitation of a basic assessment is that it includes only conditions that are detected from a ground-based inspection on the day of the assessment. Internal, belowground, or upper-crown conditions as well as certain types of decay may be impossible to see or difficult to assess and may remain undetected.

Level 3: Advanced Assessment

Advanced assessments are performed to provide detailed information about specific tree parts, defects, targets, or site conditions. They are usually conducted in conjunction with or after a basic assessment if the tree risk assessor needs additional information and if the client approves the additional service. Specialized equipment, data collection and analysis, and/or expertise are usually required for advanced assessments. These assessments are therefore generally more time intensive and more expensive.

Many techniques can be considered for advanced risk assessment. Some situations may be assessed with multiple techniques. Examples of advanced assessment techniques include the following:

- Aerial inspection (climbing or elevated work platform)
- Detailed target analysis (site use, traffic patterns, property values, etc.)
- Detailed site or weather evaluation
- Decay testing
  - Drilling with small-diameter bit
  - Resistance-recording drilling
  - Sonic or impedance tomography
- Load testing
- Laboratory analysis
  - Microscopic analysis of causal agents
- Root inspection and evaluation
  - Root and root collar excavation

Likelihood of Impact

The likelihood of a tree failure impacting a target is the combination of the likelihood of tree failure and the likelihood of that failed tree or branch impacting a target. These factors must be assessed individually and then combined to determine the likelihood of failure and impact. To assess the **likelihood of impact**, arborists consider potential targets, their occupancy rate, the direction of fall, and potential **protection factors** (shelters, vehicles, other trees, etc.).

Targets

Targets are people, property, or activities that could be injured, damaged, or disrupted by a tree failure. The most important targets are people, even if they are not present at the time of the evaluation. Assessors should focus on areas where people are likely to be present. For example, an assessor would not identify a sidewalk or a road as a target, but the people who use the sidewalk or road would, instead, be the appropriate target.

Electric lines and facilities are targets of great importance to electrical utility risk managers and

Figure 12.5 Targets are people, property, or activities that could be injured, damaged, or disrupted by a tree or tree part failure. The most important targets are people, even if they are not present at the time of the evaluation.

Figure 12.6 The target zone is the area that the tree or branch is likely to land if it fails. When determining the target zone, the assessor considers the direction of fall, the height of the tree, crown spread, slope of land, wind, potential for dead branches shattering, or other factors that might affect spread of debris.

users of electric power. Electric lines brought down by a falling tree can be a threat to people who make contact with them; they can cause large fires; and the costs, inconvenience, and possible losses associated with disruption of service can be significant.

Known targets are those visible to the assessor and those that the assessor has been informed about by the client. It is preferable to consider targets in consultation with the client so that targets that are not present at the time of the assessment may be included.

Target Zone

The **target zone** is the area that the tree or branch is likely to strike if it fails. When determining the target zone, the assessor considers the direction of fall, the height of the tree, crown spread, the slope of the land, wind, the potential for dead branches shattering, or other factors that might affect spread of debris. The target zone for dead trees or trees with dead or brittle branches is generally larger than those with live, flexible branches because dead and brittle branches are more likely to shatter upon hitting the ground and may spread debris some distance beyond the tree. It is also important to consider the location of the target within the target zone when assessing the likelihood of impact.

Occupancy Rate

Targets can be categorized by the amount of time that they are within the target zone—their occupancy rate. **Occupancy rate** is a primary component in an assessment of the likelihood of impacting a target. Some targets, such as buildings and power lines, have constant occupancy rates. Other targets, such as vehicles and pedestrians, are moving, so the assessor must consider the time spent in the target zone and the number of vehicles or pedestrians using the area. The greatest potential for injury from tree failures occurs in places where people are often

Figure 12.7 To assess the likelihood of impact, arborists consider potential targets, their occupancy rate, the direction of fall, and potential protection factors (shelters, vehicles, other trees, etc.).

examining structural conditions, defects, response growth, and anticipated loads.

Time Frame

Before assessing the likelihood of failure, a time frame must be specified to put the likelihood rating in context. The time frame is the time period for which an assessment is defined. Without a stated time frame, the rating for likelihood of failure is meaningless. The length of the time frame is directly related to the likelihood of failure rating because there are more instances of time in which failure can occur within a longer period of time and fewer instances within a shorter period of time. Time frames of one to three years are common; time frames greater than five years are not generally used because there can be too much uncertainty associated with an increased length of time.

Tree Structural Conditions

Certain structural defects or conditions are more likely to lead to failure than others. Individual defects or conditions may or may not indicate a serious structural problem, but in combination, and under additional loads, they may contribute to failure. However, through **response growth**, trees can strengthen weak areas and support loads, thereby reducing the likelihood of failure. A visual assessment includes looking for and determining the significance of each of the structural conditions, individually and in combination, that increase and decrease the likelihood of failure. More information about assessing tree

present and are unprotected from above. Examples are bus stops and locations with tree canopies where people congregate.

Likelihood of Failure

The other component of likelihood is the likelihood of a tree failure occurring within a specified time frame. The **likelihood of failure** is determined by

Figure 12.8 The second component of likelihood is the likelihood of a tree failure occurring within a specified time frame. The likelihood of failure is determined by examining structural conditions, defects, response growth, and anticipated loads.

structural defects and response growth is presented later in this chapter.

Site Factors

Site factors have significant influence on both the likelihood and consequences of tree failure. It is especially important to consider the significance of site changes such as forest clearing, trenching for underground utilities, soil excavation or filling, groundwater lowering or raising, infrastructure repair, or other construction.

As trees grow and develop, they adapt to the forces and conditions that they are exposed to. Assessors should take note of site changes that any trees present may not have had sufficient time or resources to respond to. For example, recently exposed forest-edge trees may be more susceptible to failure due to increased wind exposure, damaged or shallow root systems, or changes in soil hydrology.

Loads

Load is a term used to describe the various forces acting on a structure. The two natural forces that exert loads on trees are gravity and wind. Gravity acts as a constant pull on the mass of the tree, generating static load from the weight of the tree part and the weight of water (condensation, rain, snow, and ice) on leaves and branches.

Most tree failures occur during storms when strong wind, rain, snow, or ice loads exceed the tree's capacity to withstand the load. Tree failures

Figure 12.9 Most tree failures occur during storms when strong wind, rain, snow, or ice loads exceed the tree's capacity to withstand the load.

in normal wind speeds are usually associated with serious structural defects or other conditions, alone or in combination. When wind speeds are extreme (strong gale force in many areas), even defect-free trees can fail. When evaluating wind loads during tree risk assessment, the assessor should consider the normal range of wind and weather conditions for the region.

Species

Some species of trees and specific locations have recurring patterns of tree failure. **Species failure profiles** may provide information on the most likely failure patterns in a tree species.

Tree Health

Tree risk assessors should not confuse tree health and tree stability. High-risk trees can appear healthy and have a dense, green canopy. There may be sufficient vascular transport in sapwood or adventitious rooting to maintain tree health, but there may not be enough of a strong root system for structural support. Conversely, trees in poor health may or may not be structurally stable. For example, tree decline due to certain types of root disease may cause the tree to become structurally unstable over time, while injury due to foliar insect attack may not affect stability much.

Consequences of Failure

The **consequences of failure** (and impacting a target) are a function of the value of the target and the amount of injury, damage, or disruption (harm) that could be caused by the impact of the failure. The amount of damage depends on the part size, fall characteristics, fall distance, and any factors that may protect the target from harm.

> **Part size.** When evaluating consequences, consider the size of the tree or branch that could fail and how it could impact a target. Generally, a small branch has less potential to cause serious damage than a large branch.

> **Distance of fall.** In estimating how much damage could occur from a tree failure, one must consider the relative amount of force with which it is likely to strike the target. A falling tree or branch will gain speed as it accelerates toward the ground.

> **Protection factors.** Protection may be provided by structures that surround the people in the target zone. If the protective factor is not strong enough to stop the impact, then the assessor should judge whether it will reduce the consequences.

Tree Assessment

The ability to accurately predict tree failure is limited, but with training, arborists can learn to identify characteristics that have been associated with tree failure. **Structural defects** or conditions

Figure 12.10 Protection may be provided by structures that surround the people in the target zone. If the protective factor is not strong enough to stop the impact, then the assessor should judge whether it will reduce the consequences.

that can lead to failure are not always visible, especially those inside the tree or beneath the ground, and the forces of nature that will act upon them are unpredictable. Inspecting trees for **failure potential** involves more effort than just a casual observation. A visual assessment includes looking for and determining the significance of each of the structural conditions, individually and in combination, that could increase or decrease the likelihood of failure.

When performing an inspection, it is important to stick to a systematic and consistent process to learn as much as possible about the tree, its history, and the site. Most risk assessors use a form that ensures methodical inspection and record keeping of the conditions that they observe.

Defects and Conditions

Tree failure is the breakage of stems, branches, or roots, or the loss of mechanical support from the root system. Structural failures occur when the forces acting on a tree exceed the strength of the tree structure or soil supporting the tree. Even a structurally strong tree will fail when a load or force is applied that exceeds the strength of one or more of its parts.

Most tree failures involve a combination of structural defects and/or conditions, such as the presence of decay or poor structure, and a loading event, such as a strong wind. Some structural defects or conditions are more likely to lead to failure than others. Individual defects or conditions may or may not indicate a serious structural problem, but in combination, and under additional loads, they may contribute to failure.

Dead or Dying Parts

When a target is present, dead trees are an obvious concern, as are dead or dying branches and roots. However, some species hold dead branches for

Major Defects and Conditions That Can Increase the Potential for Tree Failure

- Dead parts
- Broken and/or hanging branches
- Cracks
- Weakly attached branches or codominant stems with included bark
- Unusual tree architecture (lean, balance, branch distribution, or lack of taper)
- Loss of root support
- Decayed or missing wood (mechanical damage or cankers)

Figure 12.11 Dead, dying, broken, and hanging branches must be assessed for their likelihood of failure if there is a potential target.

many years, so familiarity with tree species and local climate is important.

Broken or Hanging Branches

Broken branches remain partially attached at the point of breakage, or they may have completely detached and started to fall. Branches that are broken and lodged in the crown of the tree are called hangers or lodged branches. The likelihood that the branch will continue its fall depends upon how it is being held in place and if it is decaying.

Weakly Attached Branches

Certain branch arrangement and attachment configurations are associated with higher rates of failure.

Figure 12.12 Codominant stems sometimes contain included bark—bark that is embedded within the branch union. Included bark can decrease the strength of the attachment.

Strong branch attachments form when branch and trunk wood develop together over time and when the size of the branch is less than one-half the diameter of the parent stem.

Adventitious branches, which often are produced after storm-related branch breakage, lion-tail pruning, or topping, can be structurally weak until sufficient wood has formed to strengthen their attachment. If these branches are attached near a cut or broken branch end, decay developing within the wound may reduce the strength of the attachment over time.

Codominant Stems and Included Bark

Codominant stems (codominant branches) are forked branches of nearly the same diameter, arising from a common union and lacking a branch collar. Codominant stems vary in their structural stability. Though not always, they can have an increased likelihood of failure, especially when associated with decay or a crack. Codominant stems sometimes contain **included bark**—bark that is embedded within the branch union. Included bark can decrease the strength of the attachment. Response growth (described later in this chapter) can strengthen these branch attachments.

Cracks

Cracks are separations in the wood in either a longitudinal (along the length of the trunk, stem, or root), radial (along the radius of the trunk, stem, or root), or transverse (across the short axis of the trunk, stem, or root) direction. Longitudinal branch cracks (also called **shear plane cracks** or neutral plane failures) most commonly occur when branches are overloaded by freezing rain, snow, and/or wind. Although they often appear severe because the crack extends all the way through a branch, they are not necessarily prone to further failure. Cracks associated with internal decay are more likely to lead to failure. **Transverse cracks** indicate that the fibers in the wood have pulled apart or buckled. Transverse cracks often indicate that failure is imminent. Depending on the species, type of crack, and the

Figure 12.13 Cracks are separations in the wood in either a longitudinal (radial, in the plane of ray cells) or transverse (across the stem) direction. The likelihood of failure depends on the type and severity of crack, loads, presence of decay, and response growth.

severity, trees can sometimes strengthen the weakened area with response growth.

Leans

Lean is the angle of the trunk measured from vertical. Lean sometimes develops as the tree grows away from neighboring trees or structures, toward light. This type of leaning tree may be stable for long periods of time. Trees may also lean because of a partial failure of the lower stem or roots or because of soil conditions that allow excessive root movement. Trees that increase in lean over a short time period are of greater concern and should be examined promptly. Lifting of the soil on the side of the tree opposite the direction of lean, or a depression on the side of the lean, may indicate that the roots or soil are failing.

Corrected leans (sweeps) are characterized by a leaning lower trunk and a top that is more upright. Trees may develop this form when growing at the edge of a group. Such trees can be stable under normal conditions but may be less stable than straight trees under additional loads.

Root Problems

When roots are severed, decayed, broken, diseased, or restricted, they may provide less anchorage. Other root-related problems, including buried buttress roots and stem-girdling roots, can affect root decay or response growth, which, in turn, affects tree stability. In addition to structural problems with roots, some soils allow excessive movement of structurally strong roots, especially when saturated, which also can affect stability. Diagnosing root conditions must often be performed using inferences instead of direct observations because most of a tree's root system is hidden from a visual inspection from the surface. For example, roots may be hidden below the soil surface, but a newly paved section of sidewalk next to the trunk of a tree may indicate recent root cutting. A declining crown of a tree growing in poorly drained soil may indicate root problems, possibly root rot.

Wood Decay

Wood decay is caused by specific fungi, which may or may not form **conks** or mushrooms (**fruiting bodies**). Conks (**brackets**) and mushrooms are types of fungal reproductive structures. If they are attached to a tree, it is an indication of internal decay. However, a tree can have decay without the presence of conks or mushrooms on the outside. Mushrooms on the ground near a tree may or may

Figure 12.14 White rot fungi primarily decay the lignin within and between cell walls in the wood. Lignin is the material that gives wood its compressive strength, and the loss of lignin can reduce the wood's stiffness but leave some flexibility.

Figure 12.15 Brown rot fungi primarily decay the cellulose, leaving behind the stiff lignin, and reducing the bending strength of the tree. This makes the wood more brittle, and the decayed wood is dry and crumbles easily.

not be associated with tree decay. In fact, they could be the fruiting bodies of beneficial mycorrhizal fungi or associated with wood-decay fungi in mulch.

The three basic types of fungal decay are white rot, brown rot, and soft rot. Each type is characterized by the cellular and/or intercellular components of wood that they break down. However, like most things in nature, there are exceptions within each type, and some decay fungi exhibit the characteristics of more than one type of decay.

White rot fungi primarily decay the lignin within and between cell walls in the wood. **Lignin** is the material that gives wood its compressive strength, and the loss of lignin can reduce the wood's stiffness but leave some flexibility. Some white rot fungi can also attack the **cellulose** within the cell walls. White rot gets its name because the decayed wood appears white after the darker-colored lignin is decayed. Common white rot fungi include *Armillaria* spp. and *Ganoderma* spp.

Brown rot is most common on conifer trees and less common on deciduous trees. Brown rot fungi primarily decay the cellulose, leaving behind the stiff lignin and thereby reducing the bending strength of the tree. This makes the wood more brittle, and the decayed wood is dry and crumbles easily. Brown rot gets its name because, after the cellulose is decayed, the remaining lignin is dark or brown in color. Examples of brown rot fungi include *Laetiporus* spp. and *Phaeolus* spp.

Soft rot is similar to brown rot in that it usually first degrades the cellulose portion of the wood. Soft rot can be difficult to distinguish from other rots because it has some characteristics in common with both brown and white rots. An example of a soft rot found in living trees is *Kretzschmaria deusta*.

The loss of wood strength may be faster initially with brown rot than with white rot. Trees affected

Figure 12.16 Basal rot can lead to failure of the tree at the base.

by white rot are more likely to adapt to the loss of wood strength by producing more new wood around the decayed area, which may serve to compensate for the strength loss in the existing wood in that location.

Decay problems are commonly referred to by the part or parts of a tree that they affect. **Heartwood rot** is the term used to describe decay that starts in the heartwood (center) of the tree. **Basal rot** (butt rot) describes decay located in the lower trunk and/or base of the tree. Decay located in the roots is called root rot. Structural root decay often develops from the bottom upward, and visible symptoms of this rot may or may not develop in the crown.

If the decay is located in the sapwood, it is called **sapwood rot**. In this situation, both the bark and/or cambium might have been damaged or may be dead. An indicator of sapwood rot is the presence of numerous, small fruiting bodies along the bark's surface. By the time it is detected, sapwood rot is a serious problem because the decay progresses from the outside of the branch toward the center, and strength loss is severe. Most branches with sapwood rot require removal if targets are present.

There are many tree decay fungi throughout the world. The ways in which different decay fungi species interact with various tree species may vary by region. Certain wood-decay

Figure 12.17 Fungal fruiting bodies growing on major roots are an indication of root rot.

Figure 12.18 An indicator of sapwood rot is the presence of numerous but small fruiting bodies along the bark's surface. By the time it is detected, sapwood rot is a serious problem because the decay progresses from the outside of the branch toward the center, and strength loss is severe.

fungi are common and important in any given area, and arborists should become familiar with them.

Tree reaction to the presence of decay varies. The two fundamental strategies are compartmentalization and growth; the first seeks to contain the progress of decay, the second to outgrow it. Trees can sometimes compensate for loss of load-bearing capacity by developing response growth.

Indicators of Decay

Significant amounts of decay can reduce load-bearing capacity and increase tree failure potential. Decay is not always obvious during external inspection of a tree. A tree may appear to be solid and structurally sound, and it may have a thick, green crown, yet it can have significant decay inside. It is important to recognize common indicators of decay.

A **definite indicator** of decay means that decay is present. Examples of definite indicators include the presence of fruiting bodies (brackets, mushrooms, or conks) growing on or at the base of the tree, open cavities, and visibly decayed wood. Carpenter ants nesting inside a tree are a definite indicator because they nest only in decayed wood.

Potential indicators of decay are symptoms or signs that decay *may* be present; they signal that

Figure 12.19 Fungal fruiting body on trunk—a definite indicator of decay.

Figure 12.20 Carpenter ants are often associated with decayed wood in trees.

further investigation is warranted. Some potential decay indicators include cracks, seams, bulges, and wounds from old pruning cuts, especially topping cuts. Cavity-nesting mammals, birds, and bees potentially indicate the presence of decay in the tree being assessed. Birds will build their nests in hollows or dig out decayed wood to create a hollow. Honeybees often make their hives in tree cavities. Small cavities are often not a problem; they serve as habitat for birds and small animals. However, cavities may indicate more extensive internal decay and may require inspection.

Response Growth

As trees grow and develop, they adapt to the various loads that they experience (gravity and wind) by developing wood where it is needed to support the loads. This is a constant process throughout the life of a tree. **Compression wood**, most commonly produced on the underside of branches and leaning stems of conifers, and **tension wood**, most commonly produced on the upper side of branches in angiosperms, are known collectively as **reaction wood**. When trees sway in the wind, they produce **flexure wood** near the base to help support the trunk.

Site changes and weather events can abruptly change the loads that a tree experiences. For

Figure 12.21 This tree has extensive basal decay, but it is also exhibiting significant response growth. Tree risk assessors should look for and assess the significance of response growth when evaluating the likelihood of failure.

example, adjacent trees may be removed, exposing a tree to new wind loads. Even a large pruning cut changes the loading dynamics of a tree. The tree will respond to these changes as it develops new wood. Response growth is new wood produced in response to damage or loads. The amount, location, and even the wood characteristics (strength and chemical defenses) can be different from the wood produced before the response.

The development of trunk taper and buttress roots are examples of load-responsive growth. On stems or branches, bulges or enlarged areas may develop to compensate for higher load or loss of structural strength from internal decay or other weakness. Near a crack, obvious response growth may appear as spiraling or straight ribs.

Woundwood is produced in response to cambial damage. Woundwood is typically denser than and chemically different from other wood, and it resists decay better than normal wood. Its development also reinforces the strength of wounded areas. If a wound or cavity is not readily closed due to the size of the opening or other factors, woundwood may enlarge or curl inward at the edge of the cavity (a feature commonly called a "ram's horn"), adding strength to the opening but sometimes preventing full closure.

The amount of new wood that a tree produces depends upon species, health, energy reserves, and available resources (such as water, light, and nutrients). This makes the tree's overall vitality an important factor in the tree's ability to respond.

Tree risk assessors should look for and assess the significance of response growth when evaluating the likelihood of failure. When response growth is recognized, the tree risk assessor should try to determine the cause and evaluate its effect on the likelihood of failure. Trees can adapt to weaknesses and stand for many decades if sufficient structural compensation occurs.

Further Investigation

At times, it may be necessary to conduct a more detailed assessment of specific tree parts, defects, targets, or site conditions to determine tree risk. Tree risk assessors might recommend some specific type of Level 3 advanced assessment. As listed under the Level 3: Advanced Assessment section, there are many possible types of further investigation. The three most common are aerial inspection, assessment of internal decay, and root excavation.

Aerial Inspection

Aerial inspection (crown inspection) is the inspection of the aboveground parts of a tree not visible from a ground-based inspection, including the upper trunk and the upper surfaces of stems and branches. Aerial inspection can be performed from an aerial lift, adjacent building, ladder, unmanned aerial vehicle (UAV, drone), or by climbing the tree.

Assessment of Internal Decay

It is difficult to estimate or quantify the location and extent of internal wood decay during most basic assessments. When it is necessary to determine the location and extent of decay more accurately, these can be estimated with one of several techniques. Two of the most widely used advanced assessment technologies are resistance-recording drills and sonic tomography.

- A **resistance-recording drill** drives a small-diameter (3 mm [1/8 in]) flat-tipped spade bit into the tree. As the bit penetrates the wood, the resistance to penetration is recorded. With training and experience, an inspector can distinguish solid wood from voids and decay. Early-stage decay, effectiveness of compartmentalization, and response growth rates may be estimated from profiles created by some high-resolution resistance drills.

Figure 12.22 Resistance drilling can be used to detect decay in trees. With training and experience, an inspector can distinguish solid wood from voids and decay.

- **Sonic tomography** instruments send sound (stress) waves through the wood and measure the time for the wave to travel from the sending point to the receiving points. If a crack, cavity, or decay is present, the wave travels around the defect, increasing the transmission time (time of travel) from the sending point to the receiving point, as compared to the transmission time through wood with no defect. Sonic tomography instruments use measurements between many points to create a two- or three-dimensional picture (**tomogram**). By comparing the results of all time-of-travel measurements, it is possible to detect and map defects within the trunk. The tomogram illustrates the remaining load-carrying parts of the inspected tree section.

Root Excavation

The extent of damage or decay in tree buttresses and roots is difficult to evaluate in a basic inspection because most roots are beneath the soil surface. Arborists sometimes excavate soil or other materials covering the root collar to conduct an assessment. This process is called **root collar excavation (RCX)**. Frequently a high-pressure **air-excavation device** is used to remove soil while causing minimal damage to roots. If necessary, hand tools can be used. Care must be taken not to damage the roots or trunk during the excavation process. After excavation, roots can be inspected for evidence of cutting, injury, decay, response growth, or other conditions. Decay assessment techniques may also be used.

Analysis of Risk

Once a tree has been inspected and data is collected, the next step is to analyze the data to determine the level of risk. This typically involves identifying one or more potential **failure modes** (location or manner in which failure is likely to occur) based on any conditions of concern noted in the inspection. Targets of concern are also noted. Each combination of a failure mode and a target represents a separate risk to analyze. Many trees have more than one failure mode and may have multiple potential targets. For example, a tree with excessive root decay may also have several dead branches; the entire tree could fail from root decay, or dead branches may fail. Similarly, the whole tree may fall on a house, while the dead branches would fall only on the driveway. A large, mature tree could have many possible risks to analyze, so the assessor chooses the most significant ones for the site and client. When evaluating individual trees,

Figure 12.23 Acoustic devices are based on the principle that sound waves pass through decayed wood slower than through sound wood.

Figure 12.24 Advanced assessments are performed to provide detailed information about specific tree parts, defects, targets, or site conditions. Specialized equipment, data collection and analysis, and/or expertise are usually required for advanced assessments.

it is appropriate to evaluate each significant failure mode/target combination as an independent event and to report its risk.

It is important to define what risk is being rated. For example, "the risk of the large, dead branch in the oak tree in Mrs. Smith's backyard failing and striking the garage" may be very different from "the risk of the oak tree in Mrs. Smith's backyard failing at the base and striking people using the patio below." The assessment report should state clearly what risk(s) are rated to minimize ambiguity or confusion, and reports must always state the time frame.

Arborists often report an overall tree risk rating. The risk assessor identifies—among all the failure mode/target combinations assessed—the failure mode/target combination having the greatest risk and then reports that as the overall tree risk rating. Assigning an overall tree risk rating for a tree may be useful, especially when prioritizing mitigation of risk for a population of trees.

Risk Perception and Acceptable Risk

How people perceive risk and their need for personal safety is inherently subjective; therefore, risk tolerance and action thresholds vary among tree owners/managers. What is within the tolerance of one person may be unacceptable to another. It is impossible to maintain trees completely free of risk—some level of risk must be accepted to experience the benefits that trees provide.

Acceptable risk is the degree of risk that is within the owner's/manager's or controlling authority's tolerance, or that which is below a defined threshold. Some countries have a framework directive, which defines thresholds of risk tolerance and acceptability. Municipalities, utilities, and property managers may have a risk management plan that defines the level of acceptable risk. Safety may not be the only criterion used by the risk manager to establish acceptable levels of risk; budget, a tree's historical or environmental significance, aesthetics, and other factors may also be part of the decision-making process.

Risk Management

In areas where people, property, and activities could be injured, damaged, or disrupted, the consequences

Figure 12.25 The property owner must make the final decision about what level of risk is acceptable.

of tree conflict or failure may be of concern. Decisions on whether a tree inspection is required or what level of assessment is appropriate should be made with consideration for what is reasonable and proportionate to the specific conditions and situations. These are tree risk management issues. **Tree risk management** is the application of policies, procedures, and practices used to identify, evaluate, mitigate, monitor, and communicate tree risk. Tree owners or managers bear the responsibility for tree risk management. In some cases, such as in municipal arboriculture, the tree manager and the tree risk assessor can be the same person. In most situations, however, an arborist provides tree risk assessment services to a client. Roles of the tree risk manager include the following:

- Define and communicate tree risk management policies.
- Determine the need to inspect the trees in question.
- Establish the budget.
- Identify the geographical limits of the tree inspection.
- Specify the desired level of assessment.
- Determine or accept the scope of work (shared with the risk assessor).
- Decide the level of acceptable risk.
- Establish the inspection frequency.
- Verify target zone uses and occupancy rates.
- Prioritize work.
- Choose among risk mitigation options.

Many countries and jurisdictions in the United States have passed legislation that addresses responsibility for maintaining trees. In other places, courts base their rulings on legal precedents—similar cases that have been decided in the past. Arborists should be aware of the applicable laws in their country, state, and local jurisdiction.

Managers of tree populations, such as municipalities, park systems, golf courses, etc., are concerned with system-level risk at the policy level. These risks include not only the risk of harm but also the risk of financial loss, loss of credibility, and loss of productivity. Moreover, these risks are managed in the context of other risks in addition to those related to trees. How each is managed is a significant element of an organization's system-level, tree-related risk management. The risk of harm is the primary concern of this chapter. The remaining risks should be acknowledged and incorporated into a comprehensive tree risk management strategy.

Mitigation

Mitigation is the process of reducing risk. Measures to mitigate risk can be tree based, to reduce the likelihood of failure or the likelihood of impact, or they can be target based, to reduce the likelihood of impact or consequences of failure. Arborists can recommend mitigation measures to reduce tree risk, but it is the responsibility of the tree owner/manager to select among mitigation options and to take action, if warranted. Mitigation measures can also be combined to further reduce risk.

Examples of common mitigation options include the following:

Target management. Movable targets within the target zone may be temporarily or permanently relocated. Mobile targets such as pedestrians or vehicular traffic can be rerouted or restricted from using the space within the target zone. Fencing and signage may be erected to keep people from going underneath certain trees. These often are the solutions that will have the lowest impact on the tree and are therefore preferred if tree preservation is a primary management goal.

Pruning. Dead, dying, and weakly attached branches can be pruned in accordance with the applicable national pruning standards or ISA's Best Management Practices: *Pruning*. Wind resistance can be reduced with reduction pruning, but topping is not recommended due to the long-term problems with weak sprouts and the potential entry of wood-decay fungi. Crown

Figure 12.26 Pruning and/or the installation of tree support systems may be an option to mitigate risk.

raising can eliminate lower branches that could be interfering with structures, pedestrian or vehicle traffic, signs, or safe views. Excessive raising, however, can reduce taper development, change sway patterns, and limit the tree's ability to dissipate the effect of dynamic wind loading.

- **Installing structural support systems.** Structural support systems can be installed to limit movement of certain tree parts. Various types of hardware are used, depending upon the goals.

Tree risk managers should resist the ultimate security of risk elimination based on tree removal and should consider possibilities for retaining trees when practical. Trees offer many benefits, so removal should be considered as a last option to reduce or eliminate risk. In many cases, there are other options available to reduce risk to an acceptable level.

In some low-use locations, dead and decaying trees may be retained for wildlife habitat or other uses. Selection of suitable wildlife habitat trees must consider the nearby targets, the long-term management options, and the tree's risk as well as its value for wildlife. One management strategy is to ensure that trees retained for wildlife habitat are maintained at a height shorter than the distance to the nearest target.

The overall tree risk rating is based on the highest risk factor, but typically there are other factors that also should be considered, which may require additional mitigation actions if the tree is to be retained. Once the highest risk factor has been mitigated, the tree risk rating goes to the next highest risk factor.

Residual Risk

Residual risk is the risk remaining after mitigation. Following any mitigation action, there is a residual risk posed by the tree. With tree removal, that residual risk from failure is brought to zero. The level of residual risk needs to be acceptable to the risk manager/owner. Some countries follow the principle of "as low as reasonably practicable" (ALARP). To meet this principle, one must demonstrate that the cost involved in further reduction of risk would be grossly disproportionate to the benefit gained. This principle is sometimes applied in situations where large populations of trees are being managed. There has been a gradual shift, however, to a benefit-risk assessment approach, which acknowledges benefits and recognizes the trade-offs between risks and benefits.

In general, if the residual risk following mitigation would exceed the client's tolerance, that option might not be the best course of action. As previously noted, however, tree removal should consider the benefits that would be lost.

CHAPTER 12 WORKBOOK

1. List the two components of risk.

    a.

    b.

2. In tree risk assessment, likelihood is a combination of _____ _____ _____ and _____ _____ _____ .

3. List the three levels of assessment.

    a.

    b.

    c.

4. _____ are people, property, or activities that could be injured, damaged, or disrupted by a tree failure. The _____ _____ is the area that the tree or branch is likely to hit if it fails.

5. Targets can be categorized by the amount of time that they are within the target zone—their _____ _____ .

6. Without a stated _____ _____ , the rating for likelihood of failure is meaningless.

7. List at least three factors to consider when assessing the likelihood of impact.

    a.

    b.

    c.

8. List at least five factors to consider when assessing the likelihood of failure.

    a.

    b.

    c.

    d.

    e.

9. List at least three factors to consider when assessing the consequences of failure.

   a.

   b.

   c.

10. List at least five conditions that could increase the likelihood of failure.

    a.

    b.

    c.

    d.

    e.

11. _____ _____ fungi primarily decay the lignin within and between cell walls in the wood, reducing the wood's stiffness but leaving some flexibility.

12. _____ _____ gets its name because, after the cellulose is decayed, the remaining lignin is dark or brown in color.

13. True/False—A tree may appear to be solid and structurally sound or may have a thick, green crown yet can have significant decay inside.

14. List three definite indicators of decay and three potential indicators of decay in a tree.

Definite indicators	Potential indicators
a.	a.
b.	b.
c.	c.

15. _____ _____ is new wood produced in response to damage or loads.

16. True/False—As trees grow and develop, they adapt to the various loads that they experience (gravity and wind) by developing wood where it is needed to support the loads.

17. True/False—Woundwood is less dense than and chemically different from other wood, and it resists decay better than normal wood.

18. Two advanced assessment devices/techniques of assessing internal decay are _____ - _____ _____ and _____ _____ .

19. List three target-based and three tree-based options for mitigating tree risk.

   Target-based options   Tree-based options

   a.   a.

   b.   b.

   c.   c.

20. True/False—Each combination of a failure mode and a target represents a separate risk to analyze.

21. True/False—The overall tree risk is the risk of whole tree failure.

22. _____ _____ is the degree of risk that is within the owner's/manager's or controlling authority's tolerance, or that which is below a defined threshold.

23. The risk remaining after mitigation is the _____ _____ .

24. True/False—Once the highest risk factor has been mitigated, the tree risk rating goes to the next highest risk factor.

25. True/False—The tree risk assessor bears the responsibility for tree risk management.

Challenge Questions

1. Identify three trees in different settings. What are the potential targets for each?

2. What are the factors to consider when deciding what is the most appropriate level of tree risk assessment?

3. Select a large, mature tree in your area and perform a visual inspection. Record data for site conditions, tree health, conditions of concern, potential loads, and response growth.

Sample Test Questions

1. Which of the following is *not* a type of Level 1 assessment?

   a. climbing the tree
   b. walk-by
   c. drive-by
   d. aerial patrol

2. Which type of decay primarily affects the lignin within and between cell walls in the wood, reducing the tree's strength under compression?

   a. brown rot
   b. white rot
   c. soft rot
   d. sapwood rot

3. Following construction, forest trees on the edge of remaining stands are prone to failure due to

   a. losing the protection of the trees that used to surround them
   b. less trunk stability and poor taper
   c. increased exposure to the weather elements
   d. all of the above

4. Trees that lean because of ground failure or root injury

   a. have a high potential to fail
   b. are less of a risk than those that lean due to phototropism
   c. are not a threat unless located at the edge of a wooded area
   d. are a risk only if they begin to grow in compensation for the lean

5. Which of the following is the responsibility of a tree risk assessor?

   a. determining acceptable risk
   b. presenting mitigation options
   c. prioritizing work
   d. choosing among mitigation options

Recommended Resources

(See Recommended Resources in back of book for detailed information.)

Tree Risk Assessment Manual (Dunster et al. 2017)

Best Management Practices: *Tree Risk Assessment* (Smiley et al. 2017)

Wood Decay Fungi Common to Urban Living Trees in the Northeast and Central United States (Luley 2005)

ANSI A300 (Part 9) *Tree Risk Assessment a. Tree Failure* (American National Standards Institute 2017)

The Body Language of Trees (Mattheck and Breloer 1996)

CHAPTER 13
TREES AND CONSTRUCTION

▼ OBJECTIVES

- Describe how trees can be injured or killed as the direct or indirect result of construction or land development.

- Discuss the importance of arborists' participation in the planning and design phase of land development to ensure that trees are protected and preserved.

- Explain the steps that can be taken to protect and preserve trees during the construction phase.

- Describe some techniques that can be used to help protect trees when soils and roots must be disturbed.

- Explain the limitations for the treatment of trees that have been damaged by construction.

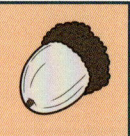

KEY TERMS

access route	soil compaction	tree well
air-excavation device	soil grade	trenching
critical root zone (CRZ)	specifications	trunk flare
drip line	tree island	tunneling
hydrology	tree protection plan	vertical mulching
radial trenching	tree protection zone (TPZ)	

Introduction

Trees can greatly enhance the environmental, social, and economic values of a property. Unfortunately, loss to development, especially from construction damage, is a common cause of tree death and decline in urban areas. Mature trees in urban sites are vulnerable to activities such as new construction or infrastructure repairs and upgrades. When a development site has existing trees that can be incorporated in the planning and design phase, they must be protected during development to survive in the long term.

When arborists are involved early in the planning process, they can help make sure trees are protected and preserved during the development process. An arborist can advise on which trees are suitable for retention, determine protection measures, and assist in designing a site layout and protection plan that maximizes tree health and minimizes tree risk during and after construction.

Information in this chapter can be applied to large multiyear development projects as well as small, routine, infrastructure repairs. Arborists should refer to the ISA's Best Management Practices: *Managing Trees During Construction* for more detailed guidelines.

Land Development and Construction Practices

Development is the process of preparing land and building structures to accommodate a variety of uses. If tree preservation is to be successful, trees must be considered in each phase of land development. For this to occur, the arborist must be an integral part of the construction team and included in each phase of development. These phases include planning, design, grading, construction, and maintenance.

Construction typically follows a sequence of events:

1. Planning, design, and permitting
2. Construction
3. Maintenance and inspection

Structures, such as buildings, roads, patios, and utility trenches, must be built in such a way that they are stable and safe. Therefore, projects must comply with specific engineering standards regarding soil compaction, footing and foundation design, and depth and separation of utilities. In many situations, building standards and local codes have

Figure 13.1 Unless protected, trees can suffer severe physical damage from construction practices, both above and below ground. If the injuries are extensive and the tree is severely stressed, the tree may never recover.

limited flexibility for modification. Arborists must work within these requirements to ensure successful tree preservation. However, the earlier arborists are engaged in the building process, the better. If arborists have input into the siting and planning process, including identifying trees that are both valuable and possible to protect, it is more likely that tree preservation will be successful. Arborists should become familiar with common terminology and work procedures involved in these processes so they can communicate effectively with engineers, planners, architects, and landscape architects.

How Trees Are Damaged by Land Development or Construction

When construction equipment such as backhoes, bulldozers, and cranes are operated near trees, irreparable damage is possible if precautions are not taken. Most construction injury occurs below ground to the root system. Symptoms above ground may include reduced shoot growth, premature autumn color, extensive watersprout (epicormic shoot) development on the trunk and main branches, dieback, and, eventually, the death of major scaffold branches. Often these symptoms may be overlooked or attributed to other causes. Further, unless the damage is extreme, trees might not die immediately but may decline over several years or even decades. Because of this delay in symptom development and the hidden injuries to the root system, tree owners and managers may not associate the decline and loss of the tree with damage incurred during construction or development. The impacts of construction are cumulative and can place a tree into a spiral of decline.

Physical Injury to Trunk and Crown

Construction equipment can injure the aboveground portion of a tree by breaking branches, tearing the bark, or creating wounds. People not trained in arboriculture may prune trees in ways that create unnecessary wounds or damage tree structure. These injuries are permanent, even if they are compartmentalized and eventually covered with new growth. If the injuries are extensive and the tree is severely stressed, the tree may never recover. Open wounds make trees vulnerable to disease and decay. Decay that starts in the wounds can shorten the useful life of a tree in the landscape.

Cutting of Roots

The digging and **trenching** necessary to construct a building or paved area or to install underground utilities will likely sever a portion of tree roots in the area. Understanding where roots grow is essential for identifying where damage from construction may occur. The roots of a mature tree can extend far from the trunk of the tree and well beyond the spread of the branches.

The amount of damage a tree suffers from root loss depends, in part, on how close to the **trunk flare** the cuts are made. Severing one major root can cause the loss of 15 to 25 percent of the root system.

Figure 13.2 Unfortunately, the effect of belowground construction damage may not be seen above ground for several years. Dieback in the top of the tree is often the first sign of root damage.

Figure 13.3 To expand this roadway, the roots have been severed to lower the grade, then the backfill has been compacted to meet construction specifications.

Figure 13.4 Clearing, grading, and compacting the land as part of the development process can cause significant changes to the soil.

Recovery from significant root loss may be difficult for stressed or older trees.

Root loss due to digging or trenching may increase the potential for trees to fall over. If major structural roots are cut on one or more sides of a tree, the anchorage may be compromised.

Soil Compaction

Soils may be compacted intentionally, such as when soils are worked to create a stable base for buildings or pavement, or unintentionally, when vehicles and/or heavy machinery are used on the site. **Soil compaction** can be very harmful to trees (see Chapter 3, Soil Science).

In addition to the damage to soil from compaction, the fine, absorbing roots near the surface can also easily be damaged, crushed, or killed by heavy construction equipment driving across the site, creating ruts or scraping the soil surface.

Grade Changes Around Trees

In construction, the **soil grade** refers to the elevation of the ground's surface. If the grade is lowered near a tree (e.g., soil is scraped away and removed), a large percentage of the root system may be removed. As much as 90 percent of the fine roots grow within a few centimeters or inches of the soil surface.

When the grade is increased, and additional soil is added over the root system, soil conditions may change dramatically, especially if the fill soil is fine textured and compacted. In some cases, this additional soil can restrict water and air movement to the roots, adversely affecting tree health. The

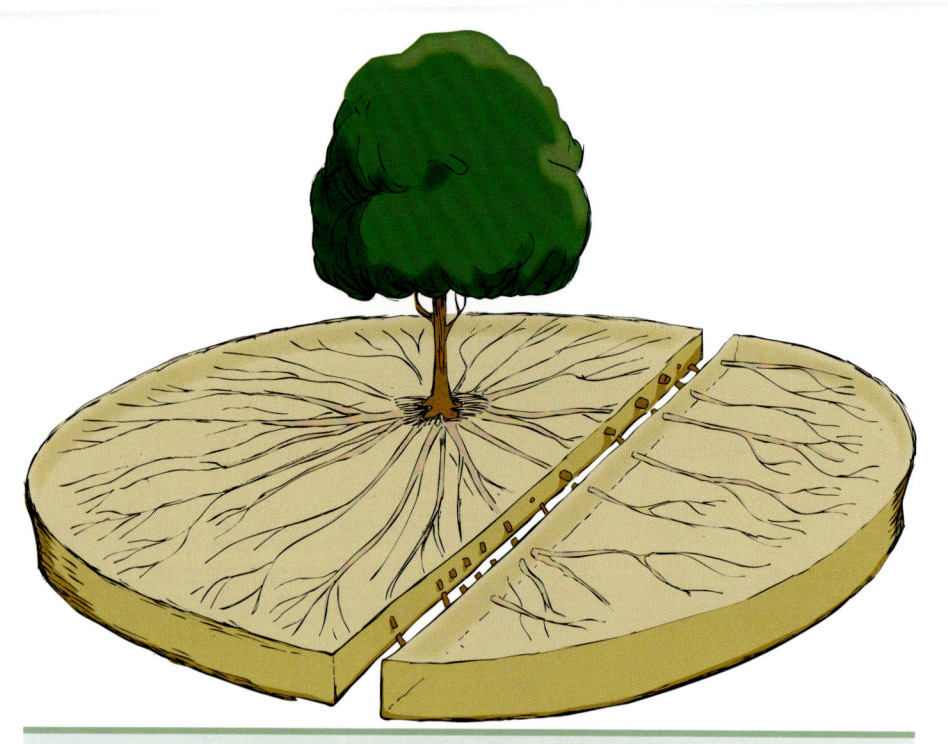

Figure 13.5 The trenching necessary to install underground utilities will likely sever a portion of tree roots in the area. The amount of damage a tree suffers from root loss depends, in part, on how close to the trunk flare the cuts are made.

degree to which added soil damages root systems depends upon the depth and type of the new soil, whether the original soil was excavated and roots were cut before the new soil was put in place, and even the tree species. Further problems may be caused if the soil is placed against the trunk.

The effects of fill soil over tree root systems may not be seen for many years, because it takes time for the roots to respond to changes in soil conditions. Even if the grade change is not in the immediate vicinity of the root zone, the water table or drainage pattern may be changed, with adverse effects to trees.

Chemical Damage

Certain chemicals used in construction, such as solvents, paint thinners, oils, and fuels, may be toxic to the foliage and roots of trees. Death of foliage and roots may place added stress on trees. Contaminated water or runoff from concrete trucks or other equipment can alter soil chemistry and cause root damage or disruption of nutrient uptake.

Exposure to the Elements

In natural settings, such as forests, trees grow in communities and protect each other. When a development project occurs in a wooded area, the removal of neighboring trees or the opening of the collective canopy exposes the remaining trees to more sunlight, frost, and wind than they were accustomed to. Exposure to higher light levels may cause sunscald on the trunks and branches of trees with thin bark. Many interior forest trees have tall, thin trunks with little taper. If neighboring trees that provided protection for many years are removed, the remaining trees will be more prone to failure, windthrow, or ice loading.

Tree Preservation During the Planning and Design Phase

The decision to preserve trees must occur before construction begins. In fact, the earlier in the process that an arborist becomes involved, the greater the chances for success.

Historically, arborists were rarely involved in the development process. In recent decades, however, arborists are much more likely to be an integral part of the process and may have opportunities to provide their expertise as a part of site planning. Arborists must communicate and work with other professionals such as developers, architects, planners, engineers, builders, and landscape architects. Becoming familiar with the standards and specifications for the design and construction of

Figure 13.6 Arborists must work with developers early in the process. If arborists have input into the siting and planning process, including identifying trees that are both valuable and possible to protect, it is more likely that tree preservation will be successful.

usually survive and better adapt to the stresses of construction. In some cases, local tree rules and regulations or local planning controls may also apply to preservation requirements.

Trees should be inventoried, mapped, and considered along with other landscape features during the preliminary design phase. The arborist should create or contribute to a map that details the location of trees and other features in relation to the buildings, roads, and infrastructure. The species, size, and condition should also be assessed prior to construction. Changes in tree condition over the time frame buildings, paved surfaces, and utilities, along with other aspects of development, is useful.

Assessing and Evaluating Trees for Preservation

Evaluating suitability of individual trees or groups of trees for preservation is one of the most important tasks for the arborist. This analysis is needed very early in the planning process. Designing projects around trees not suitable for preservation is a waste of time and money. Trees that are structurally unsound, in poor health, or unable to survive construction impacts are a liability to a project rather than an asset.

The decision to preserve a tree should be made based on its value to the site and its ability to tolerate construction activities due to its health and structure. To determine the tree's suitability for preservation, health and structure can be assessed for tolerance to construction processes. More vigorous trees can

Figure 13.7 The decision to preserve trees should be made based on their value to the site and ability to tolerate construction activities. It is often preferable to retain groups of trees rather than individual trees.

of the project should be documented, including taking pictures.

Where possible, it is preferable to retain groups of trees rather than individual trees. Retaining a mix of ages, species, and groundcover plantings will contribute to a sustainable landscape.

Sometimes design plans or construction procedures can be modified to better accommodate trees and their root systems. Small changes in the placement or design of a building can make a difference in whether a specific tree will survive. For example, bridging over the roots may be substituted for a conventional walkway. Or instead of trenching beside a tree for utility installation, **tunneling** under the root system could be considered.

Creating the Tree Protection Plan

To achieve the ultimate goal of tree preservation beyond project completion, trees must be sufficiently protected through protection requirements that are documented and enforced. After the trees have been assessed for suitability and value and decisions have been made with project leaders and stakeholders on which trees to retain, the arborist should create a **tree protection plan**.

The plan should comply with applicable regulations and standards and should document protection zones and details of all mitigation and preservation measures for the project. Tree protection plans must include details on required fencing and protection of soils. Additionally, because many individuals will

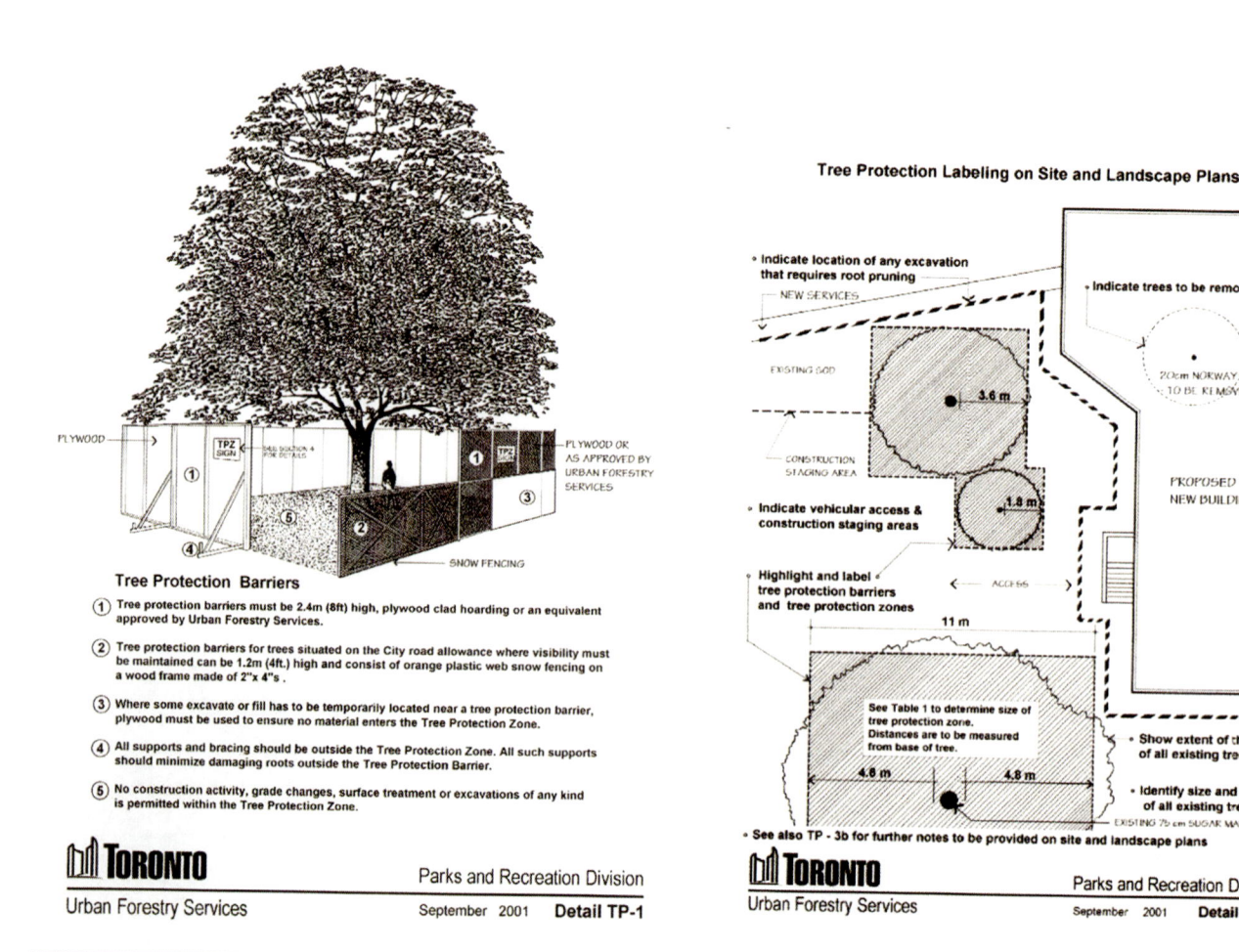

Figure 13.8 The tree protection plan must include details on required fencing and protection of soils. The arborist will work with the developer to put in place effective measures to minimize damage and maximize tree health and structure.

have access to the construction site, providing ways to communicate the tree protection plan with everyone is essential.

Two important parts of the tree protection plan are determining the **critical root zone** and delineating the **tree protection zone**.

Critical Root Zone (CRZ)

The CRZ is the area around a tree where the minimum amount of roots that are biologically essential to the structural stability and health of the tree are located. In an urban environment, it is rarely in the shape of a circle, particularly in an urban site. Factors including structures and pavement, soil characteristics and grades, drainage, and adjacent trees can influence where the roots are. For example, a tree may have more roots on the upwind side of the tree. Open grown trees will also have different root architecture than trees growing in the middle of a forest. The tree's age, size, condition, and species also influence the extent of the CRZ. Arborists must use their understanding of tree biology and observational skills to create a reasonable assessment as to where the roots are located. There are no universally accepted methods to calculate the CRZ.

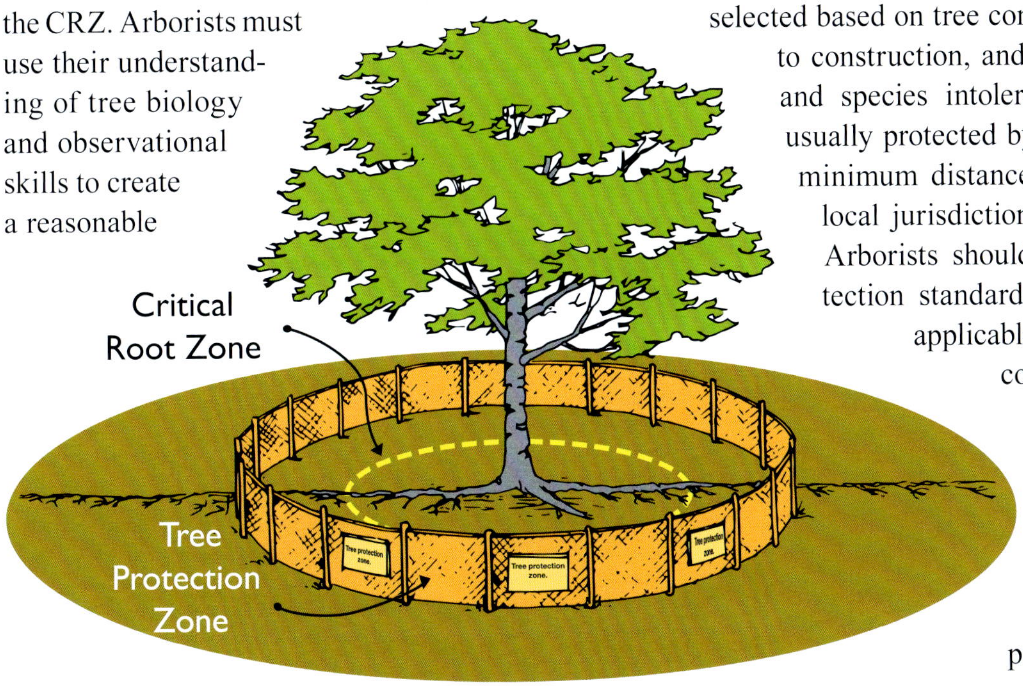

Tree Protection Zone (TPZ)

The TPZ is an area defined during site development, where construction activities and access are limited to protect the tree(s) and soil from damage, and to sustain tree health and stability. This zone is intended to protect root systems, soils, and the aboveground parts of trees. Keep in mind that the root systems can extend much farther out than the **drip line** of the crown and much farther out than the CRZ.

The boundary of the TPZ should be located as far from the tree trunk as necessary and preferably well beyond the CRZ. The minimum distance from the trunk of protected trees where the fencing for the TPZ should be placed is usually calculated with a chart or a formula. It is typically 6 to 18 times greater than the trunk diameter, and the multiple is selected based on tree condition, species tolerance to construction, and site factors. Older trees and species intolerant of construction are usually protected by greater distances. This minimum distance may be established by local jurisdictions or national standards. Arborists should seek out the tree protection standards or guidelines that are applicable to their site but should consider providing greater protection distances if possible.

Unfortunately, due to competition for space and resources in development projects, the TPZ is often much smaller than desired. This is where the arborist must use negotiating skills to find a path that allows the project goals to be

Figure 13.9 The tree protection zone (TPZ) is a defined area where construction activities and access are restricted to protect trees from damage. These zones are intended to protect root systems and soils as well as the aboveground parts of trees. The TPZ can extend farther out than the drip line of the crown and the critical root zone (CRZ).

achieved and still preserve the trees. Tree-friendly construction techniques can be used to accommodate activity within the TPZ. Knowledge of engineering drawings is helpful to determine where changes could be made. Approaching project leaders with skills that offer alternative solutions to achieve the tree protection and project goals will make negotiations more successful.

Specifications

Arborists must help ensure that all of the measures intended to protect trees are written into construction **specifications**. Written specifications should detail exactly what can and cannot be done to and around the trees. Each subcontractor and equipment operator must understand the purpose of the barriers, limitations, and specified work zones. Signs should be posted to reinforce the importance of the TPZ. Fines and penalties for violations should be built into the specifications. The severity of the fines should be proportional to the potential damage to protected trees and should increase for multiple infractions. Some municipalities have ordinances or bylaws that establish minimum tree protection specifications.

Protecting Trees During the Construction Phase

After the tree preservation plans have been made during the planning and design phase, measures must be taken to protect trees selected for preservation. Because many of the potential damages from construction are permanent, such as the severing of roots, the arborist's ability to treat trees damaged by construction is limited. Preventing injuries is more effective than attempting to treat them later.

Depending on the scale of the project, the site is first cleared, and any existing structures and vegetation that were not expressly identified to remain are removed. Rough grading is undertaken to create the site contours required for drainage and construction. Roads are made, and water, storm, and sanitary sewer systems are installed. Final grading is completed to prepare for construction of roads and buildings. Planned construction of roads, buildings, and/or installation of utilities occurs. Fine grading and installation of landscapes may occur after construction activities are completed.

When possible, measures should be taken to increase tree vitality prior to construction. Depending on site conditions, this may include water management, mulching, fertilization to address a specific nutrient deficiency, pest management, or application of tree growth regulators. A healthy tree will be able to recover more quickly from minor to moderate stresses that result from construction.

Though the following activities should be undertaken during the construction phase, they should be planned for and discussed during the planning and design phase.

Erecting Barriers

The single most important action that can be taken at the start of the construction process is to set up the TPZ previously agreed upon in the tree protection plan. This is done by erecting construction fences around each tree, or group of trees, that will remain.

Arborists should work with construction personnel to prevent access to the protected area. A sturdy

Figure 13.10 Erect signs to caution contractors not to encroach on the tree protection zone.

fence with signs indicating that this is a tree protection zone will help. Gates are not recommended. The protected area should be completely clear of building materials, waste, and excess soil. Digging, trenching, compaction, or other soil disturbance should not be allowed inside the fenced area. Exceptions should be approved in writing in accordance with the tree protection plan and conducted under the immediate supervision of the project arborist. Intact natural leaf litter or desirable understory plants should not be disturbed or covered within the fenced area. Weeds and invasive species should be removed. If appropriate, cover the entire area with mulch.

Limiting Access to the Site

An effective strategy for reducing damage to soil and tree roots is to confine traffic to the smallest area possible. Limiting the number of **access routes** on and off the property can also be effective. All contractors should be informed on vehicle access and parking. This same access route may later serve as the route for utility wires, water lines, the driveway, or other paved surfaces. Storage areas for equipment, soil, and construction materials must be planned for. These areas may result in soil compaction during the construction phase. Specify areas for cement washout pits, construction work zones, dumping (of gravel or unused cement, for example), or any other expected construction activities. Cement and some gravel types (for example, limestone gravel) can raise soil pH significantly, which can be harmful to some tree species. All areas designated for these purposes should be located away from protected trees.

Figure 13.11 Storage areas for equipment, soil, and construction materials must be planned for. Specify areas for cement washout pits, construction work zones, dumping (of gravel or unused cement, for example), or other expected construction activities.

Reducing Compaction

A thick layer of mulch can be used to protect the soil surface and reduce the potential for soil compaction on construction sites. Spreading a 20 to 30 cm (8 to 12 in) layer of coarse mulch (such as wood chips) around the trees helps to disperse the weight of construction equipment. Adding a layer of geogrid or geofabric over the ground before the mulch is spread improves effectiveness and makes mulch removal easier with less damage to the roots. Additional weight dispersal can be obtained by placing large sheets of plywood or steel over the mulch or by using trackway or other protection systems.

Construction mulching is a temporary measure, and the mulch should not be left in place at this thickness for a prolonged period if within the root zone. If the mulch has not reduced to an acceptable thickness over the course of the project (e.g., less than 10 cm or 4 in), it must be carefully removed or distributed so as not to damage the trees and roots in the process. Reducing soil compaction is extremely challenging in areas where roots are present. Therefore, preventing soil compaction in the first place is the best strategy.

Minimizing the Effects of Grade Changes

Grade changes can be devastating to trees, even if the change is not severe, because grade changes tend to cut or damage roots, compact soil, and alter patterns of water movement. If the grade must be lowered, the ability of a tree to survive depends on several factors. The most important consideration is the amount of root system that will remain after grading. Other important considerations are tree condition and species tolerance to root loss, the degree to which the grade is lowered, soil conditions, and whether irrigation will be applied. Changes in the **hydrology** and soil-water relations of the site can also affect tree survivability. Even temporary grade changes cause damage and must be planned for.

If the grade must be lowered on all sides of a tree, a **tree island** can be constructed. The tree and surrounding soil will remain at the original grade and are enclosed by a retaining wall. This practice severs most roots on the tree that extend beyond the edge of the island. Therefore, the greater the percentage of root system that remains intact and at the original grade, the greater the chance of tree survival. If retaining walls are constructed as part of the grade change, drainage must also be planned for.

Sometimes the grade must be raised near a tree. Increases in grade may require major efforts to protect the tree. On some sites, a vertical retaining wall can be constructed. On other sites, **tree wells** may be used. Large-diameter tree wells may keep

Figure 13.12 It may be possible to preserve trees when the grade must be lowered (left) or raised (right) by creating islands or pits.

fill soil far away from tree trunks and minimize the percentage of the root system that is disturbed or covered. Small-diameter wells built around the trunks of trees are rarely adequate to protect the tree and ensure survival. In addition, the compaction from construction of small wells may do more damage than the fill soil would have.

Some specifications call for the use of gravel or stone below the fill soil to increase water and oxygen penetration. However, the opposite will occur: if soil is placed over gravel, water will not drain out of the soil layer until it is extremely wet. Furthermore, roots may grow upward into gravel layers when rainfall is plentiful only to later be stressed or killed during dry periods because gravel retains little water. It is better to place the fill soil in direct contact with the original grade. When possible, use a fill soil that is similar in texture to the existing soil.

Engineered fill materials that will support structures or pavement are compacted to create a stable building base. Thus, where engineered fills occur near trees, roots may be removed, and a condition is created in which new roots are unlikely to develop and trees may become less stable.

If no structures or pavement will be placed on the fill, the grading activity can still cause significant damage to adjacent trees. Heavy equipment used to install a fill soil can injure roots near the soil surface and cause soil compaction. Using small equipment and hand tools can reduce the potential for damage. After fill is installed, it is important to ensure that the original root zone remains hydrated, as rainfall may not readily percolate through the fill to the original root zone. Soil moisture status in the root zone should be monitored, and irrigation may be required.

Maintaining Good Communications

It is important for all those involved in the development project to work together as a team. The best-laid plans between arborist and builder can be destroyed by one uninformed subcontractor. The arborist should arrange a preconstruction meeting at the site, if possible, to ensure that all subcontractors understand the importance of the TPZ. During the construction project, the arborist should visit the site frequently and maintain clear communication with the project superintendent. Vigilance will pay off as workers learn to take the arborist's recommendations seriously.

It is a good idea to take photos at every stage of construction. If any infraction of the specifications does occur, it may be important to link cause and effect. Arborists should plan and work with project managers to arrange continued maintenance and monitoring of the trees' health, structure, and

---

When Roots Must Be Severed

There are several steps that can be taken to reduce the impact of root cutting on tree health.

1. Reduce tree stress before cutting roots. During drought conditions, trees should be irrigated prior to cutting roots.
2. If roots must be severed, consider cutting them while trees are dormant.
3. Cut roots cleanly using sharp pruning tools instead of ripping them out with excavating equipment. An air-excavation device can be used to remove soil around roots before cutting roots cleanly with a saw or loppers. For large projects, mechanical root pruning tools are more efficient than hand digging and provide moderately clean root cuts. Mechanical root pruning is preferred over other forms of mechanical excavation.
4. Protect cut root ends from drying out. New roots will grow from the cut root ends, so keep the remaining roots protected by covering them with soil, or if that is not possible, protect them with other suitable materials (mulches, compost, geotextiles, burlap, or a combination of these).

Figure 13.13 Maintaining good communications with all of the contractors involved is essential.

overall condition in the postconstruction maintenance and inspection phase.

Treatment of Trees Damaged by Construction

Despite the best intentions and most stringent tree protection measures, some trees may still be injured during the construction process. Remedial treatments are available to help reduce stress and improve growing conditions around the trees. Unfortunately, the ability to restore trees damaged by construction is very limited, and this is why emphasis should be placed on protection.

Inspection and Assessment

Because construction can affect the structure and stability of a tree, and because new targets are often created, it is important to perform a tree condition and risk assessment after construction. Additional assessments may be warranted when the arborist needs more information to assess specific tree parts or the risk to high-value targets. See Chapter 12, Tree Risk Assessment and Management, for more information.

Treating Trunk and Crown Injuries: Pruning

Branches that are split, broken, diseased, or dead should be removed. Outdated recommendations suggest that tree crowns should be thinned or reduced to compensate for root loss. Such pruning is likely to reduce tree vitality because it reduces the tree's energy-making capability and may further stress the tree. It is better to limit pruning to only what is necessary for risk reduction in the first few years after construction, especially in large, mature trees.

Repairing Bark and Trunk Wounds

Sometimes the bark may be damaged along the trunk or major branches. If this happens, the loose bark can be carefully removed, leaving the attached bark intact. Jagged edges can be cut away with a sharp knife, taking care not to cut into living tissues. This procedure is called bark tracing. A common practice used to be to cut the perimeter of the wound into a smooth shape. This has been shown to be of little or no value and can cause further injury to the tree by increasing the size of the resulting wound. Wound dressing should not be applied, unless recommended for disease prevention (see Chapter 8, Pruning).

Providing Irrigation and Drainage

One of the most important treatments following construction damage is to maintain an adequate, but not excessive, supply of water to the root zone. If soil drainage is satisfactory, the trees must be kept well watered, especially during the dry summer months. A long, slow soak over the entire root zone is the preferred method of watering. Frequent, shallow watering should be avoided, and water should not be directed at or near the trunks of trees. If drainage is poor or has been altered by construction, the limitations need to be identified and corrected.

Mulching

Mulching is not only an effective measure to reduce soil compaction during construction but also helps recovery after damage. Applying a thin, 2 to 4 cm (1 to 2 in) layer of organic mulch over the root system of a tree can moderate soil temperature and moisture while also reducing competition from

Figure 13.14 Measures to remediate construction damage are limited but mulching and root zone aeration can be beneficial.

weeds and grass. The mulched area should extend as far out from the tree as practical for the landscape site. When it comes to mulch, deeper is *not* better, and the mulch should not touch the trunk. Piling it up against the trunk can create conditions for disease.

Reducing Soil Compaction Around Trees

Soil compaction and grade increases can result in poor rooting environments where air and water movement are restricted and roots have difficulty penetrating. However, soil remediation techniques must be used with care around established trees to prevent additional root disturbance. There are several techniques that are designed to reduce root disturbance, although each has limitations. Care should be taken when applying any treatment to construction-impacted trees. Excavating soil may damage roots, even when the goal is to improve conditions for root growth. The arborist might consider postponing soil remediation treatments to allow construction-damaged trees to recover.

In **radial trenching**, trenches are made in a radial pattern throughout the root zone and should extend at least as far as the drip line. Use of **air-excavation devices** has proven effective for radial trenching and can reduce root severance. In radial trenching, these devices cause much less root injury than mechanical excavation equipment, which is no longer recommended. While air-excavation devices leave most roots intact, there can still be root damage that may not be visible. Some species are

Figure 13.15 In radial trenching, trenches are made in a radial pattern throughout the root zone. Trenches are then backfilled with native soil, sometimes mixed with compost or other amendments.

Figure 13.16 An air-excavation device can be used to relieve soil compaction of the tree root zone with minimal damage to the roots.

more susceptible to injury than others. Air excavation is sometimes applied to pie-shaped wedges of the root zone rather than to the entire root zone as insurance against root damage. Radial trenches or wedges are then backfilled with native soil, sometimes mixed with compost or other amendments.

Vertical mulching involves making holes in the ground with a small auger or an air tool. The holes are typically 30 to 90 cm (1 to 3 ft) deep, depending upon at what depth the soil is compacted, and may be filled with organic material such as compost, loosened soil, or other materials or mixes of materials that could provide a loose rooting environment. Research has shown that this treatment has a limited effect, possibly because only a small percentage of the root zone is affected.

What About Fertilization?

Most experts recommend that trees not be fertilized the first year after construction damage. Water and mineral uptake may be reduced due to root damage. It is a common misconception that applying fertilizer gives a stressed tree a much-needed boost. Fertilization should be based on specific nutrient deficiencies shown by soil and/or foliar analysis (see Chapter 5, Tree Nutrition and Fertilization).

Monitoring

Despite everyone's best efforts, some trees may never recover from the stresses caused by land development and construction activities. If a tree dies as a result of root damage, it may require immediate removal. It is possible for trees to have a healthy appearance after construction while having extensive root decay. Arborists must inspect and monitor trees affected by construction to evaluate changes in health and structural condition. It is important to look for changes in drainage patterns, soil condition, and sun and wind exposure.

Other conditions of concern include lean, cracks, and indicators of internal decay in all parts of the tree. Stressed trees are more prone to attack by certain insects and pathogens and should be monitored for several years following construction.

Figure 13.17 Arborists must inspect and monitor trees affected by construction to evaluate changes in health and structural condition. Dieback and extensive watersprout production are two common symptoms of decline following construction.

CHAPTER 13 WORKBOOK

1. Name five ways that trees can be adversely affected by construction.

   a.

   b.

   c.

   d.

   e.

2. True/False—Evaluating suitability of individual trees or groups of trees for preservation is an important task for the arborist.

3. True/False—The goal of an arborist involved in a development project is to save every tree on the site.

4. True/False—The largest, most mature trees are not always the best candidates for preservation.

5. The _____ _____ _____ is the area around a tree where the minimum amount of roots that are biologically essential to the structural stability and health of the tree are located.

6. The _____ _____ _____ is an area defined during site development, where construction activities and access are limited to protect the tree(s) and soil from damage, and to sustain tree health and stability.

7. True/False—Less injury is caused by tunneling directly under a tree than by cutting directly across the root system of a tree when excavating for utility lines.

8. True/False—Preferably, the critical root zone will be much larger than the tree protection zone.

9. True/False—The purpose of the barriers, limitations, and specified work zones should be clearly communicated to each person on the jobsite.

10. True/False—It is easier for an arborist to treat trees that have been damaged by construction than to prevent the damage.

11. Written _____ should detail exactly what can and cannot be done to and around the trees.

12. Use of an _____ - _____ _____ has proven effective for soil aeration and radial trenching, causing much less root injury than mechanical excavation equipment.

13. True/False—Small-diameter wells built around the trunks of trees are usually adequate to protect the tree and ensure survival.

14. True/False—If roots must be severed, they should be cut cleanly with sharp tools and prevented from drying out.

15. In _____ _____ , trenches are made in a radial pattern throughout the root zone and should extend at least as far as the drip line.

Challenge Questions

1. Write a sample set of specifications to preserve trees on a construction site. Include specifications for protecting trees before and during construction, as well as maintenance instructions to help preserve the health of the trees.

2. What actions can be taken if a tree is damaged by construction in violation of the written specifications?

3. Why may tree death and decline due to construction occur several years after construction is complete? What are some of the signs and symptoms of construction damage that an arborist can look for following construction?

Sample Test Questions

1. Measures to reduce compaction on building sites are not always an option, because

    a. there is no way to effectively reduce compaction of soils with high clay content
    b. if the site has a high water table, compaction reduction efforts will be ineffective
    c. projects must comply with specific engineering standards regarding soil compaction
    d. soil structure and aggregate types prevent changes to bulk density

2. Arborists should be involved early in the construction planning process because

    a. tree preservation measures should be incorporated into the project specifications
    b. once construction has begun, it may be too late to save the trees
    c. there is often little arborists can do to treat construction damage
    d. all of the above

3. A measure that can be taken to minimize compaction on a construction site is

    a. watering the site thoroughly before equipment is brought in
    b. permanently raising the soil grade to protect tree roots
    c. spreading a temporary, thick layer of mulch over the site
    d. root pruning the trees in advance

4. A common strategy for tree preservation that can retain more trees and promote sustainability is

    a. retaining groups of trees with a shared root space and a protected perimeter
    b. preserving all of the largest, most mature trees on the site
    c. preserving only the youngest trees on the site
    d. selecting for preservation only the species known to tolerate soil compaction

5. What flexible guideline for a multiple of tree diameter is commonly used when determining where to establish a tree protection zone?

    a. 3 to 5
    b. 5 to 10
    c. 6 to 18
    d. 18 to 24

Recommended Resources

(See Recommended Resources in back of book for detailed information.)

Best Management Practices: *Managing Trees During Construction* (Fite and Smiley 2016)

Trees and Development (Matheny and Clark 1998)

Introduction to Arboriculture: Trees and Construction online course (ISA)

CHAPTER 14
URBAN FORESTRY

▼ OBJECTIVES

- Discuss what urban forests are and how urban foresters manage urban trees and green spaces.

- Describe the benefits and costs of urban forests to society, and present ways to quantify the values of urban forests and trees.

- List common urban forest management strategies and tools, such as urban forest management plans and tree inventories.

- Describe common legal and regulatory issues in urban forestry.

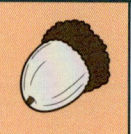

KEY TERMS

appraisal
arboriculture
best management practice (BMP)
biodiversity
canopy cover assessment
carbon sequestration
city forester
climate change
composting
cost approach
ecosystem
habitat
hardscape

income approach
i-Tree
municipal arborist
permit
resilience
risk management plan
sales comparison approach
size diversity
species diversity
specifications
stakeholder
standard
succession
sustainability

tree bylaws
tree inventory
tree officer
tree ordinance
tree preservation order (TPO)
tree warden
trunk formula technique
urban forest
urban forester
urban forest management plan
urban forestry
wildlife

Introduction

Urban forests are often defined as the trees and associated vegetation in and around urban areas. What people think of as "urban" or "a city" can change depending on where they live. Sometimes it is easier to describe urban forests as the trees and associated plants where people live, work, learn, and play.

Urban forestry involves the planning and management of urban forests, while **arboriculture** focuses on the care of individual trees and other woody plants in the landscape. Both fields may include work on both private and public land, including trees along streets, in parks and green spaces, and within commercial, industrial, and residential areas.

Urban foresters serve urban communities by improving the environment using trees and other plants while managing associated risks and costs. They work with arborists, urban planners, civil engineers, landscape architects, public works officials, government agencies, and the public to be effective and achieve their goals. Urban foresters working for cities may have job titles such as **municipal arborist**, **city forester**, **tree warden**, or municipal **tree officer**.

Figure 14.1 An urban forester must work with urban planners, civil engineers, public works officials, government agencies, and the public to effectively manage the health, risk, and sustainability of the urban forest.

This chapter is a basic overview of urban forestry. It is not intended to provide significant depth of instruction. ISA Certified Arborists® who work in urban forestry or municipal arboriculture are encouraged to pursue the ISA Certified Arborist Municipal Specialist® certification.

Benefits, and Costs of Trees

Trees provide many benefits—both direct and indirect—in urban areas. The benefits of trees include environmental improvements, economic savings, and social and health benefits. On the other hand, trees have costs associated with planting, maintaining, and managing the trees in urban areas. Those costs must be taken into consideration when quantifying the net benefits. Although the trees of urban forests are assets, when not properly cared for and managed, they can become liabilities.

The perceived benefits and costs of urban forests to cities depend heavily on human factors. For example, many people appreciate the aesthetic beauty of trees and enjoy the shade of a large tree on a hot day. However, benefits such as stormwater mitigation are dispersed across landscapes and can be harder for individuals to realize. Appreciation of trees depends on education and awareness, as well as social and cultural values.

Benefits
Environmental
- Air quality
  Trees absorb airborne pollutants and collect and filter particulates on their leaves.
- Stormwater
  The leaves and branches of trees catch and slow rainwater, preventing the water from overwhelming drains and streams. They can reduce the impact of floods.
- Soil erosion
  Tree roots hold soil in place to prevent it from eroding away. Trees also prevent soil erosion by mitigating the eroding effects of events like storms.
- Local cooling effect
  Shade and leaf transpiration reduce localized air temperature. This has ecological benefits as well as human health benefits.
- Support wildlife
  Trees provide food and shelter for many types of wildlife.
- Mitigate and adapt to climate change
  Trees can play a role in mitigating climate change (see Climate Change sidebar) by sequestering and storing carbon. Trees can also help communities adapt to the impacts of climate change, for example, by providing shade to streets and homes.

Economic
- Reduced energy costs
  Trees strategically located near and around buildings can significantly reduce heating and cooling costs. Well-placed trees can also serve as windbreaks, which can be beneficial in cold winter climates.

- Urban wood and biomass reutilization
    After urban trees are removed, wood waste can be processed and utilized as products, such as mulch and lumber.
- Increased property values
    Residential properties with trees typically have higher appraisal and sales values than similar properties without trees. Local governments may gain more sales and property tax revenue.
- Retail and commercial revenues
    Tree canopy in retail districts promote positive experiences that encourage shoppers to travel greater distances and spend more time shopping. Consumers may be willing to pay up to 12 percent more for goods and services.
- Reduced infrastructure damage
    Trees shade and protect pavements and other **hardscapes** from the effects of solar exposure and weather.
- Health services savings
    Many factors influence the health of people across communities. Having trees nearby may reduce incidence of disease and promote wellness, generating cost savings for individuals, households, and entire cities.

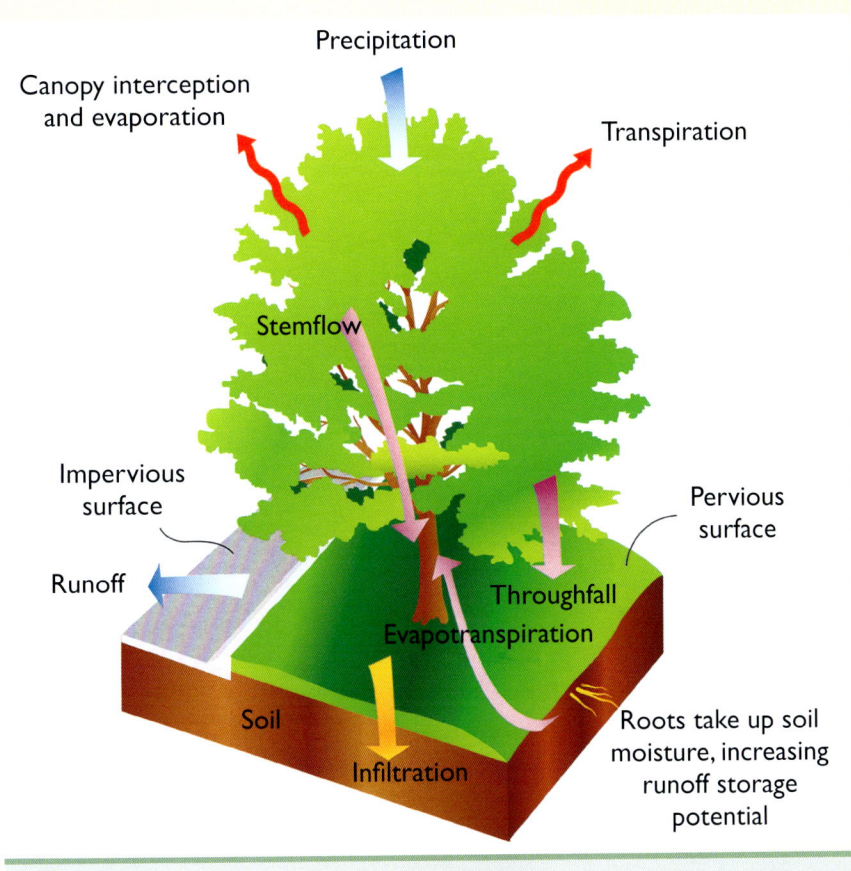

Figure 14.2 Trees reduce stormwater runoff by intercepting rainfall. The permeable surfaces that trees thrive in also intercept water and reduce the amount of waste through runoff.

Social and Health
- Strengthening communities
    Trees provide important spaces for communities to gather, meet, and work together. Having trees nearby has also been linked to less aggression in households and reduced crime in neighborhoods.
- Mental health and function
    Experiences with trees are associated with reduced stress and anxiety and improved symptoms of depression. Time in nature provides restorative experiences, leading to reduced anger and frustration and greater creativity, for children and adults.
- Physical health and wellness
    Trees in streetscapes and parks promote more physical activity and recreation, such as walking and cycling. Nature-based playgrounds are spaces for children to develop physical and social skills.

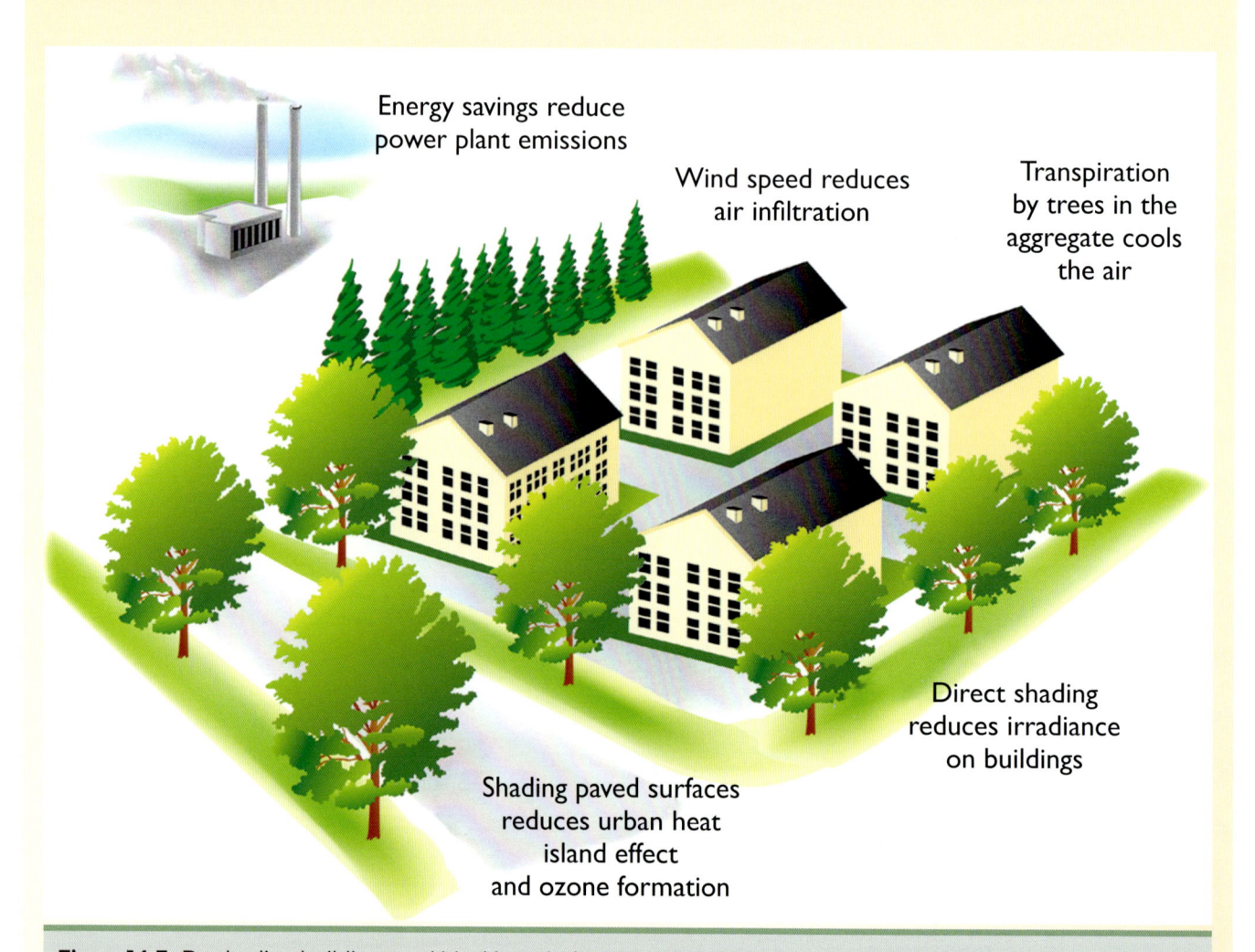

Figure 14.3 By shading buildings and blocking wind, trees reduce energy needs, which also can reduce pollutant emissions by power plants.

- Therapy and healing
  Forest therapy provides many benefits; doctors suggest spending time in parks as prescriptions for their patients. Hospital patients recover more quickly from surgery when having a view of trees.
- Cultural connections
  The cultural heritage of trees and people can be traced back thousands of years. They are considered witnesses of history and connect societies to events in their past, present, and future. Trees are often planted for memorials and celebrations.

Costs
Environmental
- Air quality
  Though trees have an overall positive effect on air quality by capturing pollutants, certain species aggravate conditions like asthma and allergies through the release of pollen.

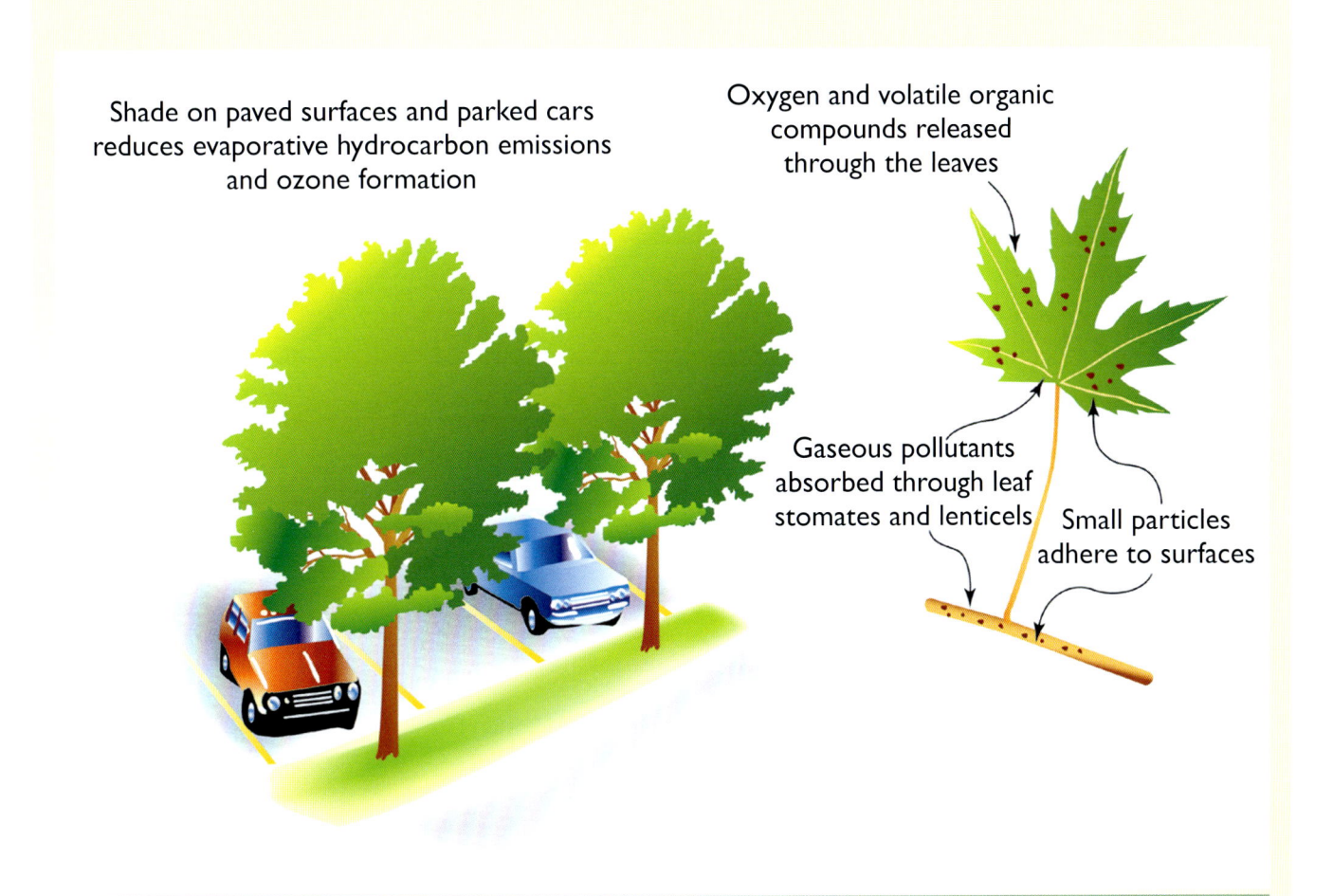

Figure 14.4 Trees collect particulate pollutants on their foliage. By cooling parking lot heat islands, they also can reduce ozone.

- Invasive species

    Many nonnative species have been introduced into cities throughout history, some of which have become invasive. Invasive species can cause significant damage to valuable natural areas, for example, by crowding out native species.

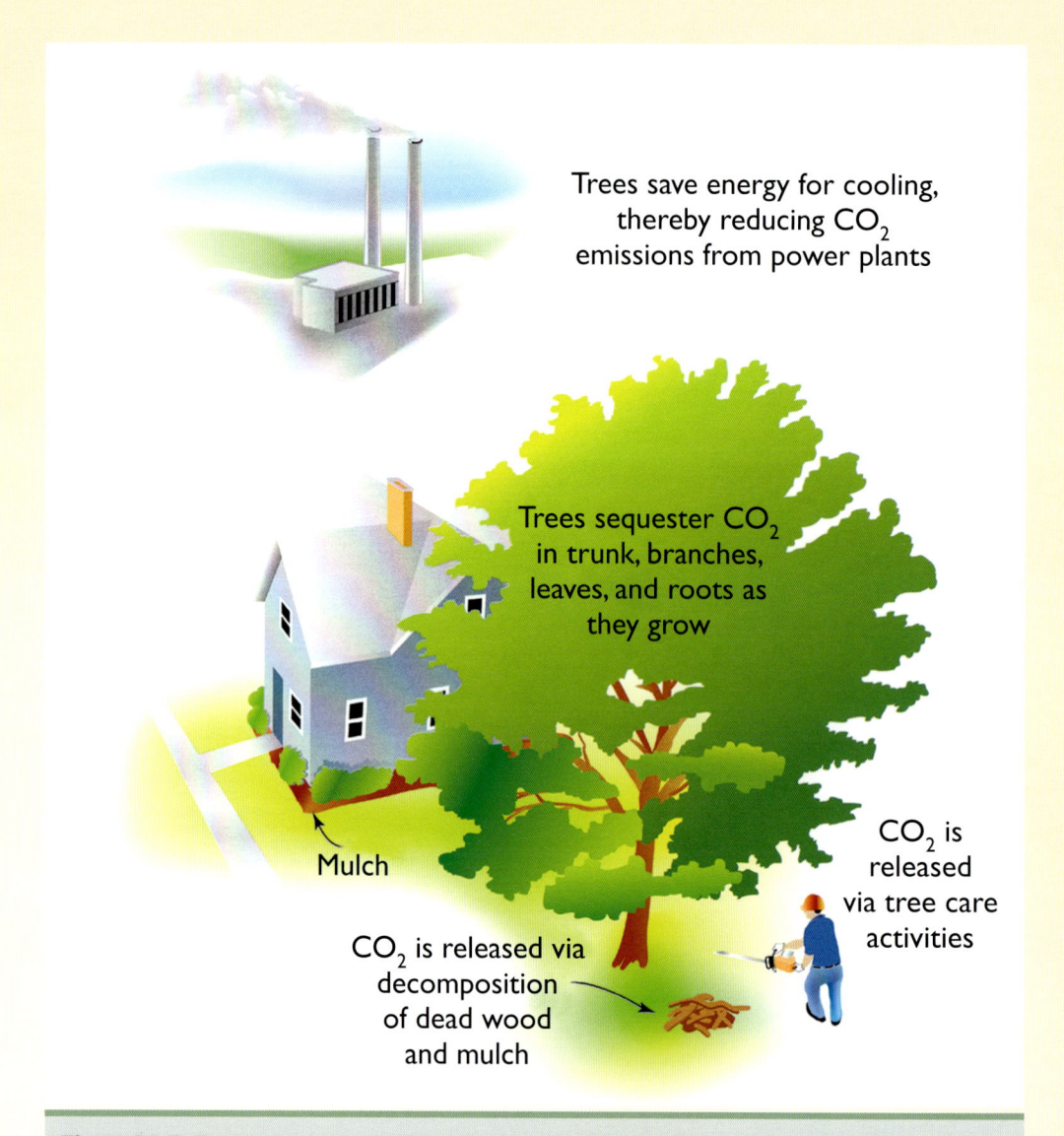

Trees save energy for cooling, thereby reducing CO_2 emissions from power plants

Trees sequester CO_2 in trunk, branches, leaves, and roots as they grow

Mulch

CO_2 is released via decomposition of dead wood and mulch

CO_2 is released via tree care activities

Figure 14.5 Trees sequester carbon as they grow. However, maintenance practices and decomposition release carbon dioxide into the atmosphere.

Economic

- Maintenance costs

    Planting and maintaining trees requires upfront costs to gain benefits in the long term. Urban foresters must maintain the health and safety of the trees in their jurisdiction. Maintenance costs include irrigation, pruning, plant health care, and risk management.
    Hiring and training qualified personnel to maintain trees and run stewardship programs is also a cost.

- Tree risk

    Urban trees can come into conflict with, and sometimes damage, infrastructure such as sidewalks, utilities, and buildings. Beyond costs associated with damage, the damaged infrastructure can also pose significant liability costs to the tree owner.

Social and Health

- Tree risk

    In rare circumstances, falling trees and branches can harm people and property. Tree risk managers must incur costs of routine risk assessments.

- Blocking lines of sight

    Trees can block lines of sight. Visual obstruction can create costs associated with public safety, such as when trees near street corners prevent drivers from seeing oncoming vehicles.

- Nuisances

    Tree parts or structures may provide benefits or be liabilities, depending on how they are viewed. Fruits, for example, may be food for wildlife or can be nuisances that must be cleaned up.

Figure 14.6 Trees in streetscapes and parks promote more physical activity and recreation, such as walking and cycling. Having trees nearby has also been linked to less aggression in households and reduced crime in neighborhoods.

> ### Climate Change
>
> **Climate change** is a long-term shift in global or regional climate patterns. In modern times, climate change is being caused by the increase in heat-trapping gases such as carbon dioxide (CO_2) in the Earth's atmosphere. As trees grow, they absorb carbon from CO_2 in the atmosphere and store it in wood and other carbon-based tissues. This is called **carbon sequestration**, and it reduces excess carbon in the atmosphere. Trees have a finite life span, and when they die or are removed, some of the carbon is returned to the atmosphere through normal decomposition processes or through burning. Climate change mitigation refers to actions that reduce emissions that lead to climate change. Climate change adaptation refers to actions to reduce the effects of climate change impacts.

Valuation and Appraisal

Many tools exist for urban foresters to assess the value of urban trees, for example, by determining their environmental, economic, or social and health benefits. Other tools help urban foresters estimate the cost to repair or replace trees in the landscape.

Appraisal is the act or process of developing an opinion of value or cost. There are three primary approaches for appraising a tree or group of trees: the sales comparison approach, the income approach, and the cost approach.

The **sales comparison approach** compares a buyer's willingness to pay with a seller's willingness to sell. The most common indicator of the market value of the subject of an appraisal is the comparison with recent similar transactions. The market value of a tree could refer to the market value of a living tree, the value of usable lumber the tree can be converted to, or the market value added by a tree to a piece of real property. Even when amenity trees are not directly bought and sold, their contributory value can be appraised by taking the difference in value between two similar parcels of land differing only by the presence or absence of trees.

The **income approach** quantifies the present value of future benefits expected to be generated by the subject of an appraisal. Some trees provide direct benefits such as products that may be sold to generate income. Others provide indirect benefits such as cooling cost savings. Future expected benefits are estimated and then discounted based on their uncertainty and how far into the future they will be realized. Some valuations of ecosystem services, such as the benefits we receive from nature, are possible using online applications such as **i-Tree**, a tool created by the USDA Forest Service.

The **cost approach** determines the amount of resources necessary to reproduce, replace, or repair the subject of an appraisal. The reproduction method appraises the cost to recreate an identical copy of the subject tree of the same size and species. The functional replacement method appraises the cost to reproduce the benefits provided by the subject of an appraisal, which may involve one or more replacement trees that are smaller or are a different species. The repair method determines the cost to correct damage to the subject tree.

Figure 14.7 Data can be used to assess the structure, function, and value of the urban forest, using software such as i-Tree, to communicate the benefits of urban trees and forests.

It is difficult to replace very large trees with trees of equal size. To address this appraisal challenge, various techniques of extrapolating costs have been developed and adopted around the world. These valuation techniques involve scoring systems that consider tree size, species, condition, and location. One of the most widely used reference guides for tree appraisal, especially in North America, is the *Guide for Plant Appraisal*, developed by the Council of Tree and Landscape Appraisers (CTLA) in the United States. Among other techniques, it outlines the **trunk formula technique**, which extrapolates the cost to reproduce nursery stock per unit of trunk cross-sectional area. The cost of reproducing a tree will often exceed its contribution to the market value of the real estate on which it is growing or the present value of its future expected benefits.

One of the roles of a tree appraiser is to select the most appropriate approach and methods of appraisal for each unique assignment. It is also important for a tree appraiser to be clear about the appraisal methods used. Individuals and communities will value trees differently and may react strongly when a tree is appraised at a higher or lower value than they feel it should be.

Planning and Management

Urban forests can help to make our cities more sustainable and resilient for both current and future generations. **Sustainability** is the ability to maintain environmental, social, and economic benefits over time. **Resilience** is the ability to respond and recover from disturbances and stresses.

Figure 14.8 The infrastructure of a city includes the streets, sidewalks, and utilities that support urban life. The "green" infrastructure includes the green space, trees, and other natural resources.

City infrastructure is often divided into several types: (1) "gray" infrastructure, such as streets, sidewalks, and sewer systems, (2) "green" infrastructure, including parks, green spaces, greenways, and urban forests, (3) "blue" infrastructure, including urban streams, lakes, rivers, and sea fronts, and (4) "brown" infrastructure, including former industrial areas, harbor areas, etc. Often, such infrastructures are mixed or linked.

To make cities and their urban forests more sustainable and resilient, planning and management are required. Planning is the process of identifying the activities required to achieve a desired goal. Management is the process of acting strategically to achieve certain goals and is performed on different levels within an organization. At the policy level, long-term goals establish the direction. These may include overall tree policies and tree preservation strategies. At the tactical level, more specific plans and guidelines are created. These may include developing budgets and a tree inventory. At the operational level, the actual maintenance activities such as planting and pruning are carried out.

To be effective in planning and managing urban forests, urban foresters must work collaboratively with professionals in many disciplines and with members of the public, including private residents and business owners. For example, city planners, engineers, and architects often decide where and how urban trees fit into new urban spaces. Urban foresters must be able to communicate the specific measurements of the space, above and below ground, that trees require for growth and survival.

To collaborate successfully, it is important to communicate in terms that the audience understands. Different professions use different terminology, which can pose a challenge. To promote and protect trees, urban foresters must work at the policy level, to build a shared understanding of the benefits and requirements of urban forests, and at the operational level, to ensure proper tree maintenance.

Urban Forest Assessment

Urban foresters rely on several methods to assess the extent and condition of the urban forests they manage. Two common approaches are canopy cover assessment and tree inventories. Both approaches provide information to help urban foresters better manage the tree resources.

- A **canopy cover assessment** is a two-dimensional measure of the area of ground covered by trees and other vegetation. It is most often represented as either a percentage of a total area (e.g., "City A has a canopy cover of 20 percent") or an area measurement (e.g., "City A has 10 sq km [4 sq mi] of total canopy cover"). Canopy cover can be assessed based on tree sizes and location using remote-sensing technologies such as satellite imagery and aerial photography. This data is most often used for strategic level planning, goal setting, and understanding the spatial distribution of the urban forest across property boundaries. It has different uses and purposes for urban foresters than tree inventories.

- A **tree inventory** is a record of the location, characteristics, and assessment of

Figure 14.9 Most inventories include tree species, diameter, location, condition, maintenance information, and notes. More detailed inventories may include site information, crown dimensions, maintenance recommendations, and risk assessment.

Figure 14.10 Data analysis will provide information such as the number of trees of different species and size and condition categories. Inventory data can be used for planning, budgeting, emergency preparedness, and to provide record of maintenance actions.

individual trees and groups of trees over a well-defined area (such as streets or parks). Most inventories include tree species, diameter, location, condition, maintenance information, and notes. More detailed inventories may include data input fields for site information, crown dimensions, detailed maintenance recommendations, and risk assessment. Data analysis will provide information such as the number of trees of different species and in various sizes and condition categories. Vacant planting spaces can be inventoried for future planting efforts. Inventory data can be used for planning, budgeting, emergency preparedness, and to provide record of maintenance actions. Data can also be used to assess the structure, function, and value of the urban forest, using software such as i-Tree, to communicate the benefits of urban trees and forests.

Urban foresters decide what types of data to collect based on their needs and resources. A cost-benefit analysis will help in making decisions about what information to collect and how it will be maintained. Not only can data collection for tree inventories and canopy cover assessments be expensive, but it can also take time. The time invested and cost of the data collected vary based on the depth of information collected and the size of the area.

Most urban forest assessments use global positioning system (GPS) technology to accurately map canopy cover and tree locations. Urban forest assessment and mapping technology is rapidly advancing, by integrating data collected by field personnel on the ground and by using data collected by satellites, unmanned aerial vehicles (UAVs or drones), and other types of sensors.

Management Plans

One of the primary goals of urban forestry is to manage tree resources to sustain and protect their environmental, social, and economic benefits while managing associated risks and costs. Managing large populations of trees requires a comprehensive management plan. An **urban forest management plan** formulates and documents the strategies and procedures for managing trees within a predefined area or jurisdiction over a specified period, typically several years or decades. A management plan may contain a number of subplans:

- Planting plan
- Preservation plan

- Maintenance and operations plan
- Tree removal and replacement plan
- Risk management plan
- Storm-response/emergency plan
- Public outreach and education plan
- Community engagement plan

The management plan establishes priorities and goals within the limitations of the local authority, climate, financial constraints, and resources available. It should also involve the assessment of the current human, financial, public, and environmental resources. The plan should chart a course for meeting the stated goals. Key **stakeholders** should be involved in the development. Finally, periodic evaluation of progress and outcomes is important for making adjustments and to help achieve desired outcomes.

One component of a management plan is a **risk management plan**. It should include a risk management policy statement and state the procedure for identifying, assessing, reporting, and mitigating risks from trees posing elevated risk. Other components of the plan may include (but may not be limited to) a standard-of-care statement, identifying risk exposure, determining acceptable risk, frequency and methods of tree risk assessment, mitigation alternatives, risk response protocol, record-keeping protocols, and a statement of how the program is to be funded.

Figure 14.12 One approach to utility and municipal tree maintenance is to contract all tree work to private businesses. Companies contracted to perform municipal tree care operations must follow the city's tree ordinances and perform work according to specifications.

Figure 14.11 A tree risk management policy statement should establish policies for identifying, assessing, reporting, and mitigating risks associated with trees.

Key Issues in Management

Although there is a wide range of responsibilities associated with urban forests, those that follow are of particular importance.

Maintenance Planning

Proactive and planned maintenance of the urban forest is necessary to sustain tree benefits and minimize risks. Most urban foresters strive to keep their routine maintenance on a cycle by assessing and pruning trees on a predetermined schedule. This can help manage costs in the long term and reduce the time and expenses involved with responding to calls from the public, emergency removals, and cleanup. Deferring maintenance can lead to higher costs and risks and is likely to reduce the benefits the trees provide.

Figure 14.13 Routine maintenance cycles can help manage long-term costs and reduce the time and expenses involved with emergency removals and cleanup. Postponing maintenance can lead to higher costs and risks.

Figure 14.14 A storm-response plan is an essential part of an urban forest management plan. When a storm hits, the plan must already be in place so that the response is safe, efficient, and effective.

One approach to utility and municipal tree maintenance is to contract all tree work to private businesses, and another is to perform the work using in-house staff. Many cities and utilities do both, contracting more specialized tasks but performing planting and maintenance themselves. There are advantages and disadvantages to each approach. Some small municipalities do not employ a working crew or even one arborist.

Emergency Planning and Response

From seasonal storms to natural disasters, urban foresters must manage the impact of emergency events to the trees they manage and the communities they serve. Damage and costs can be avoided in many ways. Preparation activities include regular maintenance as well as quick, vehicle-based assessments prior to storms. Response during events will require collaborating across sectors, such as with public safety and utilities, and having trained crews ready to respond. Recovery after events includes activities such as caring for damaged trees and organizing replanting programs.

Tree Diversity and Urban Biodiversity

When planting an individual tree, appropriate species selection is based on the tree and site characteristics and the goals of the property owner. When managing large populations of trees, however, there is additional concern about tree **species diversity**. Planting a single or few species in a large area may result in significant damage and loss due to a host-specific pest or disease.

Urban foresters try to maintain a diversity of species that are tolerant to urban site conditions, are hardy in the climate, and have few pest problems or unwanted characteristics. The number of species that will thrive in a given area will vary according to geography, soils, and climate. More tropical areas generally have a large selection of species, while colder or high-altitude climates may have extremely limited possibilities.

Tree species diversity is often measured by percentages of overall tree populations by species or genus. Diversity should be considered at multiple geographic scales—from the single block or neighborhood to the regional scale. Urban foresters should avoid allowing the desire for increased diversity to override the necessity to plant appropriate, noninvasive, species.

Figure 14.15 A lesson from history. Prior to the introduction of Dutch elm disease, many urban streets in the United States and other countries were lined with majestic overarching American elm (*Ulmus americana*) trees. The deadly disease spread quickly in mass plantings, or monocultures, where roots were prone to grafting.

Tree diversity also includes age and **size diversity**, which must be considered to maximize the health, sustainability, and benefits of the urban forest. Knowing that trees have a limited life span, which may be shortened in some urban environments, it is critical to plan for **succession**. An urban forest should always have trees of various ages. Because of the long time between initial planting and functional maturity, retaining and maintaining mature trees in the urban forest helps to maximize tree benefits. Small, understory trees also have important roles to play.

Biodiversity is the variety of all living organisms in an **ecosystem**, including plants, animals, bacteria, and fungi. Urban forests made up of diverse tree species will support more biodiversity by providing diverse habitats and food sources and will contribute to the total variety of species in an urban area.

Invasive Pest Management

While arborists practice plant health care at the tree and site level, it is crucial for urban foresters to work with not just arborists but also entomologists, plant pathologists, and invasive species experts to plan and manage the impacts of pests and disease on a larger scale. Early-detection and monitoring systems and educational programs can prepare communities for invasions that may impact urban tree populations. Climate change, urbanization, and global trade all contribute to an increase in the threats to urban forests from invasive species.

Reusing and Recycling Wood Waste

There are growing movements in many cities to manage urban forests from a life-cycle approach, considering the costs, benefits, and environmental impacts of every stage in a tree's life from nursery to removal. Tree maintenance and removal can generate a tremendous volume of wood waste in the form of logs, branches, and leaves. In urban landscapes where there is limited space to leave wood on-site, or when it is not appropriate to do so, arborists must find other locations for these materials. Wood and landscape waste can be deposited in landfills for a fee or used for alternative purposes.

Logs are often cut up into firewood and may be sold or given away. However, wood-burning fireplaces are now restricted in some communities due to the air pollution they generate. Converting urban wood waste into a biofuel is showing potential in some regions as an alternative energy production method. In some cases, urban wood is sold as specialty wood for use by artisans. Another means of recycling logs and tree debris is to create mulch using a tub grinder. Wood chips generated from

Figure 14.16 Wood chips generated from municipal tree work can be used as mulch or composted.

brush chippers may also be recycled as mulch and may even be run through a tub grinder to make a finer-grade, more uniform mulch, which can be sold.

Wood chips are sometimes composted with other organic waste. **Composting** is the process of decomposition of organic matter by microorganisms. Because the decomposition process generates heat, most of the pathogens and weed seed present will be killed, and the final product is a rich, organic material that is used in gardening and landscaping.

Wildlife in the Urban Forest

Wildlife is a broad term that includes all animals living in a natural, undomesticated state. Many studies have shown that urban forests provide critical **habitat** and food for a wide range of wildlife. Many jurisdictions have wildlife legislation or regulations that affect tree management. For example, certain migratory birds are protected by the North American Migratory Bird Treaty Act in the United States and by the Migratory Birds Convention Act in Canada. Harming protected birds or removing their nests can be a violation of the law and result in significant penalties.

Urban foresters should recognize that wildlife play important ecological, economic, and cultural roles. Many wildlife species are declining worldwide, in part due to pressures from urbanization. Urban foresters can support wildlife, from the management and planning side, by planting and preserving trees that wildlife depend on, and from the maintenance and training side, by minimizing the impact and disturbance to wildlife.

Figure 14.17 Urban foresters can support wildlife from the management and planning side, by planting and preserving trees that wildlife depend on, and from a maintenance and training side, by minimizing the impact and disturbance to wildlife.

Professional Collaboration, Public Outreach, and Education

Urban foresters must collaborate with government decision makers and other professionals to obtain resources and staffing to achieve objectives. Increasing awareness of the urban forestry program's benefits and goals with key policy makers is essential. When possible, demonstrating benefits in economic terms and showing community support is helpful.

Community-led tree commissions and tree boards are an integral part of most municipal programs, and urban foresters can work with these groups to better articulate and publicly share needs, vision, and goals. Partnering with community associations can help ensure that the urban forestry program fairly and equitably serves the needs of city residents, and can help support urban forestry

Figure 14.18 Community engagement and communication are key parts of successful urban forestry programs and urban forest management plans. Informed citizens can be champions for the cause as well as keen observers who can provide valuable feedback.

through volunteer steward programs and neighborhood planting campaigns.

The most effective urban foresters have found that strong communication and partnerships with the public and community organizations develop allies that can help achieve objectives.

Community engagement and communication are key parts of successful urban forestry programs and their urban forest management plans. In many cities, the majority of trees are located on private land, so urban foresters need to engage the landowners to promote good tree care to achieve sustainability and resilience. Informed citizens can be champions for the cause as well as keen observers who can provide valuable feedback.

Regulatory and Legal Issues

Tree Ordinances and Bylaws

Tree ordinances, also called **tree bylaws** or tree code in some local governments, are legal regulations enacted to protect trees within a given jurisdiction. A typical tree ordinance will define the jurisdiction's authority, describe the conditions and requirements of the ordinance, establish penalties for noncompliance, and specify the responsibility for enforcement. The urban forester should be involved with drafting and reviewing tree ordinances to ensure that tree preservation and maintenance requirements are consistent with professional best management practices and that applicable standards are referenced.

Ordinances that affect trees and their management do not always have the word "tree" in their title. Zoning ordinances and comprehensive plans are examples of legal regulations that impact urban forests. Regulations may also be in the form of Covenants, Conditions, and Restrictions (CCRs), which are rules that attach to one or more individual parcels of land.

Examples of common provisions include the following:

- Requirements for property owners to care for trees in right-of-way zones adjacent to their property
- Guidelines and requirements for obtaining

permits for tree planting, maintenance, or removal
- Authorization for public workers to enter private property for tree inspections or to perform required maintenance
- Lists of acceptable species to be planted on properties within the jurisdiction
- Regulations prohibiting topping of trees
- Special protections given to trees of a certain species, size, historical or cultural importance, or growing location

Arborists should be familiar with all relevant sources of law and rules applicable to trees on a given property before working on any of them. Some ordinances carry substantial penalties for violations.

Permits

It is common for government agencies such as municipalities to require permits for planting, pruning, tree maintenance, or any activity, such as construction, that might affect trees. Tree ordinances and bylaws will often define the requirements and sometimes the procedures for obtaining permits. The tree ordinance or bylaw gives the municipality the authority to review and approve all tree-related permits to ensure the protection and best care of the trees involved. Approval might be conditional upon establishing certain work procedures and clearances, requiring that all applicable standards and best management practices are followed, and requiring monitoring by a qualified arborist. Some jurisdictions require that tree care companies working in that jurisdiction have at least one ISA Certified Arborist® on staff. Some jurisdictions may have a tree service licensing requirement for companies engaged in tree care.

Tree Preservation Orders (TPOs)

Trees may be subject to the protection of **tree preservation orders (TPOs)**. Common in the United Kingdom and being adopted elsewhere, a TPO is

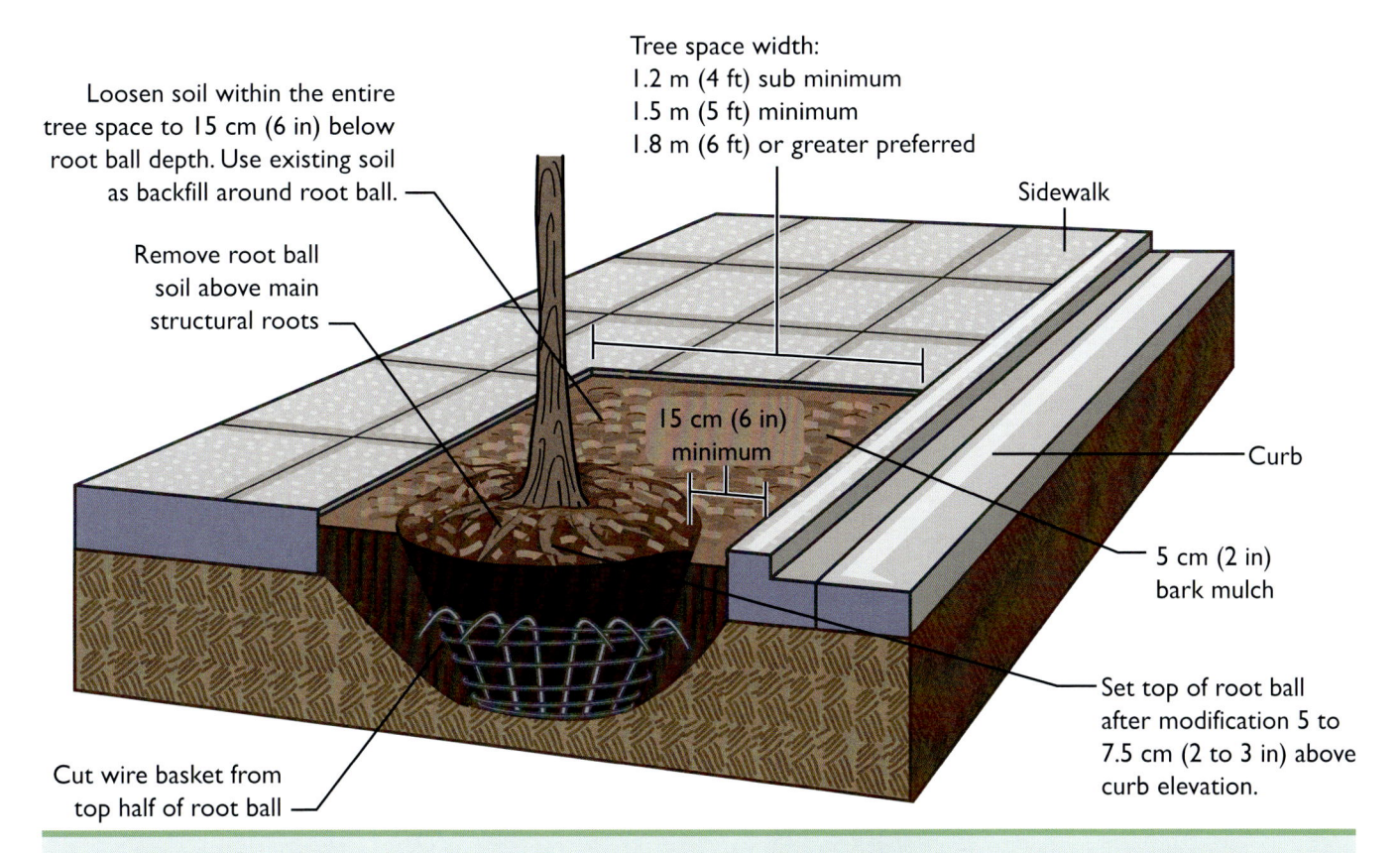

Figure 14.19 Urban foresters work with plans and specifications, which detail how work is to be performed.

Figure 14.20 ISA publishes a set of best management practices (BMPs) for arboriculture and urban forestry. Pictured are examples of some of the topics covered.

order must be specified. A TPO can be established for a single tree, groups of trees, all trees within a defined area, or even woodland areas.

Standards and Specifications

Many countries have established national **standards** for tree work. A standard is an established or widely recognized authority of acceptable performance. The legal authority and enforcement of standards may vary from one jurisdiction to another. In some cases, failure to comply with standards can lead to fines or other penalties. However, even where standards do not carry direct legislative authority, they may be recognized in a court of law. Failure to act in compliance with recognized standards has resulted in findings for liability and resultant damages in many cases.

Specifications are detailed plans, requirements, and statements of particular procedures and/or standards used to define and guide. Urban foresters should establish detailed specifications for all tree work, including planting, pruning, fertilizing, pest control and monitoring, installation of support or protection systems, construction near trees, and removals. Specifications should apply not just to contracted work but also to internal work of the agency or jurisdiction. In the United States, the American National Standards Institute (ANSI) A300 standards for tree care operations are designed to be used as a tool for writing specifications. However, specifications should be explicit and contain much more detail than simply a statement of compliance with applicable A300 standards.

The International Society of Arboriculture (ISA) publishes a set of **best management practices (BMPs)** for arboriculture and urban forestry, which cover a wide array of topics. Where national standards draw the lines, best management practices color in the picture and provide a more descriptive how-to for tree care and management procedures. Many of the BMPs are recommended in this study guide as other sources of information.

a legal regulation, established by the local authority, that protects a tree or multiple trees. Tree preservation orders are often put in place by local planning authorities to preserve trees during land developments. Other TPOs protect trees in already developed areas from removal or unapproved pruning or other forms of maintenance. Some TPOs are in place for a fixed period; others may be permanent. They do not necessarily apply to all trees within a jurisdiction; trees that are the subject of an

CHAPTER 14 WORKBOOK

1. Urban forestry is the management of naturally occurring and planted trees and associated plants in urban areas. Whereas arboriculture focuses on the_____ , urban forestry focuses on the _____ .

2. List five professionals or groups that an urban forester should learn to communicate with.

    a.

    b.

    c.

    d.

    e.

3. True/False—The leaves and branches of trees catch and slow rainwater to reduce soil erosion from runoff.

4. _____ _____ occurs when trees absorb carbon from CO_2 in the atmosphere and store it in the form of wood and other carbon-based tissues.

5. List three economic benefits of trees.

    a.

    b.

    c.

6. List three environmental benefits of trees.

    a.

    b.

    c.

7. List three social and health benefits of trees.

    a.

    b.

    c.

8. _____ is the ability to maintain ecological, social, and economic benefits over time.

9. List four types of data that are typically collected in a tree inventory.

   a.

   b.

   c.

   d.

10. A _____ _____ policy statement should set out the policies for identifying, assessing, reporting, and mitigating risks.

11. _____ _____ are legal regulations drafted and instituted to protect trees within a given jurisdiction.

12. List three component plans that are commonly part of an urban forest management plan.

    a.

    b.

    c.

13. A _____ _____ _____ is a legal regulation, established by the local authority, that protects a tree or multiple trees.

14. True/False—Even where standards do not carry direct legislative authority, they may be recognized in a court of law.

15. Urban foresters should establish detailed _____ for all tree work, including planting, pruning, fertilizing, pest control and monitoring, installation of support or protection systems, construction near trees, and removals.

16. The _____ _____ is a document laying out how a municipality will balance the maintenance of its large population of trees within the common urban pressures and financial restraints of a municipality.

17. The urban forest provides _____ and food for a wide range of wildlife.

18. True/False—Deferring maintenance can lead to higher costs and risks and is likely to reduce the benefits trees provide.

19. True/False—Widespread planting of a single species is recommended to bring a uniform appearance to the urban forest.

20. True/False—Community engagement and communication are key parts of successful urban forestry programs and their urban forest management plans.

Challenge Questions

1. Explain why the ability to communicate the benefits and costs of trees is important to arborists and urban foresters in all sectors of practice.

2. Discuss the differences between standards, specifications, and best management practices.

3. List the components of an urban forest management plan, and state why each should be included.

Sample Test Questions

1. A social benefit of trees and natural areas that has been identified through research is

    a. stress reduction from settings with trees
    b. faster healing of patients in hospitals
    c. crime reduction in communities
    d. all of the above

2. A commonly used set of methods for appraising trees was developed by the

    a. Council of Tree and Landscape Appraisers
    b. Society of Consulting Tree Workers
    c. Society of Urban Foresters
    d. Consortium of Landscape Professionals

3. A typical tree ordinance or bylaw will define the jurisdiction's authority and

    a. describe the conditions and requirements of the ordinance
    b. establish penalties for noncompliance
    c. specify the responsibility for enforcement
    d. all of the above

4. Detailed plans, requirements, and statements of particular procedures and/or standards used to define and guide are called

    a. laws
    b. best management practices
    c. specifications
    d. ordinances

5. A problem associated with overplanting of a single species or a few species is

   a. all of the trees maturing and dying within a short period of time
   b. increased biodiversity of associated pest populations
   c. unsustainable management due to uniform and consistent maintenance needs
   d. the risk of catastrophic loss due to an insect or disease outbreak

Recommended Resources

(See Recommended Resources in back of book for detailed information.)

Municipal Specialist Certification Study Guide (Matheny and Clark 2008)

Urban Forestry (Miller et al. 2015)

Best Management Practices: *Tree Inventories* (Bond 2013)

Guide for Plant Appraisal (Council of Tree and Landscape Appraisers 2019)

CHAPTER 15
TREE WORKER SAFETY

▼ OBJECTIVES

- Identify appropriate safety standards for tree care operations.

- Select necessary personal protective equipment (PPE).

- Describe the steps of establishing a safe work site, including tree and site inspection and establishing the work zone and drop zone.

- List the elements of a thorough job briefing, and explain the importance of good work-site communication.

- Identify the potential hazards associated with working in proximity to electrical conductors, along with the fundamental measures necessary to avoid direct or indirect contact.

- List the proper procedures for operating chain saws and chippers and the PPE required for their operation.

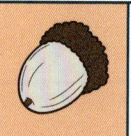

KEY TERMS

aerial lift
ANSI Z133
approved
back cut
barber chair
cardiopulmonary resuscitation (CPR)
command-and-response system
conventional notch
direct contact
drop zone
electrical conductor
emergency response

feller
first aid
hinge
indirect contact
job briefing
kickback
kickback quadrant
landing zone
leg protection
minimum approach distance (MAD)
mobile elevating work platform (MEWP)

open-face notch
palm skirt
personal protective equipment (PPE)
reactive force
retreat path
safety standards
shall
should
tagline
work plan
work zone

Introduction

Working in and around trees can present a significant risk of personal injury if safety measures are not followed. Safety must always be the first concern. Safety is more than using special equipment, wearing appropriate gear, or attending meetings. Safety is an attitude and should be a part of the organization's culture. It is an ongoing commitment at every level, from the top down. Safety requires a conscious recognition of potential risks and hazards and the development of a program designed to prevent accidents. Safety precautions must be built into every task performed by tree workers. A small investment of time in safety education and training can help avoid injuries and save a great deal in lost production, insurance costs, negligence claims, and damages. Training can help reduce the emotional and psychological stress on all employees when a coworker is seriously injured or killed.

This study guide is based primarily on the **ANSI Z133** standard used in the United States. **Safety standards** in many other countries are similar. Organizations in several countries have adopted the ANSI Z133 in part or altogether to use as the safety standard.

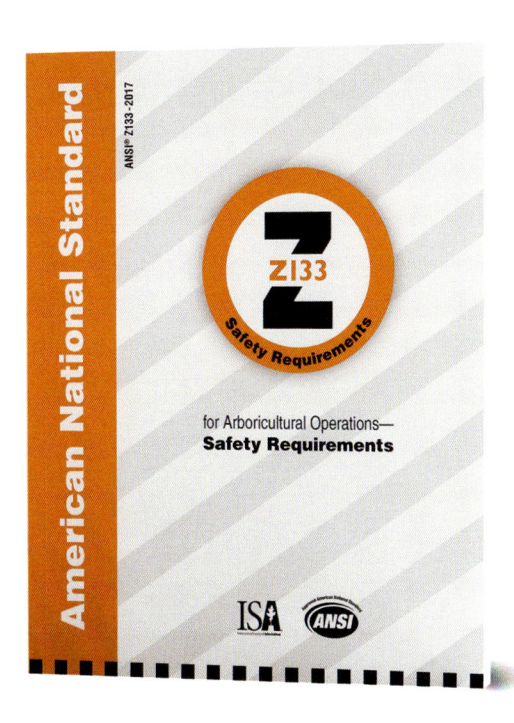

Figure 15.2 Always follow all applicable safety regulations and standards for your region. ANSI Z133 is the applicable safety standard in the United States. Many other countries have similar national standards for safety.

Figure 15.1 A small investment of time in safety education and training can help avoid injuries and save a great deal in lost production, insurance costs, negligence claims, and damages. Training can help avoid the emotional stress that occurs when a coworker is seriously injured or killed.

This study guide is an educational tool for introductory-level arboriculture. Arborists should use it as part of, but not as a replacement for, comprehensive education and training. Although some equipment and techniques are explained and illustrated in this chapter, simply reading about them is neither equivalent to being trained in their safe use, nor is it enough to be able to implement them on a job. Proper training in the safe use of equipment and techniques is necessary before using them in an actual work environment.

Safety Standards

Most countries, and many states and provinces, have standards that guide safety in tree care operations. The term "standard" is used generically to

Standards in the United States and Canada

In the United States, private employers are subject to occupational safety and health regulations developed under the auspices of the federal Occupational Safety and Health (OSH) Act of 1970. The Act created a federal agency, the Occupational Safety and Health Administration (OSHA), which regulates worker safety and health. [In Canada, the occupational health and safety acts are administered by federal, provincial, and territorial governments.] Many states have what are known as "state plan" OSHAs, which may enforce the federal rules, their own unique occupational safety and health rules, or some combination of the two.

In the United States, ANSI Z133 is the standard for arboricultural operations developed in accordance with the American National Standards Institute (ANSI). It is intended to provide safety standards for workers engaged in pruning, repairing, maintaining, or removing trees or cutting brush. In Canada, workers must comply with the standards established by the Canadian Standards Association (CSA).

All tree care workers in the United States must be familiar with and comply with the ANSI Z133 standard and all applicable OSHA standards. Standards enforced at the state or local levels, or company policies, may be more restrictive than ANSI Z133 or applicable OSHA standards. In Canada, standards governing tree care operations are very similar and are sometimes more stringent.

It is the responsibility of employers to ensure that employees comply with all applicable safety standards and policies.

---

indicate any applicable law, regulation, or standard that pertains to safety in the arboriculture industry. The purpose of standards is to reduce occupational injury, illness, and death through the establishment and enforcement of safe work practices and the provision of mandatory education and training. In some regions, safety standards have been developed by industry consensus—when industry representatives work together to agree on safe work practices for a region—but they have also been developed by government agencies or as a collaboration between industry and government. Industry-based standards are not legally enforceable by themselves, but regulatory agencies often refer to industry standards when issuing a citation or fine to an employer for unsafe work practices.

Employers should provide copies of applicable standards, which can be placed prominently in the office, shop, and vehicles. Employers should emphasize the importance of safety standards and regulations to all employees and make it clear to employees that the company they work for strictly follows safety standards that are applicable in their area. Employers may also develop safety policies to supplement industry or government standards. Safety is a partnership between employer and employee, and everyone is responsible for ensuring safety.

There are specific words or terms that are consistently used in most safety regulations. **Approved** means acceptable to the federal, state, provincial, or local enforcing authority having jurisdiction. In many cases, the word "approved" applies to specific equipment. References to applicable standards sometimes appear on equipment labels. Become familiar with the words "shall" and "should" when reading standards: **shall** defines a mandatory requirement; **should** refers to an advisory recommendation. *Because of inconsistencies among standards, and for readability, the use of the terms "shall," "should," and "must" in this study guide does not necessarily parallel the requirements of any specific standard.*

This study guide is not a substitute for any safety standard and cannot reference all of the pertinent standards and regulations for tree work performed in all countries or regions. The content of those applicable standards takes precedence over this and other educational resources.

Personal Protective Equipment (PPE)

Personal protective equipment (PPE) consists, at a minimum, of clothing and footwear appropriate for the work and weather; head, eye, and ear protection; and, if working with a chain saw, cut-resistant pants or chaps. When working with potentially hazardous materials like pesticides, arborists must follow applicable PPE recommendations from the manufacturer and government regulations of the material such as face masks or protective gloves. Personal protective equipment specific to first aid is discussed separately in a later section.

Clothing should be made of durable fabric that also allows for free movement. Avoid loose-fitting clothing and wearing jewelry that may get caught in equipment or tools and become a hazard. Tree workers should wear durable work boots to provide good support, traction, and feet protection. Boots should offer ankle protection to minimize injuries due to slips, trips, and falls. If climbing often on spikes (spurs), workers might choose boots with a deep, square heel to brace the stirrup of the climbing spur as well as a steel or polymer shank for arch support and comfort. Some boots are designed with flat, rubberized soles to facilitate better contact with tree bark and climbing lines. Check applicable standards for the use of steel- or composite-toed boots.

Tree workers must always wear head and eye protection on a jobsite. Head protection can be a hard hat or helmet that meets applicable impact and penetration standards. Chin straps or ratcheting devices can help keep a hard hat or helmet in place, which is especially important when climbing through a dense crown. Some hard hats and helmets are designed to accept add-ons such as a face shield or earmuffs. For working near electrical conductors,

Figure 15.3 Personal protective equipment (PPE) for tree work includes head protection, hearing protection, eye protection, and leg protection, when needed. Tree workers should also wear appropriate clothing and footwear.

head protection that is approved and tested for working near high voltage is required.

Eye protection must also conform to applicable local standards. A pair of sunglasses does not offer the same protection as a pair of approved safety glasses. Helmets with a built-in or add-on face shield or face mask must comply with local standards to provide proper eye protection. A mesh face mask attached to a hard hat or helmet provides protection, but small particles and dust, common in tree work, can easily pass through the mesh and cause eye injuries. Some arborists prefer to wear a face mask or face shield as additional protection with a pair of safety glasses.

Prolonged exposure to the noise of chain saws and brush chippers causes permanent hearing loss, so hearing protection is required for workers exposed to loud machinery. All workers on a jobsite must wear hearing protection, especially when operating or working close to a chain saw or chipper. It is personal preference whether to use earmuff- or earplug-type hearing protection, provided they meet applicable standards.

Some standards or company policies require workers to wear gloves on a jobsite. Gloves are strongly recommended for certain operations such as sharpening saw chain and chipping brush, but gauntlet-type gloves must be avoided while chipping brush because they have an open cuff that can get caught up in the brush, pulling the worker into the chipper. Some climbers prefer to wear latex-dipped "gripper gloves," which reduce the exertion of hand and forearm muscles to grip a climbing line. Gripper gloves are not appropriate when handling a rigging line; instead, wear gloves with a thick palm to protect hands from heat and abrasion associated with friction when lowering a load.

Good Communication

Good communication among workers is an integral part of working safely. This is true for all tree care operations. Each crew member, whether on the ground or aloft, must always be aware of what the others are doing, and each must take measures to manage risk and to prevent accidents. Using good communication from the start and consistently on every jobsite helps to minimize the likelihood of property damage or injury.

Communication begins at the start of the workday when a crew is assigned to a particular job and they decide what equipment and tools are needed to complete the job. Once on the jobsite, each job should begin with a **job briefing**, which coordinates the activities of every worker. The job briefing summarizes what has to be done and who will be doing each task, potential hazards and how to prevent or minimize the potential, and what special PPE may be required. All workers must have a clear understanding of the communication system being used. The crew leader should formulate and communicate the **work plan**. There should be no question about assignments—teamwork is essential on a tree crew. The written job briefing sheet should include emergency phone numbers (which also should be posted in vehicles and programmed into workers' phones). The crew should also note the location of

Figure 15.4 Every job should begin with a job briefing, which coordinates the activities of every worker. The job briefing summarizes what has to be done and who will be doing each task, the potential hazards and how to prevent or minimize the potential, and what special PPE may be required.

the nearest medical facility and know how to get there.

Setting up a work zone comes after the job briefing. The **work zone** includes the entire area where work will occur throughout a job. A smaller **landing zone** or **drop zone** within the work zone is where the crew expects cut branches or logs to be dropped or lowered from above. The climber or aerial lift operator making cuts must clearly communicate with ground workers to ensure that the drop zone is clear before any cutting occurs. Depending on the circumstances, the crew will use a vocal or visual **command-and-response system**. In a vocal system, the worker aloft loudly and clearly issues the command such as, "Stand clear!" Before cutting, the worker aloft waits for the response from the ground, "All clear!" When there are multiple workers on the job, confusion can be reduced by assigning one person to respond to the climber after ensuring that the area is clear and safe. Some companies have instituted a three-step communication system to include cross acknowledgment of commands.

In some situations, a vocal system might be problematic, so visual hand signals may be used instead.

Chapter 15: Tree Worker Safety

Figure 15.5 The climber or aerial lift operator making cuts must clearly communicate with ground workers to ensure that the drop zone is clear before any cutting occurs.

culture of safety in the company. Every employee shares in the responsibility for safe operations. It is a team effort.

Fire and Vehicle Safety

Trucks should be equipped with a fire extinguisher, and all workers should be trained in its use. Gasoline-powered equipment must be refueled only after the engine has been turned off. Any spilled fuel should be removed before starting. Equipment must not be started or operated within 3 m (10 ft) of the refueling site. Smoking is prohibited when handling or working around any flammable liquid. Flammable liquids must be stored, handled, and dispensed from approved safety containers and kept separate from all ropes and equipment.

When work requires equipment, vehicles, or personnel to be in or near a road or sidewalk, pedestrian and vehicular traffic control is critical. Most governments regulate traffic control measures, and workers must be familiar with them. At a minimum, the work zone must include necessary buffer zones around equipment, vehicles, and crew members in the road. Workers must also wear appropriate high-visibility clothing or vests. As long as they conform to applicable governmental regulations, traffic control devices such as safety cones, warning signs, barriers, or flags may be used to expand the work zone. In most jurisdictions, crews have a legal responsibility to secure the work zone so that no individuals or vehicles pass underneath trees where tree work is in progress. A secure work zone also ensures crew safety.

Another alternative, which can be especially useful on jobs involving cranes, long distances between workers, or any loud machinery, is a Bluetooth communications system that connects to workers' helmets and other workers on the site.

Training

In addition to good communication, training is essential to tree worker safety. Workers must be trained prior to using equipment and tools on a job, and they must be trained to complete the tasks they are expected to undertake. The employer is responsible for providing and documenting training and for documenting worker competency. In addition to training for the use of specific equipment, tools, and techniques, workers must become familiar with applicable safety standards. Safety starts at the top, and it is the employer's responsibility to foster a

Electrical Hazards

During the job briefing, the crew must inspect the site to determine whether any electrical hazards exist. An electrical hazard exists when there is a risk of injury or death associated with direct or indirect

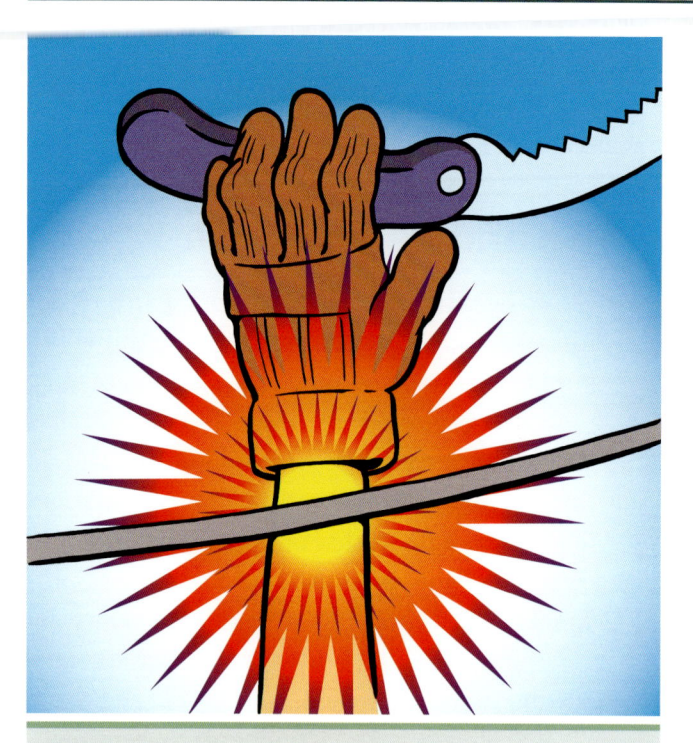

Figure 15.6 Direct contact with an energized conductor could be fatal.

Figure 15.7 When an aerial lift truck is in use and the platform is near electrical conductors, the truck could become energized.

contact with an electrical conductor. An **electrical conductor** is any overhead or underground electrical device, including communication wires and cables, power lines, and related components and facilities. Workers should consider all such lines to be energized with a potentially fatal voltage. **Direct contact** is made when any part of the body contacts an energized conductor or other energized electrical fixture or apparatus. **Indirect contact** is made when any part of the body touches any conductive object that is in contact with an energized conductor. Indirect contact can occur through tools, tree branches, trucks, equipment, or other conductive objects or as a result of communication wires or cables, fences, or guy wires becoming energized.

Direct or indirect contact will result in an electric shock that can cause serious injury or death. The shock is a result of electricity from a conductor flowing through the body to a grounded object (such as a tree) or to the ground itself. Simultaneous contact with two energized conductors also causes electric shock that may result in serious injury or death.

Workers must also be aware of potential underground hazards. Stump grinders, augers (for fertilizing or aerating), and tree spades all have the potential to come into contact with underground utility lines such as power and communication lines, gas lines, and water lines. All underground utilities should be marked before any type of digging.

Neither footwear, including those with electrical-resistant soles and lineman's overshoes, nor rubber gloves, with or without leather or other protective covering, can be considered as providing any measure of protection from electrical hazards.

Hand tools powered by AC current—that is, those with a power cord—must never be used in trees near an energized electrical conductor when there is a possibility of the power cord contacting the conductor. This is less of a concern when using battery-powered hand tools because there is no cord to contact a conductor. When using electric tools aloft, climbers should support them with an independent line or lanyard and prevent cords from becoming entangled or coming in contact with water. Only use clean, nonconductive tools and ladders when there is any possibility of contacting an energized conductor.

All tree workers should receive appropriate and documented training in electrical hazard safety awareness. Workers must receive proper training in electrical hazard tree work procedures to perform tree work in proximity to electrical conductors. In the United States, employers are required to certify this training.

Chain Saw Safety

Chain saws are frequently used in tree work, but they are also one of the most hazardous pieces of equipment arborists use and may cause serious injury or death. Whether used aloft or on the ground, safe chain saw operation requires proper training and adherence to manufacturer's instructions for safe use and proper maintenance. Using a chain saw in poor working condition or cutting with dull or improperly sharpened cutters can lead to injury.

When operating a chain saw, workers should wear **leg protection** in the form of cut-resistant pants or chaps. In the United States, leg protection is only required when using a chain saw on the ground, but in many countries, leg protection is required any time a chain saw is used (on the ground or aloft). Many tree care companies in the United States have adopted a similar policy. The material used in the construction of cut-resistant pants and chaps will jam and slow the movement of the chain in the sprocket if contact is made with the saw chain. However, "cut-resistant" does not mean "cut-proof." Leg protection has been shown to reduce the severity of chain saw injuries. Newer designs of chain saw pants are lightweight and less bulky than earlier models. Cut-resistant materials have been incorporated into other clothing such as shirts, jackets, gloves, bib-style pants, and work boots.

The chain saw operator must have secure footing when starting and operating the saw, and the immediate area must be clear of debris. Always engage the chain brake before starting the saw. Larger saws should be started on the ground or otherwise be firmly supported. *Never operate a chain saw with one hand.* Always grip both handles firmly, with the left hand and thumb wrapped around the front handle and the right hand and thumb wrapped around the rear handle, activating the throttle trigger.

When operating a chain saw, be aware of all surroundings and coworkers (or if a passerby unknowingly enters the work or drop zones). If a coworker is operating a chain saw, do not approach them from the rear. If two workers are operating saws at the same time, they should be at least 3 m (10 ft) apart and should not be cutting on the same piece of wood. Never operate a chain saw above shoulder level.

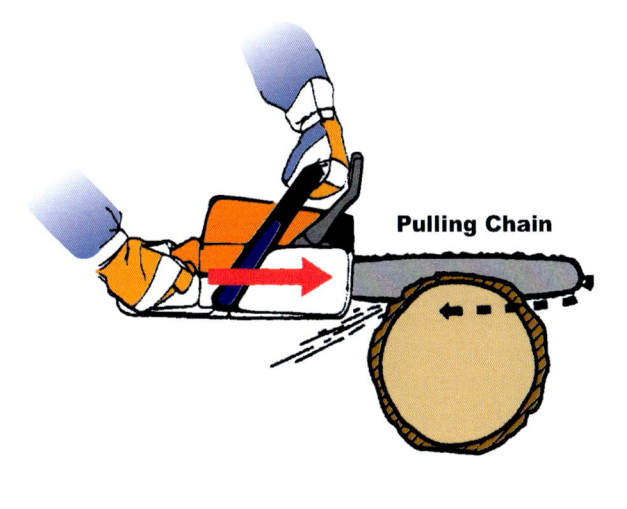

Figure 15.8 When cutting with the bottom of the bar, the saw has a tendency to pull into the cut. When cutting with the top of the bar, the saw pushes back toward the operator.

Release the throttle and engage the chain brake before removing one hand from, or taking more than two steps with, a running chain saw. Always turn off the engine to clean, refuel, or adjust a chain saw, according to the manufacturer's instructions. Wear gloves when sharpening or tensioning the chain.

Chain saw operators should understand the saw's **reactive forces**. When cutting with the bottom of the guide bar, the rotation of the chain will pull the saw into the cut. When cutting with the top of the bar, the rotation of the chain will push the saw back toward the operator. A dangerous reactive force that frequently causes injury is **kickback**, a violent backward and upward movement of the chain saw toward the operator. Kickback occurs when the upper portion of the tip of the guide bar (**kickback quadrant**) contacts a log or other object. Always be aware of where the tip of the guide bar is, and avoid contact between the kickback quadrant and objects. Kickback occurs at a speed many times faster than a human can react. Maintain a firm grip on both handles of the chain saw, and operate the saw to the right side of the body to reduce the likelihood of injury in the event of kickback. Keeping the chain saw engine close to the body increases control and reduces operator fatigue.

Climbers must take extra precautions when using a chain saw aloft. Only experienced climbers with proper chain saw training should use a chain saw in a tree. Before cutting with a chain saw, the climber must use a second means of securing. Typically, this means being tied in with the climbing line and securing with the work-positioning lanyard. Make sure to have three points of contact to provide a stable and secure work position; this will help maintain control when cutting with a chain saw aloft. Never cut with only one hand on the chain saw, and never cut above shoulder level. Also, proper work positioning when cutting aloft means taking care to avoid (1) kickback, (2) follow-through with the chain saw (potentially cutting an object or body part after the saw passes through the wood), (3) being hit by the limb being cut, and (4) cutting the climbing line, lanyard, or other ropes in the tree.

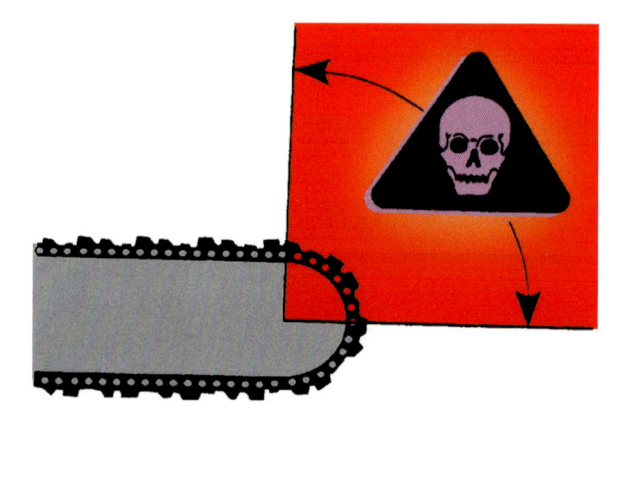

Figure 15.9 A common cause of chain saw injury is kickback. Kickback occurs when the upper portion of the tip of the guide bar (kickback quadrant) contacts a log or other object.

Tree Felling and Removal

When removing trees, it is critical to set up and clearly mark the work and drop zones and to establish appropriate crew communication. These precautions will help avoid injuries to the crew and passersby. Inspect the tree and site, and identify obstacles and hazards to avoid during the work. Tree removals, especially when rigging is involved, can subject the tree to much greater forces than other aspects of tree work such as cabling and pruning. Tree defects such as decay or cracks reduce the load-bearing capacity of the tree, and large forces might cause it to fail, resulting in serious injury, death, or property damage. The crew should decide whether defects or other conditions make it unsafe to perform the removal without additional planning, equipment, or gear.

It may be possible to fell trees without having to climb and rig them. When felling an entire tree or a section of trunk that remains after rigging the crown, workers should install a **tagline** (pull line) for added control. Wedges can also be used when felling; they can be particularly useful to initiate the fall, compensate for lean, and avoid getting the bar of the chain saw pinched.

Tree felling should be performed only after assessing the tree and site and developing a plan. The assessment considers the terrain, obstacles, hazards, wind, and other site factors that could affect the felling operation. The condition of the tree should also be factored in, including defects, lean, height and crown spread, potential hazards, and the amount of solid wood available for cutting. The plan includes the direction for felling, equipment needed, type and placement of felling cuts, property protection needed, and the position and role of each worker involved and their path for retreat.

At the start of a felling operation, workers not directly involved in the operation and any bystanders must be at least two tree lengths away from the base of the tree. Workers involved in the removal, including those on the tagline, must be at least one

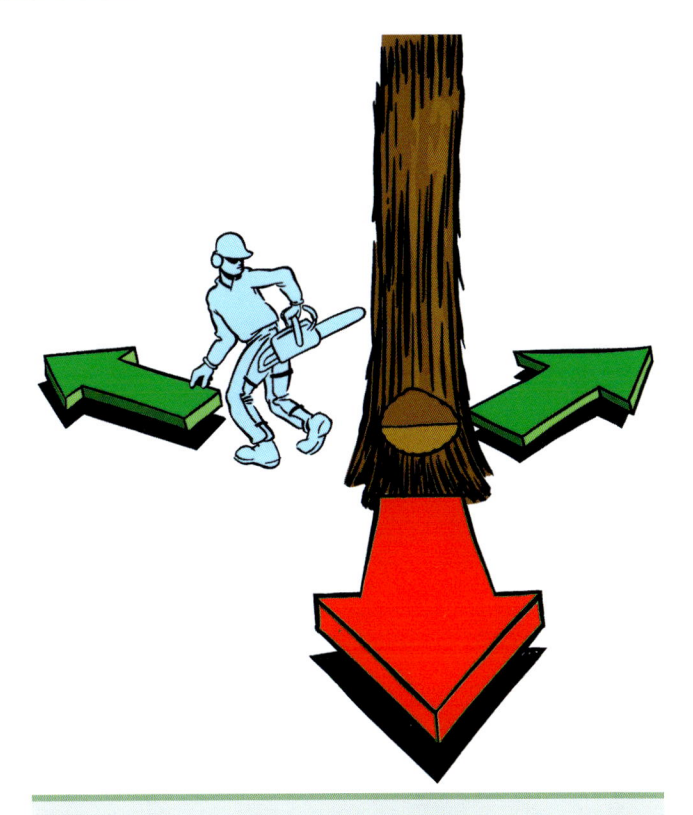

Figure 15.10 The preferred retreat path for the chain saw operator in a felling operation is 45° on either side of a line drawn opposite the intended direction of fall.

and a half tree lengths away, have an established means of communication with the **feller** (the person felling the tree), and have a planned **retreat path** (escape route). The preferred retreat path for the feller is 45° on either side of a line drawn opposite the intended direction of fall. If the tree has a lean, the retreat path should be 45° to the rear on the side away from the lean, if practical. As soon as the tree starts to fall, the feller should engage the chain brake, shut off the chain saw, and follow the retreat path while keeping an eye on the tree.

To fell the tree, the feller makes a notch in the intended direction of fall. It is easier to make a good notch by making the top cut of the notch first. When making the bottom cut, the feller can then observe the guide bar through the opening of the top cut to avoid making a bypass cut. A bypass cut occurs when the bottom cut goes farther into the trunk, past the end of the top cut. Making a bypass cut can

Figure 15.11 The hinge is critical in controlling the direction of fall.

Figure 15.12 An open-face notch of 70° or more allows the hinge to control the tree longer.

reduce the effectiveness of the hinge, reducing the feller's control of the tree as it falls.

In the past, it was common to use a **conventional notch** of 45°, but the preferred notch today is an **open-face notch** of 70° or more. An open-face notch is preferred because the wood fibers of the **hinge**, which direct the tree to the ground as the notch closes, do not break until the notch closes. If the notch is 70° or more, the trunk will usually be on the ground before the notch closes. A conventional notch closes when the trunk is at an angle of 45° to the ground; once the notch closes and the fibers in the hinge break, control is lost.

The depth of the notch should be about 20 to 25 percent of the diameter of the tree. Remember: a small notch can be enlarged, but it is impossible to correct a notch that is too deep. The length of the hinge should be about 80 percent of the tree's diameter. If possible, the notch should not be placed in the vicinity of cracks or decay; it is important to have solid fibers to form the hinge. Create a hinge on small-to-medium-diameter trees with a thickness of 5 to 10 percent of the tree's diameter. Large-diameter trees may require a hinge thickness of 5 percent or less. Never cut into the hinge when making the back cut.

The traditional, straight **back cut** is made from the back of the tree toward the notch. The hinge is formed as the back cut approaches the notch. It is easy to cut through the hinge while making the back cut, especially if the feller is looking toward the top of the tree. When fellers made conventional notches, they made the back cut slightly higher than the apex of the notch to reduce the possibility of the tree kicking back toward them as the hinge broke. But felling with an open-face notch minimizes

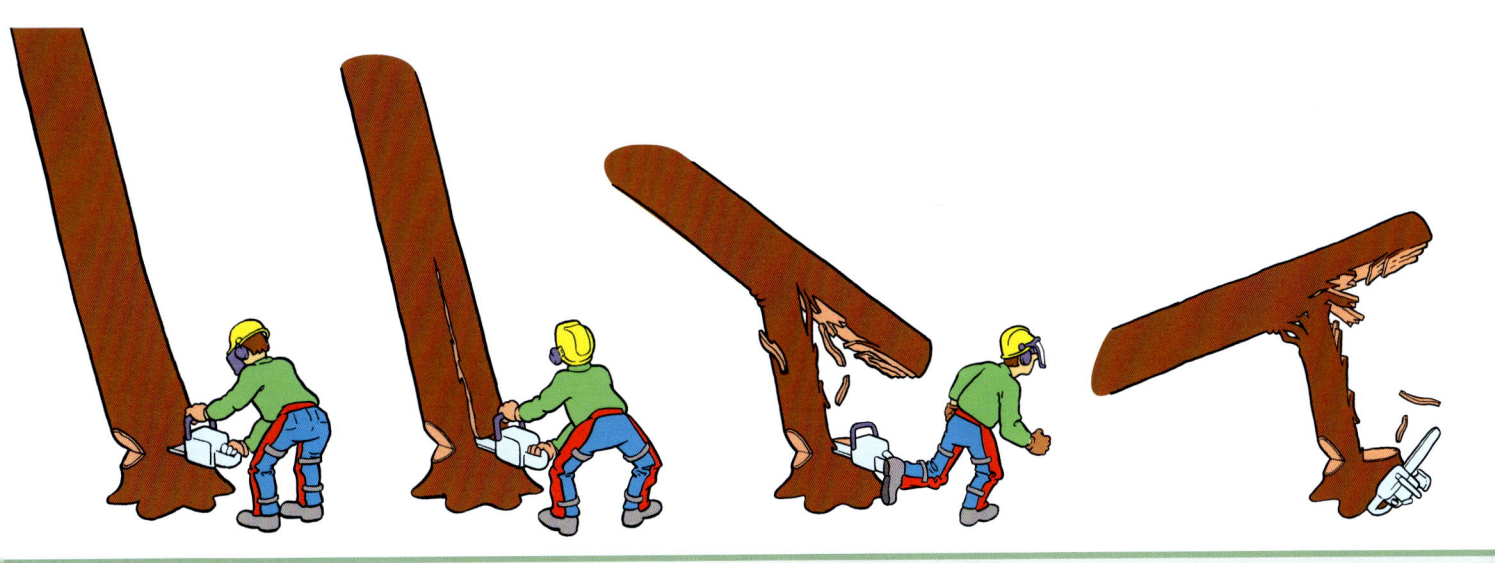

Figure 15.13 Sometimes, if the tree is leaning in the direction of fall, has internal faults, or if there is too much tension on a pull line, the tree can split upward from the back cut. This is called a barber chair, and it can be very dangerous. The split trunk can hit the person felling the tree, which is often fatal.

the likelihood of the trunk leaving the stump before the notch closes. The back cut should be made at the same height on the tree as the apex of the notch.

If the tree is leaning heavily (or is pulled) in the direction of fall, or if it has internal defects, it may split upward from the back cut. This is called a **barber chair**, and it can be very dangerous to the tree feller. The split trunk can hit the feller, often causing serious or fatal injuries. There are cutting and felling techniques that can reduce this danger. Workers should be trained to recognize such trees and to fell them only if they have been trained in advanced felling techniques.

Chipper Safety

Like chain saws, brush chippers have made tree work much more efficient, but they can also be very hazardous. Good communication, jobsite awareness, proper training, and safe work practices are essential when operating a chipper. Training should include instruction on daily inspection and maintenance, steps that are necessary to connect and disconnect it from the chip truck, towing procedures, starting and stopping the chipper (including emergency stopping procedures), feeding brush, and safety hazards involved with operation. All instructional and warning stickers and labels on the chipper must be in place and legible; always follow the manufacturer's instructions for safe use.

Proper PPE is required when operating a chipper. Additionally, to avoid getting pulled into

Figure 15.14 Brush should always be fed from the side of the chipper, and the worker feeding the brush should move away after the brush is fed. The larger butt end of each branch should be fed in first. Note the tear-away vest worn by the worker when operating a chipper.

the chipper while operating it, do not wear loose clothing or loose-fitting chaps, jewelry, climbing saddles, harnesses or body belts, and gauntlet-type gloves.

Feed brush into the chipper and move to the side and away from the intake chute after the chipper grabs the brush. Feed brush by inserting the larger butt end of a branch first and using larger branches to push smaller ones into the chipper. No part of the operator's body should ever reach beyond the back edge of the infeed chute. Avoid placing foreign material such as rocks, wires, or other debris into the chipper. Such material can damage the knives or cause projectiles to be thrown from the machine.

Many accidents have occurred when workers attempt to perform maintenance while the chipper disk or drum is still moving. Never work on a chipper unless the engine is turned off, the ignition key is removed, and the cutter wheel is completely stopped (with the lockpin in place, if applicable) and prevented from moving.

Aerial Lifts (Mobile Elevating Work Platforms)

Aerial lifts, known in some parts of the world as **mobile elevating work platforms (MEWPs)**, dramatically improve tree worker efficiency; however, operators must receive proper training in their use to avoid injury or death. Aerial lifts require training and familiarity with the specific piece of equipment. Always follow manufacturer's instructions for safe use. Training should cover applicable regulations, inspection, PPE, fall protection, emergency lowering procedures, aerial rescue, and routine maintenance. At the start of each day, before leaving for the jobsite, the operator should perform a routine inspection and make sure the aerial lift functions properly. All crew members should be familiar with the capabilities and limitations of the equipment, including safe working loads and the maximum number of people who can be carried aloft.

As part of the job briefing, the operator should inspect the site to locate above- and belowground obstacles to work around or those that could be damaged. It is especially important to be aware of overhead wires that the lift might come into contact with, as well as belowground features that might indicate unstable ground, which can cause the lift to overturn. Traffic control is also critical when the lift will be operated in or near a street or sidewalk.

Trained and experienced aerial lift operators should be able to demonstrate smooth and controlled operation of the equipment to perform the job without unnecessary damage to the tree, the equipment, or the site. Operators must always look

Figure 15.15 Aerial lift operators must always look in the direction the basket/bucket is moving.

Figure 15.16 Aerial lift operators must always be aware of the location of all parts of the device in relation to its surroundings. No part of the aerial device should ever make contact with or violate minimum approach distances (MAD) with energized electrical conductors.

in the direction the basket/bucket is moving but must also be aware of the location of all other parts of the device in relation to its surroundings. No part of the aerial device should ever make contact with or violate **minimum approach distances (MAD)** for energized electrical conductors.

Palm Safety

Some species of palms retain a full **palm skirt**, consisting of dead fronds that may remain attached and hanging along some or all of the length of the trunk. Birds, rats, raccoons, centipedes, scorpions, and bats are among the species that may nest within the skirt. Large gaps in the skirt can indicate sloughing, when large sections of frond rings detach and settle on a lower section of fronds. Sloughing of fronds is a leading cause of fatalities among climbers in palms, as they can pose a risk of suffocation. Aerial rescue of a climber trapped underneath or within a sloughed skirt is particularly challenging and is why climbing and pruning under the skirt must be avoided.

Fan palms can contain a significant amount of dry, dead material within these skirts. Arborists should be aware of the fire hazard potential. Chain saws used should be properly operated and maintained, and workers should not smoke while working to avoid starting a fire.

Palms require a larger drop zone for pruning or removal operations, as fronds can often glide down and land a significant distance from the trunk. Precautions should be taken to avoid dropping fronds onto or allowing fronds to blow into electrical lines.

Emergency Response

All crew members must be trained in **emergency response** procedures, and each individual must know what to do in an emergency situation. Emergency response includes training in how to contact emergency personnel and what information to provide. It also includes **first aid**, **cardiopulmonary resuscitation (CPR)**, and aerial rescue. In an emergency, it is easy for panic and confusion to take hold. Regularly practicing the appropriate responses to a variety of emergencies will help the crew remain calm and efficient in the event of a real emergency.

It is recommended—and required in some places—that all tree workers receive training in first aid and CPR. In addition, an approved and adequately stocked first aid kit must be provided on each vehicle and must remain on the work site at all times. All employees must be familiar with the use of first aid kits and with emergency procedures.

First aid often involves responses to cuts, exposure to poisonous plants, and insect stings or bites. Workers should be familiar with local hazards; they should avoid them and be competent in providing first aid. Weather-related injuries such as heat stress, frostbite, and hypothermia may require first aid. Tree workers should be familiar with symptoms related to temperature extremes and how to provide first aid. Wearing appropriate clothing for the weather (and having additional clothes in the vehicle, in case of a change in the weather) and properly hydrating can help workers avoid weather-related injuries.

Figure 15.17 Sloughing of a "palm skirt" (dead fronds that can detach all at once) is a leading cause of fatalities among climbers in palms, as they can pose a risk of suffocation (left). Climbers must work from the top down and avoid working under the skirt (right).

First Aid Procedures

This section reflects current practices at publication date and focuses on traumatic injuries and CPR. It does not replace the need for skills training and certification in first aid or CPR.

Directions for All First Aid

- Assess the situation, including the person's injuries and site hazards.
- Call 911 (in North America) or emergency medical services (EMS). Situations in which to call EMS include unconsciousness, difficulty breathing, severe bleeding, broken bones, coughing up blood, change in mental status, electric shock, seizures, or chest pain.

- A conscious adult has the right to refuse first aid or to stop your assistance at any time. Always ask permission to assist.
- Always put on PPE before treating an injured person. At a minimum, PPE includes nitrile gloves. If there is any possibility of being splattered with bodily fluids (blood or vomit), wear safety glasses.
- Have a biohazard bag or plastic bag to dispose of gloves and dressing waste after applying first aid.
- Wash your hands with warm water and soap after applying first aid.

Severe Bleeding
Severe Bleeding from a Laceration
- Have the person lie down if they might faint or go into shock.
- If they are able, have them apply direct pressure on the wound with a dressing while you put on nitrile gloves.
- Once your gloves are on, apply direct pressure firmly on the wound. Place a dressing between the flat palm of your hand and the wound to slow the bleeding.
- If bleeding continues, apply another layer of dressing over the wound. Do not remove the first dressing. Use a bandage to keep pressure on the wound.

Figure 15.18 Always put on PPE before treating an injured person. At a minimum, PPE is nitrile gloves, which should be in the first aid kit.

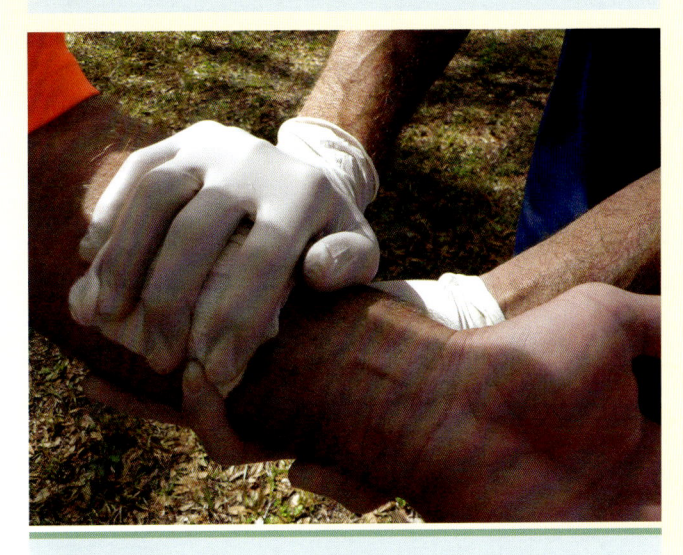

Figure 15.19 For severe bleeding, put on gloves and apply direct pressure firmly on the wound. Place a dressing between your hand and the wound to slow the bleeding.

Deep Laceration from Chain Saw Cuts
- If there is severe bleeding from a deep chain saw cut on an arm or leg, use a soft (tactical) tourniquet. Do not use a rope or wire as a substitute!
- Place the cut arm or leg through the tourniquet band. Position the tourniquet high on the extremity.
- Tighten the windlass on the tourniquet.
- Note the time the soft tourniquet is applied and let EMS know.

- Never release the windlass. Leave it in place for EMS.

Amputations

If a finger or toe is severed, it may be reattached in surgery if prompt action is taken.

- Cover the wound with dressing and a bandage to stop the bleeding.
- Rinse the amputated finger or toe with clean water, and cover it with sterile dressing.
- Place the amputated finger or toe and dressing in a clean, watertight plastic bag.
- Place the bag on ice, not in ice.
- Note the time, and provide this information to EMS.

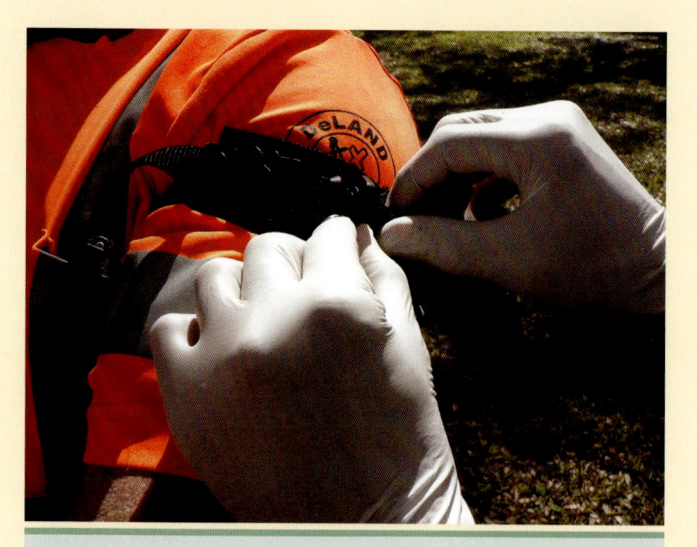

Figure 15.20 Soft tourniquets are valuable for managing bleeding from deep lacerations such as those caused by chain saws. Tourniquets should be applied only by people who have been trained to use them.

Head, Neck, or Back Injuries

Suspect head, neck, or back injuries if the person fell from height or was struck in the head, neck, or back. Symptoms include unconsciousness, confusion, numbness, vomiting, headache, unequal pupil size, and seizures.

- Minimize movement. Do not twist or bend the head, neck, or spine.
- Lay the injured worker on their back, and support their head and neck. Place them on their left side in case they might vomit.
- Do not clean deep scalp wounds. This may cause severe bleeding or contamination.
- Place a dressing on any head wound, but do not apply excessive pressure.
- Do not give the injured worker food or drink.

Fractures

Suspect fractures (broken bones) if the person fell from height or was struck by an object. Treat possible sprains (joint injury where bone ends are dislocated) and fractures the same way.

Closed Fracture or Sprain
- Place a towel over the injury, and place a plastic bag filled with ice and water on the towel to reduce swelling. Cooling packs may be used if ice is not available.
- Do not move the injured body part. Support it at the joints in its present position.
- A splint can be used to help support the body part and protect it from further injury. The splint must support the joints above and below the injury.

Open Fracture (a Broken Bone That Has Pierced the Skin)
- Do not attempt to push the fragmented bone back into place. Any movement might result in severing a blood vessel.
- Cover the injured area with dressing and support it in place.

Burns
Superficial Burns: First Degree
These burns present as redness and minor swelling. They are usually due to friction heat from a rope running through the hands or thermal heat from touching a hot object such as a chain saw muffler.

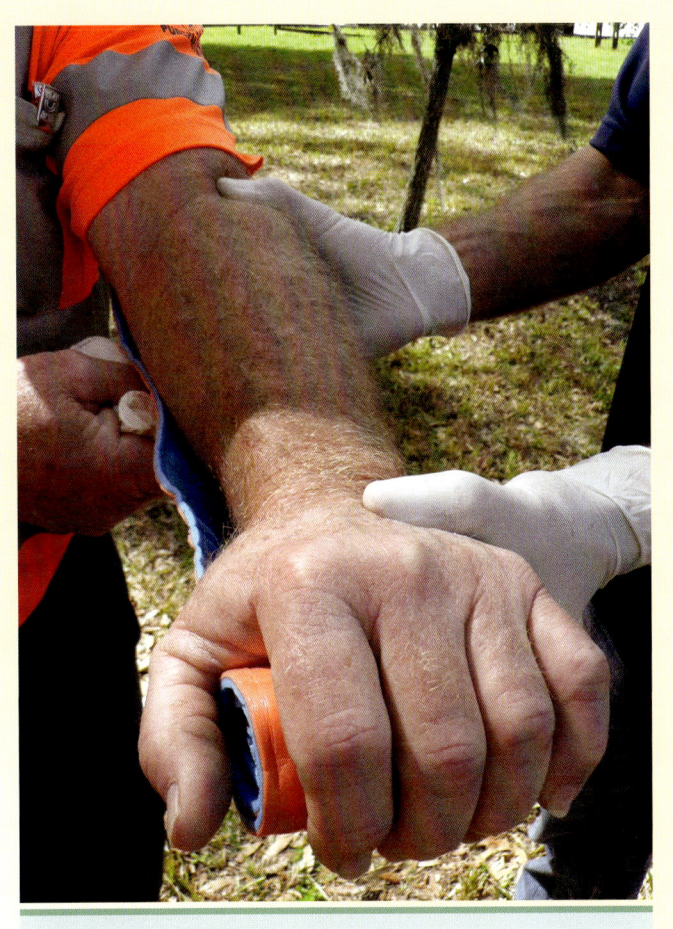

Figure 15.21 For a suspected bone fracture, a splint can be used to help support the body part and protect it from further injury. The splint must support the joints above and below the injury.

- Apply cold water, and cover the burn with burn ointment and a dressing.

Partial-Thickness (Second Degree) and Full-Thickness (Third Degree) Burns
Partial-thickness burns present as redness, mottling, and blisters. Full-thickness burns present with a dry, charred, glossy white appearance. Both are caused by direct contact with open flame, hot objects, or electric contact.

- Do not apply cold water or burn ointment.

- Do not attempt to remove clothing that is stuck to the wound unless it is still burning. Remove any jewelry or watch bands.
- Cover the burned surface with a dry dressing. Do not use any material that may stick in the wound, such as cotton.

Heat Injuries

Heat exhaustion can lead to heat stroke. Heat stroke can lead to death.

Heat Exhaustion

Heat exhaustion presents as pale, clammy skin, sweating, cramping, dizziness, nausea, or vomiting.

- Place the person in a shaded spot or building.
- Remove or loosen their clothing, and provide moving air with a fan.
- Cool the person with damp towels placed on their neck and armpit.
- Rehydrate them with water or a commercial sports drink.

Heat Stroke

Heat stroke presents as hot, red, dry skin and confusion or unconsciousness.

- Call for EMS. While waiting, follow the same first aid as for heat exhaustion. Do not give liquids unless the person is conscious and able to swallow.

Venomous Insect Bites

Minor Reaction to Bites or Stings

- For a bee sting, remove the stinger with tweezers. Be careful not to squeeze the attached venom sac, as doing so injects more venom. A credit card can also be used to scrape the skin's surface to remove the stinger. Wasps and hornets do not leave a stinger.
- Apply ice to the area.

Severe Reactions to Bites or Stings

- Call 911 or EMS if the person has a known allergy to stings or is

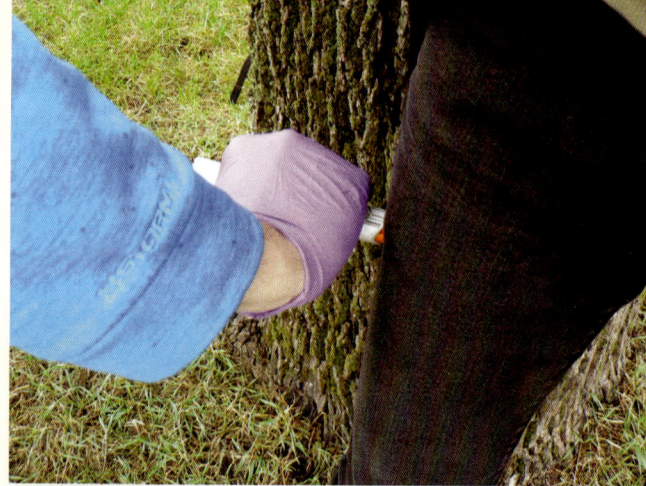

Figure 15.22 People with a known allergy to bee stings should always carry an epinephrine auto-injector. Follow directions and use immediately. It must be the person's auto-injector; this is a prescription medication.

developing hives, has flushed or pale skin, or is having trouble breathing.
- If the person has an epinephrine auto-injector, use immediately, following directions. It must be the person's auto-injector because this is a prescription medication.

Cardiopulmonary Resuscitation (CPR)

There are two CPR versions: hands-only CPR and CPR with breaths. Both versions begin with these steps:

- Assess the person's responsiveness. Tap on their shoulder while talking to them. If there is no response, rub the sternum (chest bone) with your knuckles. If there is still no response, call 911.
- Check for breathing. Place your ear near their nose and mouth to listen for breathing. If there are only gasps, or no breaths, CPR is needed.
- Lay the person face up on a firm surface. You may not be able to rule out spinal injury in an unconscious person, so move them carefully.
- Pull their clothes away from their chest.
- Have a coworker locate and bring the nearest AED (automated external defibrillator).

Hands-Only CPR

Perform hands-only CPR if you are not sure how to give breaths, do not have a mask, or are concerned about contamination with bodily fluids (e.g., saliva, vomit), *and* if the event just happened and EMS will be responding within four minutes. (CPR with breaths is preferred for longer time periods because the amount of oxygen in the blood decreases over time.)

- Place the heel of one of your hands on the person's lower breastbone—in line with the nipples—and put the heel of your other hand on top of the first.
- Position your shoulders over their chest and push straight down, depressing the chest about 5 cm (2 in) and allowing it to spring back. Repeat this at the rate of 100 to 120 times a minute.

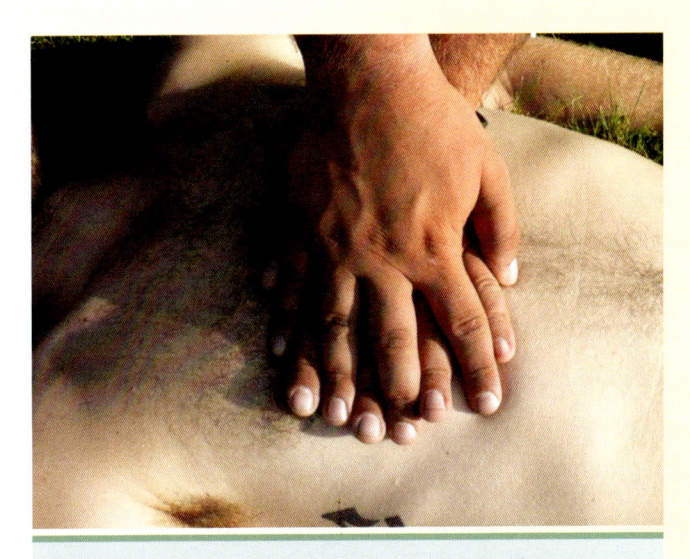

Figure 15.23 To perform CPR chest compressions, place the heel of one hand on the lower breastbone—in line with the nipples—and put the heel of the other hand on top of the first. Position your shoulders over their chest and push straight down, depressing the chest about 5 cm (2 in), and allowing it to spring back.

- If a coworker is available, alternate every two minutes.
- Continue compressions until EMS arrives.

CPR with Breaths

Perform if you are confident in your ability to give breaths, if you have a mask, and if you do not know how long the person may have been unconscious or if EMS response may be longer than four minutes.

- Position your hands on their chest and perform compressions at the same depth and rate as hands-only CPR.
- After 30 compressions, stop and tilt the person's head back with one hand on the forehead. Place your fingers of the other hand on the bony part of their chin to simultaneously tilt their head back and lift their chin.
- Take a deep breath. Create a tight seal around the person's mouth with your mouth or with a mask. Blow air into the person's mouth giving two quick, full breaths. Watch for their chest to rise.
- If there is no chest rise, return their head to normal position, tilt it back again, and try another breath.
- If the chest does not rise, return to compressions. Compressions should not be interrupted for more than 10 seconds.
- Repeat the process of 30 compressions alternating with two breaths for two minutes.
- If a coworker is available, alternate every two minutes.

Use of an AED

- Have a coworker turn the AED on, either by opening the lid or pressing the "on" button.
- Follow the AED's provided instructions. It will guide you through the steps.

CHAPTER 15 WORKBOOK

1. As used in many standards and regulations, the term _____ denotes a mandatory requirement, and the term _____ denotes an advisory recommendation.

2. Name four pieces of personal protective equipment (PPE) that are generally required for all tree workers.

   a.

   b.

   c.

   d.

3. True/False—All communications wires and cables shall be considered to be energized with potentially fatal voltages and shall never be touched directly or indirectly.

4. True/False—Industry-based standards are not, by themselves, legally enforceable, but regulatory agencies often refer to industry standards when citing or fining an employer for unsafe work practices.

5. True/False—The ANSI Z133 is the standard used in the United States, and arborists in other countries have adopted the ANSI Z133 in part or have similar standards.

6. True/False—Head protection need only be worn while there are climbers in the trees.

7. True/False—Eye protection is not required for tree work.

8. True/False—Hearing protection may be in the form of earplugs or earmuff-type devices.

9. True/False—Workers must not wear gauntlet-type gloves while chipping brush.

10. The _____ _____ summarizes what has to be done and who will be doing each task, the potential hazards and how to prevent or minimize them, and what special PPE may be required.

11. The voice _____ -and- _____ system ensures that warning signals are heard, acknowledged, and acted on.

12. All workers should receive some education and training in _____ _____ procedures, including CPR, first aid, and aerial rescue.

13. True/False—Any tree workers who work in proximity to electrical conductors must receive training in electrical hazards.

14. True/False—"Electrical conductor" is defined as any overhead or underground electrical device, including wires and cables, power lines, and other such facilities.

15. True/False—Rubber footwear and gloves provide absolute protection from electrical hazards.

16. Always engage the _____ _____ before starting a chain saw.

17. True/False—A chain saw engine does not need to be stopped for refueling.

18. True/False—Kickback can occur when the upper tip of the chain saw guide bar contacts an object.

19. True/False—A well-trained climber, in good condition, should be able to dodge the kickback of a chain saw.

20. True/False—An open-face notch is preferred for felling trees because the wider notch allows the hinge to work until the tree is almost on the ground.

21. True/False—If using a conventional notch to fell a tree, the back cut does not need to be stepped up higher than the hinge.

22. The _____ is critical in controlling the direction of fall of a tree.

23. Sometimes, if the tree is leaning in the direction of fall or has internal faults, it can split upward from the back cut. This is called a _____ _____ , and it can be very dangerous.

24. Workers should feed brush into chippers from the _____ and not allow any part of the body to cross the plane of the infeed chute.

25. True/False—Safety is the responsibility of all employees from the owner to the ground worker.

Matching

___ shall   A. leg protection for chain saw use

___ approved   B. advisory recommendation

___ CPR   C. cardiopulmonary resuscitation

___ direct contact   D. body touches energized conductor

___ should   E. mobile elevating work platform

___ indirect contact   F. mandatory requirement

___ MEWP   G. meets applicable standards

___ chaps   H. touching an object in contact with an energized conductor

Challenge Questions

1. Describe the steps that should take place before any work begins on a jobsite.

2. Simulate a job briefing that includes all the required elements and communication.

3. What are some of the most common causes of injuries in tree care, and how can they be prevented?

Sample Test Questions

1. According to many standards, the term "shall" denotes

    a. an advisory recommendation
    b. a mandatory requirement
    c. a safety suggestion by ISA
    d. a legal requirement

2. The area within the work zone where the crew expects cut branches or logs to be dropped or lowered from above is called the

    a. landing pad
    b. danger zone
    c. drop zone
    d. barrier zone

3. Head protection is required for tree workers

    a. whenever performing tree care operations
    b. when specified by the supervisor
    c. whenever there are climbers working aloft
    d. only if chain saws or chippers are in use

4. The most common situation that can cause chain saw kickback is

    a. failure to maintain adequate chain tension
    b. a worn sprocket or guide bar
    c. when the upper quadrant of the guide bar tip contacts an object
    d. uneven sharpening of the cutter teeth

5. A leading cause of fatalities among climbers in palms is

    a. insect bites
    b. small mammal attack
    c. chain saw cuts
    d. sloughing of fronds

Recommended Resources

(See Recommended Resources in back of book for detailed information.)

ANSI Z133-2017 *American National Standard for Arboricultural Operations—Safety Requirements* (American National Standards Institute 2017)

Tailgate Safety (Tree Care Industry Association 2018)

Z133 Online Course (ISA)

Introduction to Arboriculture: Safety online course (ISA)

CHAPTER 16
CLIMBING AND WORKING IN TREES

▼ OBJECTIVES

- Explain the functions of various climbing and rigging equipment, including ropes.
- List common knots according to their uses as friction hitches, attachment knots, and other applications.
- Compare and contrast moving rope systems (MRS) and stationary rope systems (SRS).
- Describe techniques for ascending and working in trees.
- Discuss emergency response procedures and precautions associated with aerial rescue.
- Explain the decision-making process for selecting rigging techniques and equipment.

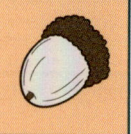

KEY TERMS

3-strand rope
12-strand rope
16-strand rope
24-strand rope
aerial rescue
arborist block
ascender
balance
basal anchor
bend
body-thrust
butt-tie
canopy anchor
carabiner
climbing harness (saddle)
climbing hitch
climbing line
climbing spikes (spurs)
connecting link
cordage
cycles to failure
design factor
double braid rope
drop cut
dynamic load
emergency response

footlock
friction
friction device
friction hitch
friction-saving device
hinge cut
hitch
hitch cord
job briefing
kerf
kernmantle rope
knot
lanyard
life-support equipment
limb walking
load line
mechanical friction device
micropulley
minimum approach distances (MAD)
moving rope system (MRS)
personal protective equipment (PPE)
primary suspension point
quick link
redirect

rescue pulley
rigging
rigging line
rigging point
rope angle
rope snap
rope walking
scabbard
self-double locking
shock-loading
sling
snap cut
spliced eye
split-tail
static load
stationary rope system (SRS)
tagline
tensile strength
throwline
tie-in point
tip-tying
triple-action
working-load limit (WLL)
work-positioning lanyard

Introduction

A great deal of arboricultural work involves climbing trees. Tree climbing is a physical and potentially hazardous profession, and rigging to lower wood and branches adds even more complexity to the work. Arborists, even if they are not professional climbers, should have a basic knowledge of how tree work is performed and of the equipment and techniques used.

This chapter is an educational tool for introductory-level arboriculture. It may be used as part of, but not as a replacement for, a comprehensive training program. Knowledge and experience are essential with all equipment and techniques described prior to use in the actual work environment.

Figure 16.1 A climber's personal protective equipment (PPE) includes head, eye, face, hearing, and foot protection. Climbing helmets offer many advantages over traditional hard hats. In many countries, PPE for climbers includes leg protection. Lightweight, flexible chain saw pants are widely available.

Climbing Equipment

The safety of tree climbers depends on their skills, knowledge, and the reliability of their gear. A climber must be able to select, inspect, and use their equipment correctly. Recognizing when it is time to retire and replace a piece of equipment is just as important as the selection and use. All equipment used by tree workers, including climbing gear and tools, must conform to applicable safety standards and should not be altered.

Personal Protective Equipment (PPE)

A worker's safety gear is called **personal protective equipment (PPE)**. This equipment includes head, eye, face, hearing, and foot protection. It also includes apparel specifically designed to prevent or reduce the severity of chain saw cuts, often in the form of cut-resistant pants or chaps. In many regions, PPE also refers to fall protection, including harnesses, hardware, and ropes. For the purposes of this publication, PPE refers to the protective items worn on the body.

Fall Protection and Work Positioning

Fall protection includes any components that keep a climber safely positioned above the ground. These components are referred to as **life-support equipment**. Many components of a climber's life-support equipment are for work positioning, to facilitate working safely in trees.

A **climbing harness** (sometimes referred to as a **climbing saddle**) suspends a climber on a **climbing line** and provides attachment points for work-positioning equipment. Harnesses also include a variety of attachment loops and accessory options.

A **work-positioning lanyard** is a relatively short length of rope, usually adjustable in length, equipped with carabiners, rope snaps, and/or eye splices, used to temporarily secure a climber in place or to provide additional security when using a chain saw or other tools in a tree. Work-positioning lanyards must meet the same strength requirements as ropes and hardware.

Copyright © 2022 International Society of Arboriculture

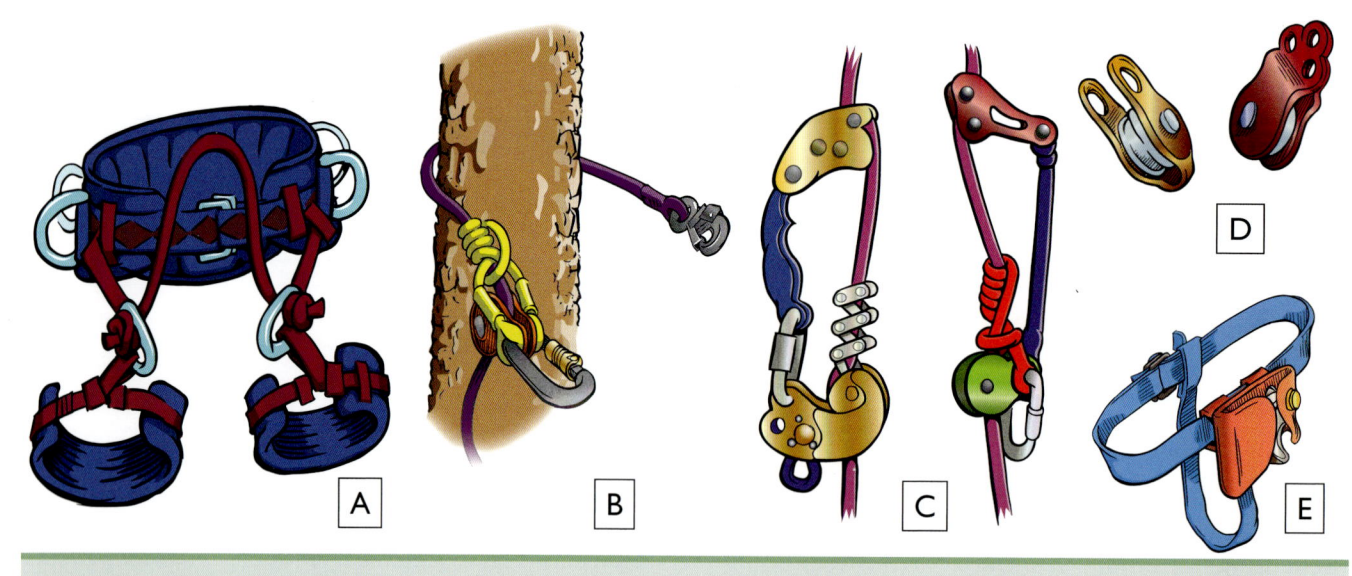

Figure 16.2 Fall-protection and work-positioning equipment and climbing aids include (A) climbing harness, (B) work-positioning lanyard with connecting links, (C) mechanical friction devices, (D) micropulleys, and (E) foot ascender. A variety of other equipment is also used.

Connecting Links

A **connecting link** is a component of a climbing or rigging system that connects other components. **Rope snaps** are one type of connecting link. Rope snaps used in securing the climbing line or work-positioning lanyard must be self-closing and self-locking. **Carabiners** are available in a wide variety of shapes, sizes, and constructions, with several types of locking mechanisms. Carabiners used for climbing must be self-closing and **self-double locking (triple-action)**. Carabiners must always be loaded along their major axis. **Quick links** are semipermanent connecting links; they should be tightened with a wrench. All connecting links must meet the minimum strength standards when used to support a climber.

Hitch Cords and Split-Tails

Tree climbers employ various "rope tools" as part of their climbing gear, including **hitch cords** and **split-tails**. A split-tail is a separate, short length of **cordage** (rope) used to tie a climbing hitch on the climbing line. Hitch cord is typically a smaller-diameter cordage than the climbing line and is also used to tie the friction hitch. Hitch cords frequently have eyes in both ends, either stitched or spliced. Rope components used in a climbing system must meet the minimum strength standards for climbing lines.

Mechanical Friction Devices

Mechanical friction devices function in place of, or in addition to, a friction hitch. They provide friction on the climbing line for ascending, descending, and work positioning. Many have been introduced to provide the additional friction necessary to secure a climber using a stationary rope system (SRS).

Climbing Aids

Climbing aids may or may not serve as life support, depending on their intended use. They are largely incorporated to create mechanical advantage, to support a more ergonomic body position, to reduce fatigue on the body and equipment, and in some cases, to minimize damage to the tree.

Pulleys

Two primary types (sizes) of pulleys are sometimes incorporated into climbing systems: **rescue pulleys** and **micropulleys**. A rescue pulley is a light-duty pulley that is sometimes set as part of an in-tree anchor suspension point or for establishing mechanical advantage. Micropulleys are smaller, light-duty pulleys, often used to tend slack and assist advancement of the friction hitch. Pulleys used in life support must meet minimum strength standards for climbing hardware.

Ascenders

An **ascender** is a mechanical device that enables a climber to ascend a rope. Attached to the rope, an ascender will grip in one direction (down) and slide in the other (up). Ascenders are used for ascent. They are not designed for working in a tree, for fall arrest, or for descending. Toothed ascenders must only be used on compatible ropes. In most configurations, two attachment points on the rope and/or a backup are required.

Friction-Saving Devices

Friction-saving devices, also referred to as friction savers or cambium savers, are devices that are installed to reduce friction at the tie-in point in the tree. The most common designs employ rope or webbing with rings at either end—double-ring type. Reducing friction at the tie-in point reduces wear on both the tree (branch union) and the climbing line. While this provides more consistent friction (and performance) at the friction hitch, it also creates additional friction at the friction hitch, so the climber must be sure to monitor the condition of their hitch cord or split-tail.

Throwlines

A **throwline** is a thin, lightweight cord with a throw weight (bag) attached to the end(s). It serves as a pilot line to set climbing or **rigging lines** in trees. The throwline is first installed in the desired branch union in a tree and is then used to pull the heavier climbing or rigging line into place.

Climbing Spikes

Climbing spikes (spurs) are long shanks with sharp, angled spikes (gaffs) on the ends that are attached to the inside of a climber's lower legs to assist in climbing and positioning during removals. They are used for ascending and working in trees, but only if the tree is to be removed, or to perform an aerial rescue.

Equipment Inspection

Inspections should be performed in a clean, dry area free of debris, fuel, and oil. It is a good idea to spread a tarp out to inspect equipment. Employers should develop a protocol for crews, such as a checklist and inspection by other arborists. Some standards require documented inspections at specified intervals.

Inspections should be conducted daily, periodically, and post-incident. A daily inspection involves a check prior to each use. This is a visual inspection to look for any nicks, deformations, cracks, improperly functioning components, or other major defects that could cause the equipment to fail. All climbing gear and ropes should be inspected daily. A periodic inspection should be conducted monthly or quarterly, depending on applicable regulations or the extent of use. A post-incident inspection occurs following an incident,

Figure 16.3 All climbing gear and ropes should be inspected daily.

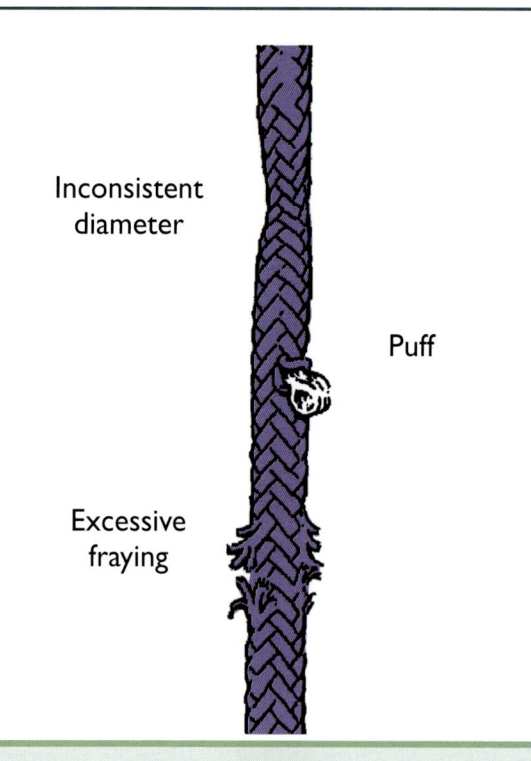

Figure 16.4 Check ropes for cuts, puffs, abrasions, pulled or herniated fibers, changes in diameter, discoloration, or glazing (melting) of the fibers.

such as an incident on a jobsite or a piece of equipment that dropped out of a tree.

Rope

The characteristics of a rope (strength, stretch, durability, etc.) are the result of the materials and techniques used to make it. Polyester and polyester blends are widely used by arborists, and many commercially available climbing and rigging lines are made from polyester, at least in part. Climbers need to use ropes that hold their shape and maintain flexibility. Arborists' ropes used for climbing or rigging should be durable, without excessive stretch. The characteristics of a rope can make it ideal for some applications yet inappropriate for others. Rock climbing and other sport ropes are not the same and should not be used for working in trees.

Most modern ropes have a braided construction. Braided ropes vary in the number of fibers and strands used, the diameter of each, and the method and tightness of the braiding. **Kernmantle ropes** are those with a cover and an inner core. Although the broad definition of kernmantle includes either braided- or parallel-core construction, the tree care industry has used kernmantle to more specifically refer to those with a parallel core. Kernmantle ropes are designed to be used with life-supporting toothed cams and with some other climbing hardware.

Double-braid ropes are a braided rope inside a braided rope. Thus, double-braid ropes fall in the kernmantle family, but they have a braided core. The core and cover are balanced and share the load almost equally. For this reason, they are not recommended for rigging through natural branch unions, in which the friction of the cover with the tree causes an imbalance in the load taken by the core and cover braids.

A **3-strand rope** has relatively low strength and high elongation, is inexpensive, and is the only nonbraided rope commonly used by arborists. It is appropriate to run a 3-strand rope through natural branch unions for climbing or rigging. A major drawback to 3-strand rope is the twisting that occurs as the line is used.

A **12-strand rope** is a braided rope that has no core (i.e., it is hollow in the center) and may be either loosely or tightly woven. Loosely woven, hollow-braid, polyester 12-strand rope is often used for rigging slings, but it would not be appropriate as a climbing or rigging line. Tightly woven, solid-braid, polyester-blend 12-strand ropes are sometimes used for climbing and rigging through natural branch unions, but, because they have no core, they tend to flatten and do not work well in certain situations.

A **16-strand rope** has relatively large, braided cover strands for strength and abrasion resistance and a small-diameter core to keep the rope round and firm under load. These ropes are commonly used for climbing and tend to be easy to tie.

A **24-strand rope** is typically double-braid construction with a core and cover but with a tighter braid than a 16-strand rope. This rope construction is well suited for use with mechanical friction

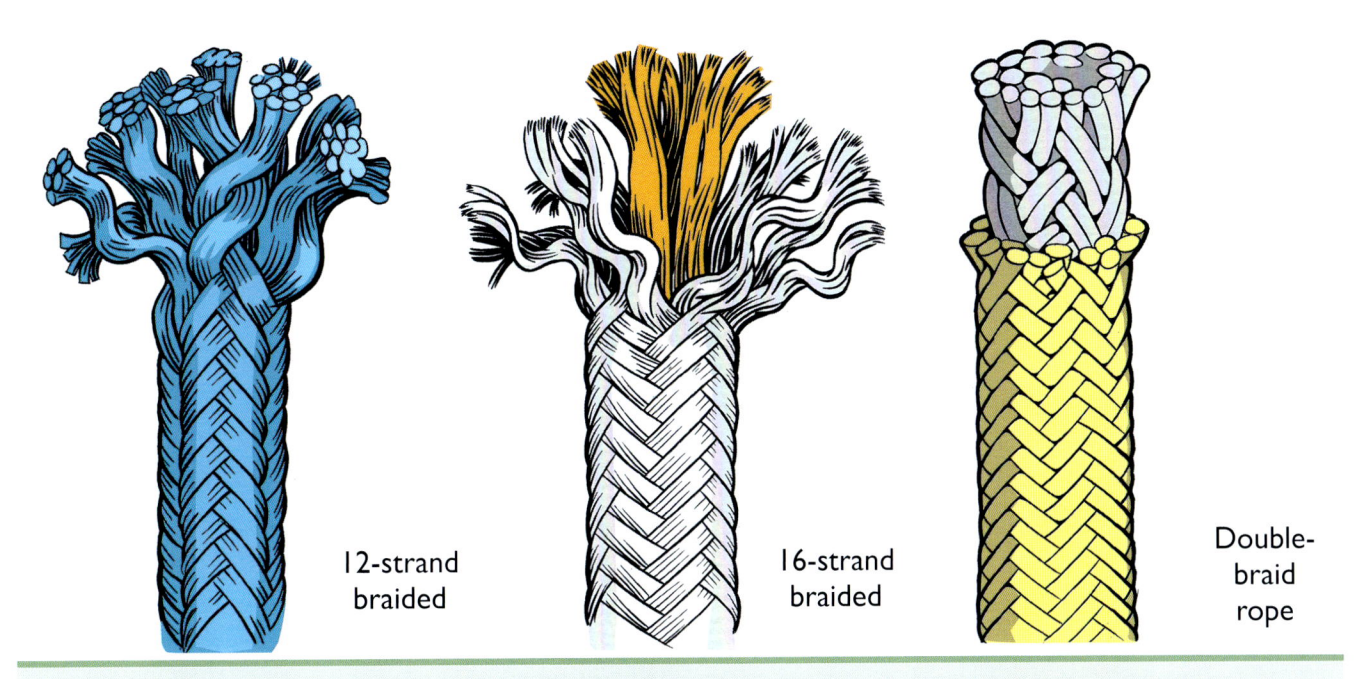

Figure 16.5 Ropes commonly used in climbing and rigging.

devices but may not be a good choice for running over bare branch unions.

Many ropes can be purchased with a **spliced eye**—interwoven strands forming a loop in one end—or a sewn, stitched eye for easy connection to hardware. The ability of a rope to be spliced is another important characteristic for arborists' ropes. Some ropes may have a spliced eye on one or both ends.

Design and Limitations

Understanding the design and limitations of each piece of equipment, as well as the tree itself, is an important part of setting up a safe and efficient climbing or rigging system. The failure of any link in the system can have disastrous effects, including catastrophic injuries, property damage, or even a fatality.

The **tensile strength** reported by the manufacturer is the breaking strength of a new rope or piece of hardware. The tensile strength of rope is diminished through use due to dirt, wear, knots, and, of course, loading. **Cycles to failure** must also be considered. One cycle means one lift, or drop, for a rigging line. Each cycle creates permanent damage in the rope, and eventually the rope will fail. At larger loads, the number of cycles to failure is reduced.

If the load in a rope is equal to its tensile strength, the rope may fail when used one time (one cycle to failure). To increase the life of a rope, a **working-load limit (WLL)** much less than the tensile strength must be established. The WLL is the highest intended load that the rope is manufactured to bear, and workers must ensure that the loads in each cycle are less than this WLL. The **design factor** reflects the relationship between the tensile strength and the WLL. A design factor of 3 or more is recommended for use with metal rigging components. For arborist rigging, with dynamic loading, high wear, and dirty conditions, a design factor of 5 or more is often chosen for synthetic rigging components such as rope, slings, and cordage. A design factor of 10 is commonly used for life-support ropes, lanyards, hitch cords, and connecting hardware. Keep in mind that the load on the equipment is often many times higher than the weight of the log or branch being lowered!

Various components of any rigging system have different WLL, and it is important to consider each component. Arborists generally design rigging systems such that the rigging line is the weakest link in the system and all loads are within the WLL of the rigging line.

Knots

All tree workers should be familiar with some of the various knots and hitches used in tree work. A climber should know how to tie and untie each and how to dress and set the knot properly. To tie a knot, first correctly form the knot in the rope. The dressing of the knot is the aligning of the parts; setting the knot tightens it in place. A climber must know how each of the common knots is used, as well as the advantages and disadvantages of each.

"Knot" is the general term given for all knots, hitches, and bends, although specifically, a **knot** is tied to itself, utilizing only one rope. A **hitch** is a type of knot used to secure a rope to an object, another rope, or the standing part of the same rope. A **bend** joins the ends of two ropes, or both ends of the same rope, together. There are several categories of knots, hitches, and bends. Tree climbers use endline knots, hitches, and bends to secure the climbing line to carabiners or rope snaps. Endline or termination knots are also used to tie off branches being rigged and lowered to the ground. A type of knot important in tree climbing is the **climbing hitch (friction hitch)**. Climbing hitches are the "climbing knots" used by climbers. They allow the climber to ascend or descend on a climbing rope or line and to tie in (secure themselves in the tree).

COMMON KNOTS FOR CLIMBING AND RIGGING

These are the basic knots that are frequently used in tree work. Many are described and demonstrated in the *Arborists' Knots for Climbing and Rigging* video listed in the bibliography.

Climbing Hitches (Friction Hitches)

Tautline Hitch

For years, the primary climbing hitch used by climbers in North America was the tautline hitch, although it is not as commonly used today.

Use and Characteristics
- Used as a friction (climbing) hitch
- Has a tendency to roll out (come undone during use), so it requires a stopper knot (such as a figure-8 knot)
- Must be adjusted (tended) frequently

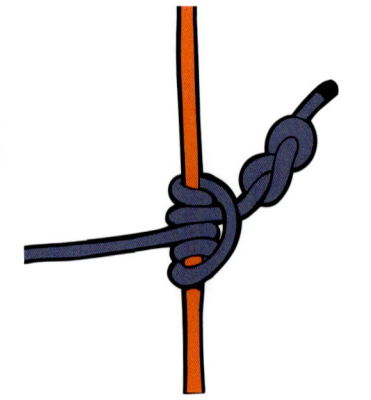

Blake's Hitch

Blake's hitch maintains more uniform friction than the tautline hitch and does not roll out.

Use and Characteristics

- Climbing (friction) hitch, often preferred over the tautline
- Stays dressed and set, less need to tend
- Doesn't tend to roll out (although a stopper knot is still recommended)
- Higher tendency to glaze on a long or rapid descent

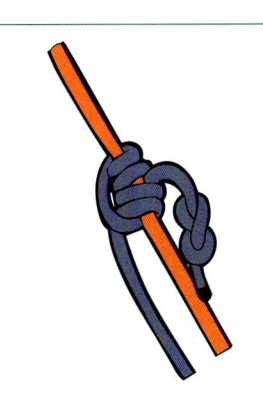

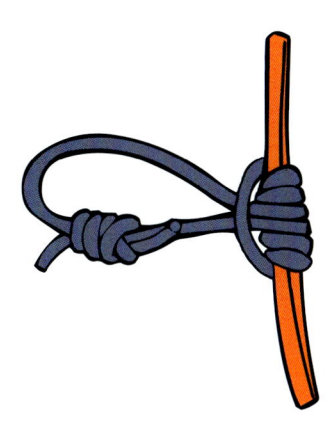

Prusik Hitch

For years, the primary climbing hitch used by climbers in Europe, Australia, and New Zealand was the Prusik hitch.

Use and Characteristics

- Friction hitch used in both climbing and rigging applications
- Bidirectional in some applications
- Can be tied with the end of a line, a length of hitch cord, or a loop of cord called a Prusik loop
- When tied using a Prusik loop (as in the secured footlocking technique), the cord used for the loop is of a smaller diameter than the working line to which it is attached; the type of rope affects how the knot will perform

Valdôtain tresse (Vt)

Use and Characteristics

- Tied with a length of cord (eye-and-eye split-tail) that is smaller in diameter than the line on which it is tied
- Tied as a closed friction hitch, with both legs of the split-tail attaching to a carabiner
- Holds securely and releases easily but may require slight adjustments to ensure that it performs properly
- Also used in rigging applications

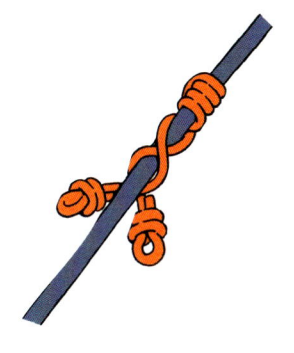

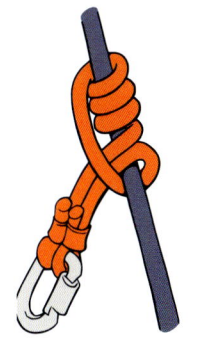

Schwäbisch

Use and Characteristics

- Essentially an asymmetric Prusik hitch
- Tied with a length of cord (eye-and-eye split-tail) that is smaller in diameter than the line on which it is tied
- Tied as a closed friction hitch, with both legs of the split-tail attaching to a carabiner
- Holds securely but can bind

Distel

Use and Characteristics

- Tied similarly to the tautline hitch but performs differently because it is a closed hitch
- Tied with a length of cord (eye-and-eye split-tail) that is smaller in diameter than the line on which it is tied
- Tied as a closed friction hitch, with both legs of the split-tail attaching to a carabiner

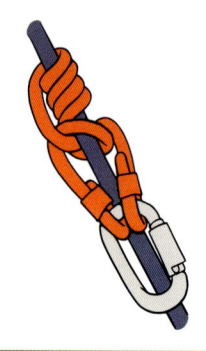

Michoacán

Use and Characteristics

- Essentially a Blake's hitch tied with an eye-and-eye split-tail
- Tied as a closed friction hitch, with both legs of the split-tail attaching to a carabiner
- Holds securely
- Grabs quickly but releases easily

Attachment Knots (Hitches)

Girth Hitch

Use and Characteristics

- Very simple hitch
- Used to attach a loop to a carabiner, ring, or another rope
- When tied with a length of rope, an eye sling, or the end of a rope, it is called a cow hitch and requires a half hitch as a backup
- Easy to untie, even after loading

Anchor Hitch

Use and Characteristics

- Used as a termination knot; snugs tightly against a carabiner or other hardware
- Does not loosen when loaded and unloaded repeatedly, though a stopper knot, such as a figure-8 knot, may be added
- Can easily be untied after loading
- Generates two loops around the hardware; may not fit in all hardware
- The loaded side of the knot should be closest to the spine of a carabiner

Bowline

Use and Characteristics
- Commonly used for forming a loop
- Easy to untie, even after loading
- The basis for other knots in the "bowline family" (running bowline, bowline on a bight, sheet bend, double bowline)
- Should *not* be used for attaching a climbing line to a carabiner
- The bowline with the Yosemite finish is shown in orange and is a more secure variant

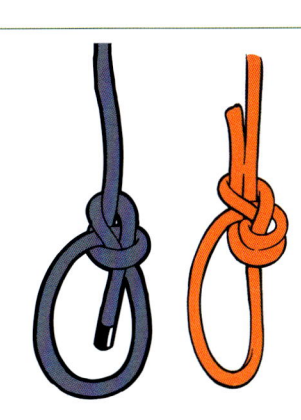

Running Bowline

Use and Characteristics
- A cinching knot, often used in tying off branches
- Can be tied around something that is far away and run up the line
- Easy to untie after loading

Midline Clove Hitch

Use and Characteristics
- Used to send equipment up to the climber
- Quick to tie in the bight (middle) of a line

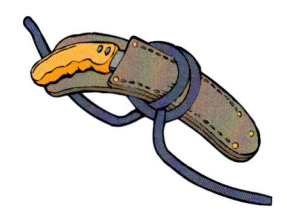

Endline Clove Hitch with Two Half Hitches

Use and Characteristics
- Used to tie off branches or sections of wood
- Requires at least two half hitches to prevent rollout of the tied piece
- Quick and easy to tie

Sheet Bend

Use and Characteristics
- Used to join two ropes of different diameters; often used to send a line up to the climber
- The smaller line should be tucked under its own standing part
- The same form as a bowline but tied with two ropes and does not form a loop

Double Fisherman's Knot (Grapevine Knot)
Use and Characteristics
- Used to form a Prusik loop
- May be difficult to untie after it has been loaded
- Essentially two double overhand knots, each tied around the standing part of the other

Cow Hitch with Half Hitch
Use and Characteristics
- Tied with an eye sling and used for securing hardware to a tree
- Variation on a simple girth hitch but formed with a line instead of a loop and backed up with a half hitch
- Cinching functionality is directional, though the half hitch minimizes movement

Timber Hitch
Use and Characteristics
- Tied with an eye sling and used for securing hardware to a tree (especially on large trees when the rope sling is not long enough to tie a cow hitch)
- Always make at least five wraps, spread around the stem
- Most secure when tied on larger pieces and when the pull is always at 90° from or against the bight, such that it tightens the hitch on the stem; loading toward the wraps can cause the timber hitch to loosen

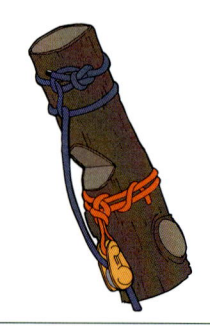

Half Hitch and Running Bowline
Use and Characteristics
- Used to tie off a section of wood when rigging from below or lowering branches or when there is no secure point for a running bowline alone
- The half hitch shares the load with the running bowline and reduces the chances of the knot slipping off the piece once cut

Other Common Arborist Knots

Figure-8
Use and Characteristics
- Used as a stopper knot
- Fast and easy-to-tie endline knot
- Unties easily

Slip Knot

Almost any knot can be "slipped." Typically, this means the final tuck of the working end is replaced by tucking a bight instead so that the knot can be rapidly untied by pulling on the working end. The knot known as the slip knot is a slipped overhand knot.

Use and Characteristics
- Easy to tie, even with one hand
- Experienced climbers find many uses for this knot
- A directional knot—it tightens when loaded one way but spills when pulled from the other side

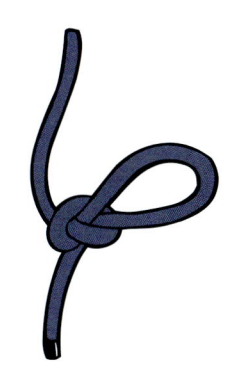

Butterfly

Use and Characteristics
- Primarily used for forming a midline loop
- Symmetrical, without extreme bends; can be loaded in either direction
- May be difficult to untie after it has been heavily loaded

Inspection of the Tree and Site

Every job must begin with a **job briefing** that covers the work plan, PPE, potential hazards, hazard mitigation, and the required gear and procedures. Before ascending a tree, a climber must always look carefully and locate any electrical conductors and other utilities. If there are electrical conductors nearby, the climber must develop a work plan to maintain a safe distance so as not to come into contact accidentally. It is important to know and abide by **minimum approach distances (MAD)** for electrical conductors as set forth by regulations and governing authorities.

Next, the climber and crew should inspect the rest of the surrounding area for other potential hazards and obstacles. Additionally, the site should be checked for signs of recent construction, grade changes, or root decay that could compromise the root system.

Figure 16.6 Climbers must perform a systematic inspection of the tree, starting at the root collar, and working up to the top. This preclimb inspection should take into consideration not only the tree's climbability, but additional loads that may be placed on the tree during operations.

Following a thorough site inspection, the climber can start a systematic inspection of the tree. Evidence of decay in the root collar area could indicate a defect that can lead to whole-tree failure. Snow, vines, debris, soil, and landscaping could hide fruiting structures of decay organisms, such as conks, which indicate decay in the tree. Dieback in the top of the tree could indicate root damage. Included bark, cankers, and other structural defects could result in trunk failure. In the crown, dead or broken branches, cracks, and insects or other animals should be evaluated and considered in the work plan. Cracks, sapwood decay in major branches, substantial dieback, and an unstable root plate should all serve as warnings to the climber. Problems that are identified may require mitigation measures to reduce or eliminate the risk. The work plan can change as the work progresses, including the use of additional equipment or personnel.

Climbing Systems

There are many systems for ascending into and working within a tree; each offers its own set of advantages and limitations. The systems are divided into two main types: **moving rope systems (MRS)** and **stationary rope systems (SRS)**. The equipment and techniques vary somewhat, but the basic principles remain the same: managing friction to move efficiently in a tree. Both systems are options for a climber but not necessarily for every tree. Many employers and trainers require that beginners learn the fundamental skills of MRS before learning SRS. Knowing these skills can be essential for self-rescue or descent if a key piece of equipment is dropped or damaged during work.

Moving Rope Systems

Moving rope systems (MRS), formerly called doubled rope technique (DdRT), are so named because the climbing line moves through the tie-in point as the climber moves up and down in the tree. Moving rope systems tend to be less gear intensive

Figure 16.7 With moving rope systems (MRS), the climbing line moves through the tie-in point when the climber moves up and down in the tree. Certain ascent techniques, such as body-thrusting, rely heavily on core and upper body strength, so long ascents can be exhausting.

and less complicated than SRS, leaving less of a chance for error for a new climber.

A limitation with MRS is the ascent, which is slower than ascending on SRS. Because the rope moves as the climber moves in MRS, the climber advances only 1 ft for every 2 ft of rope pulled. Even though ascending is slower, an advantage is that the climber's weight is split between the two parts of the line, creating a mechanical advantage with less weight to pull and less load at the tie-in point. Certain ascent techniques, such as **body-thrusting**, can only be used on an MRS. Body-thrusting relies heavily on core and upper body strength, so long ascents can be exhausting. Another limitation of MRS is associated with setting a climbing line in the tree. For MRS to work efficiently, the two parts of the climbing line must be isolated over a single branch union, meaning both ends follow the same path from the branch to the ground.

Stationary Rope Systems

When climbing using MRS, the rope runs over a branch union or through a friction-saving device, constantly moving as the climber navigates the tree. In stationary rope systems (SRS), formerly called single rope technique (SRT), one end of the rope is secured in place so that the climber can navigate up and down the single, stationary leg of rope. Stationary rope system climbing lines can run through one or more branch unions, but they are held taut at a single anchor point, which serves as a stable attachment for the climbing line. Two different kinds of anchors are used: canopy and basal. A **canopy anchor** is isolated over and secured to a single branch union in the canopy. A **basal anchor** is set at the base of the tree.

Each anchor configuration has advantages and limitations. Canopy anchors can be set to be retrievable, which allows the climber to remove the full system without being physically near the anchor point. A basal anchor can be configured in ways that make it lowerable, which could potentially aid in a rescue. A very important consideration with a basal anchor is that the load exerted at the branch union where the climbing line first passes over can be nearly twice as great as when using a canopy anchor or an MRS.

The selection of the type of anchor used is based on several factors. One factor is the tree itself and the work to be done. The branch configuration or the amount and location of the work to be performed may dictate the type of anchor used.

Whether using a canopy anchor or a basal anchor, climbing only one leg of the rope changes the dynamics of the system. As the climber ascends using SRS climbing, advancing the gear 1 ft up the rope translates into 1 ft of upward movement for the climber. In MRS, there is a 2:1 mechanical advantage; however, that corresponds to only 50 percent gain in upward movement.

Ascent with SRS is not only more efficient, but many climbers also find it to be less fatiguing. Climbers ascend in a more ergonomic, vertical position and

Figure 16.8 (left and right) In stationary rope systems (SRS), one end of the rope is anchored in place while the climber navigates up and down the single leg of rope. SRS climbing lines can run through one or more branch unions and are held taut at an anchor point, either at the base of the tree or within the canopy. The simple addition of a foot and knee ascender can shift a climber from the horizontal body-thrust to the more efficient, vertical technique "rope walking."

use the larger muscles in their legs to ascend. Also, because the rope is not moving through a branch union or friction-saving device in SRS, the reduced friction conserves energy and reduces wear on the climbing rope. Because of these advantages, many climbers have adopted SRS ascent techniques even when they work the tree using the MRS climbing method. Potential limitations are that SRS can be more complex and gear intensive, requiring more investment, training, and setup time.

Selecting a Tie-in Point

The **tie-in point** in MRS climbing is the branch or branch union through which the climber has installed the climbing line to work in the tree. With SRS climbing, the climbing line can be anchored either in the canopy or at the base of the tree, so climbers refer to the **primary suspension point** instead of a tie-in point. The primary suspension point is the position in the canopy where the climbing line is anchored or, if the climbing is anchored at the base of the tree, whichever branch union the climbing line crosses over that experiences the highest loads during the climb.

The position of the tie-in point affects the climber's ability to access various parts of the tree. Also, it is the point from which the climbing line and climber are suspended, so a strong branch attachment is required. For these reasons, the selection of a safe, sturdy tie-in point is essential for a safe climb. Generally, it is advisable to select a high, central location in the tree. This allows ease of movement and easy access to most points in the tree. The higher the tie-in point, the farther the climber can move out on the branches, maintaining safe rope angles.

The branch attachment selected for tying in should be wide enough for the climbing line to pass through easily. Wood strength, and therefore the minimum size of the branches, varies with species, but the main branch should have the diameter and strength large enough to support the loads a climber may place upon it. The branch needs to be alive and healthy, with leaves or living buds. If there are signs of sapwood decay or possible internal decay, that branch union must not be used. Preferably, the climbing line is installed by passing it through a branch union, going around the larger branch or trunk, and over the

Figure 16.9 As a rule, the higher above the workstation the tie-in point is, the greater the distance you can move out from the trunk.

smaller or lateral branch below it. This way, if the upper branch breaks, the line will drop to the next branch down, rather than out of the tree.

Rope Installation and Ascent

Climbers must be protected from falling at all times when working in a tree. The first step to climbing is safely installing a rope in the tree. Sometimes, a climber may throw the climbing line directly into the tree. To set a rope higher in the tree, a climber may choose to use a throwline instead of throwing the rope. A throwline can be thrown with amazing accuracy through branch unions 20 m (70 ft) or higher. Once the throwline is installed, it is attached to the climber's line, which is then pulled up through the branch attachment and back down to the ground.

Using Ladders

If a climbing line has been installed, then ladders are an acceptable means of entering a tree. But ladders can easily shift when not placed on level ground or when loaded unevenly. A climber can also lose their balance and fall from a ladder. To prevent this, the climber should first tie in while on the ground and then ascend the ladder, advancing the friction hitch while the ladder is held steady by a ground worker. Once the climber is off the ladder and secure in the tree, the ladder can be removed from the area.

Body-Thrusting

Once a climbing line has been set in the tree, there are several techniques for ascending. When climbing with MRS and when the climbing line is set close to the trunk of the tree or when in a tree with many horizontal branches along the trunk, body-thrusting is one way to ascend. Using the inner edges of the boot, the climber straddles the trunk of the tree, orienting their body in a horizontal position. Using leg, hip, and core muscles, the hips are thrust upward, creating slack in the line; simultaneously, the climber pulls down on the standing part of the line to take up the slack and keep the line taut. Technique is important when body-thrusting; the climber who relies solely on upper body strength to body-thrust can be exhausted after reaching the top. The addition of a micropulley below the friction hitch allows a ground worker to advance the knot and pull slack out of the climber's line while the climber ascends.

Footlocking the Standing Part of the Line

If the climbing line is installed far from the trunk, the climber can execute a single-line **footlock**, which is performed by footlocking on the fall of the line below the friction hitch. The climber pulls their legs up into a squatting position, wraps the fall of the line around one boot, and then locks it in place with the other boot. Then the climber stands on the lock, advancing the friction hitch and taking up the slack. A slightly more sophisticated variation is to add a foot ascender to grip the climbing line below the friction hitch.

Rope Walking

Another option for ascent is to incorporate cammed ascending devices for a technique known as "**rope walking**." It is the primary ascent method for SRS climbing. The ascenders grip the line and remove slack as the climber advances the friction hitch (or mechanical friction device). The addition of a chest harness can assist with keeping the climber in an upright position and help pull the system upward as the climber ascends. Though more gear intensive than some other ascent techniques, rope walking will save time and energy. Rope-walking techniques can also be incorporated into MRS climbing. The simple addition of a foot and knee ascender can shift a climber from the horizontal body-thrust to a more efficient, vertical technique. Rope walking can be a bit awkward if the climbing line is set close to the trunk, however.

Climbing with Spikes (Spurs)

Another method of ascending a tree is the use of climbing spikes or spurs. Using climbing spikes to

Figure 16.10 Because spurs can damage a tree, they are approved for use only on trees to be removed, or for aerial rescues.

climb trees for pruning is unacceptable and strongly discouraged because it can do irreparable damage to a tree, including to palms. Using spurs for tree removal, however, is common and considered a standard technique. Having the climbing line installed before ascending on spikes allows the climber to have a secure attachment while maneuvering up the tree.

If a climbing line is not installed beforehand, climbers must be secured while ascending and should always carry two means of securing, one of which must be a climbing line. The climbing system must include a method that prevents the line from sliding. Climbing line systems that can be cinched on the tree are preferred. In addition to a climbing line, a work-positioning lanyard is required. Some **lanyards** have a wire core that makes the line more rigid; these wire-core lanyards should not, however, be assumed to provide protection from cutting with a chain saw, nor should they be used in proximity to electrical conductors.

Climbing Basics

Friction is an important concept to climbing. Managing the friction of the climbing hitch (or mechanical friction device) on the climbing line is how the climber moves safely and can stay suspended in place. Safe tree climbing requires keeping the rope taut between the climber and the tie-in point or suspension point, tending any slack that can result in dangerous falls or swings. Though climbers may swing from branch to branch, they are not falling into their climbing line to be caught. Rather, they use their climbing line to support their movements throughout the tree.

Limb Walking

Limb walking is an important skill to learn to be able to reach all parts of a tree. A climber can walk out on branches to reach the tips of branches. Generally, the preferred method is to walk backward or sideways on the limb, with the climber facing the direction of the tie-in point, always keeping the climbing line taut. It is important for the climber to keep much of their weight on the climbing line; if the climber allows their full weight to be on the limb, the limb may break. The angle of tying in is important here. As a rule, the higher above the work area the tie-in point is, the greater the distance the climber can move horizontally away from the trunk. Ideally, the **rope angle** formed by the trunk and the climbing line should not exceed 45°.

Using Redirects

A **redirect** is a change in the direction of the climbing line to create a safer or more efficient rope angle and reduce the chances of an uncontrolled swing. Redirects can be achieved two ways. The easiest

Chapter 16: Climbing and Working in Trees

the climbing line can be run through the hardware. This method is preferable because the climber can set the webbing loop to optimize the rope angle relative to the work location. Some redirect devices are self-retrieving, allowing the climber to pull the device back to them.

Safe Work Practices

While in a tree, the climber may require various tools and equipment depending on the work plan or project. This may include a chain saw, a pole pruner, or a pole saw. If the tree work includes installing support systems, then cabling hardware and other tools may be required. Climbers should always climb with their handsaw and **scabbard** (sheath for the handsaw). For efficiency or reducing climber fatigue, ground workers can send up tools as requested by the climbing arborist. If pole pruners or pole saws are used in the tree, they should be hung vertically in the tree when not in use, hung in such a way that the sharp edge is away from the climber and such that they will not accidentally dislodge.

Chain saws can be equipped with a breakaway chain saw lanyard for use in a tree. The climber is to descend through a branch union above the work, which is known as a natural redirect. This type of redirect can be challenging to undo because it often requires the climber to ascend back through the union—a tricky maneuver—and is usually only performed before the climber descends out of the tree. Another type of redirect can be established by girth-hitching an appropriately rated webbing loop around a branch. A carabiner or pulley can be attached to the loop, and one or both sides of

Figure 16.11 When limb walking, the preferred method is to walk backward or sideways on the branch, with the climber's rear end opposing the direction of the tie-in point, keeping the climbing line taut at all times.

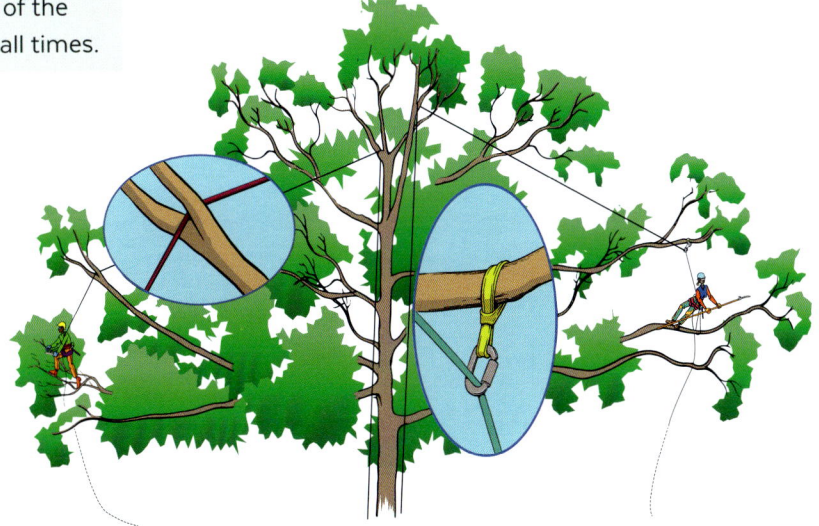

Figure 16.12 In SRS, a climber can use redirects without changing the friction in the system. Examples of redirects include natural branch unions, webbing loops, or retrievable redirects.

Copyright © 2022 International Society of Arboriculture

must be stable and secure when using a chain saw and other equipment in the tree. Chain saws should be shut off, with the chain brake engaged, when the climber moves to another position. A climber must be secured with a work-positioning lanyard or a second climbing line in addition to the main climbing line when using a chain saw in a tree.

Maintaining regular communication with the ground crew is important, including informing them when making cuts or when dropping or lowering anything from the tree.

This chapter provides a broad overview of climbing basics. For more detailed information, refer to the *Tree Climbers' Guide*, listed in the bibliography.

Emergency Response and Aerial Rescue

Arborists working in trees should be trained in **emergency response**, which includes not only first aid and cardiopulmonary resuscitation (CPR) (both covered in the previous chapter) but also training in how to contact emergency personnel and what information to provide. Emergency response training should be practiced regularly. Practicing the response will reduce the likelihood of panic and will reinforce what steps each worker should take. **Aerial rescue**, a potential component of emergency response, is the process of safely bringing an injured or unconscious worker to the ground. Ground workers should maintain a close watch on climbers and remain in voice contact during aerial operations. Although Bluetooth capabilities can help with on-site communications, a climber could be injured and lose consciousness without ever calling for help.

The first step in any emergency response is to assess the situation using the ABC's of first aid: airway, breathing, and circulation. In cases where the injured worker is aloft, ground workers should try to communicate with the victim. Once the situation has been quickly assessed, the second step is calling emergency medical services (EMS) with the information. Most countries have a universal emergency number, such as 911. Expect the dispatcher

Figure 16.13 Ground workers should maintain a close watch on climbers and remain in voice contact. Bluetooth capabilities can help with on-site communications. However, a climber could get hurt and lose consciousness without ever calling for help.

to ask the following: the nature of the emergency, the location, the telephone number in case the call is disconnected, the number of injured workers, any hazards (tree or electrical), and the injured worker's height in the tree.

If there are no electrical hazards, if the tree is safe to climb, and if a crew member has the equipment and capability to climb, the third step is to gain access to the injured worker. There the rescuer can provide first aid and either prepare to assist EMS in a rescue or, if necessary, perform the rescue. The rescuer should carry their cell phone with them so they can maintain communication. If there are additional workers or other people on-site, one should

With Any Aerial Rescue, These Fundamental Principles Apply:

- No rescue should ever be attempted unless it is safe for the rescuer and necessary for first aid to the injured worker.
- The injured worker should not be moved unless necessary, such as to control serious bleeding or to perform CPR.
- Both the injured worker and the rescuer must always remain secured.
- A rescuer must support the injured worker, if possible, which often requires being secured to the injured worker.

Figure 16.14 Upon reaching the injured worker, the rescuer should assess the injured climber's equipment and tie-in point for damage or defect, then they can more thoroughly assess the injuries. If the injured worker is conscious and breathing and is not bleeding severely, no attempt should be made to move them, especially if a neck or spinal injury is possible.

stand by the road to flag in EMS, and any other workers can clear a path to the tree and provide support to the rescue climber.

Upon reaching the injured worker, the rescuer should assess the injured climber's equipment and tie-in point for damage or defect, and then they can more thoroughly assess the injuries. If the injured worker is conscious and breathing and is not bleeding severely, no attempt should be made to move them, especially if a neck or spinal injury is possible. If CPR is required, however, it can only be performed on a flat, hard surface, requiring the rescuer to lower the victim to the ground. In many cases it will be best to wait for EMS; they have the knowledge and equipment to prevent further injuries while lowering the worker.

Once EMS arrives, they will take control of the situation. Most aerial rescues are performed by EMS professionals. These may be high-angle EMS teams that have the capabilities to climb or use aerial devices, such as towers, to reach the injured worker. There will be instances where the EMS team does not have the capability to conduct a high-angle rescue and the tree crew's assistance may be necessary. Generally, EMS will not allow someone they do not know to assist in the rescue; however, if a crew member is already in the tree with the injured worker, EMS may call upon their assistance.

Rigging

Rigging is the use of ropes and other equipment to safely take down trees or remove branches with minimal damage to the landscape. Rigging is necessary when felling the tree is not possible due to obstacles such as a house or potential hazards such as power lines. Rigging techniques provide a controlled removal of larger branches. Rigging is an advanced aspect of working in a tree, and only experienced climbers should attempt to use these techniques.

The tools and techniques used in rigging vary with the situation. If removing branches from a tree that is being pruned, it is important not to damage the remaining branches and trunk. There are more options available when rigging is used during whole tree removal.

This text is introductory in nature and presents only the most basic of rigging techniques. Arborists

> ### "Low and Slow"
>
> Tree climbers often use the expression "low and slow." This refers to the principle that new equipment, knots, and techniques should always be introduced in a low-risk environment—low to the ground and tried out slowly.
>
> The same concept is especially important in rigging. Avoid combining multiple new processes in a work environment in which the consequences of a mistake are severe. Practice using new techniques in noncritical situations.

wanting to advance their skill and knowledge of rigging should seek qualified, hands-on training and are encouraged to study *The Art and Science of Practical Rigging*, an eight-part video series and study guide published by the International Society of Arboriculture.

Forces in Rigging

Removing large, heavy sections of trees using ropes and other equipment generates large forces. These forces are significantly affected by the equipment and techniques used. The size and weight of the piece removed is the base factor in determining the force involved, but forces are also affected by the distance of fall, the type and amount of rope in the system, the rate of deceleration, and the angles involved. An especially important factor in rigging forces is whether the rigging point is above the load or below the load (sometimes confusingly referred to as negative rigging). By selecting appropriate equipment and employing rigging techniques that minimize forces, arborists can reduce the potential risks of rigging. While rigging equipment and ropes have become stronger over the years, trees have not, and anchor points in the tree can sometimes be the weakest point in the rigging system.

Friction is essential in rigging systems. Without friction, ground workers could not lower pieces that

Figure 16.15 Rigging is the use of ropes and other equipment to take down trees or remove branches.

weigh more than the workers. With more friction at the **rigging point**, such as when rigging using a natural branch union, the lead (the part of the line from the rigging point to the piece being cut) experiences significantly more of the force in this relatively short length of rope than does the rest of the rope. Over time, these forces shorten the life of the rigging line.

One important concept to understand is the difference between **static loads** and **dynamic loads**. In arborist rigging, an example of a static load is the weight of a suspended tree section or piece of wood; an example of a dynamic load is the load experienced by stopping a piece of falling wood by using ropes and rigging equipment (an example of **shock-loading**). The dynamic load can be many times the static load (the weight of the piece of wood). Failure to account for dynamic loads can lead to premature or sudden equipment failure or even tree failure.

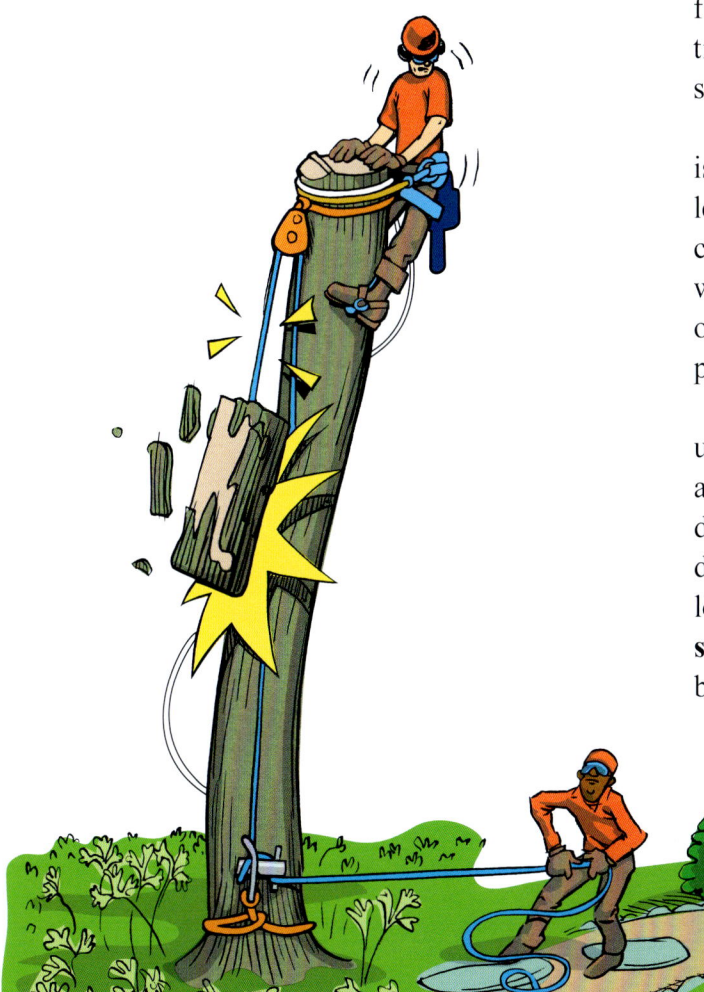

Figure 16.16 Shock-loading occurs when a dynamic, sudden force is placed on a rope or rigging apparatus when a moving load/piece is stopped.

Equipment

Choosing the most appropriate equipment for a given situation can make the job much more productive—and safer. The demands placed on arborist equipment by dynamic loading and abrasion mean that tools from other industries are not always applicable to tree care.

Friction devices (lowering devices) for lowering cut pieces have been designed so that workers can easily take wraps around the device with the **load line** instead of wrapping it around the trunk of the tree. In addition to being far more efficient, friction devices also provide more consistent, predictable friction than the trunk of the tree. Mechanical friction devices should never be used for rigging, nor should any other piece of climbing gear.

Rather than having to tie a knot each time a rope is used, connecting links such as carabiners, shackles, or screw links can speed up the process. Most connectors are *not* designed for dynamic loading, which eliminates their use in most tree removal operations. Steel connecting links are usually the preferred choice in rigging.

Compared to running lines through branch unions, using arborist blocks can increase control and decrease dynamic loading, wear on ropes, and damage to the tree. **Arborist blocks** are heavy-duty pulleys, with a large, rotating sheave for the lowering line and a smaller sheave to accept a rope **sling**, eliminating the need for a connecting link between the block and sling. Arborist blocks provide minimal, consistent friction, which (1) reduces wear on the rope, (2) minimizes damage to a branch union, and (3) allows more of the lowering line to share the load. When using a block in a tree, it is important to understand that the load at the point of attachment of the block can be double the load in the rigging line.

Through knotting or splicing, rope can be made into any number of rope tools. There are many variations of slings and other rope tools used in rigging. In addition, webbing slings

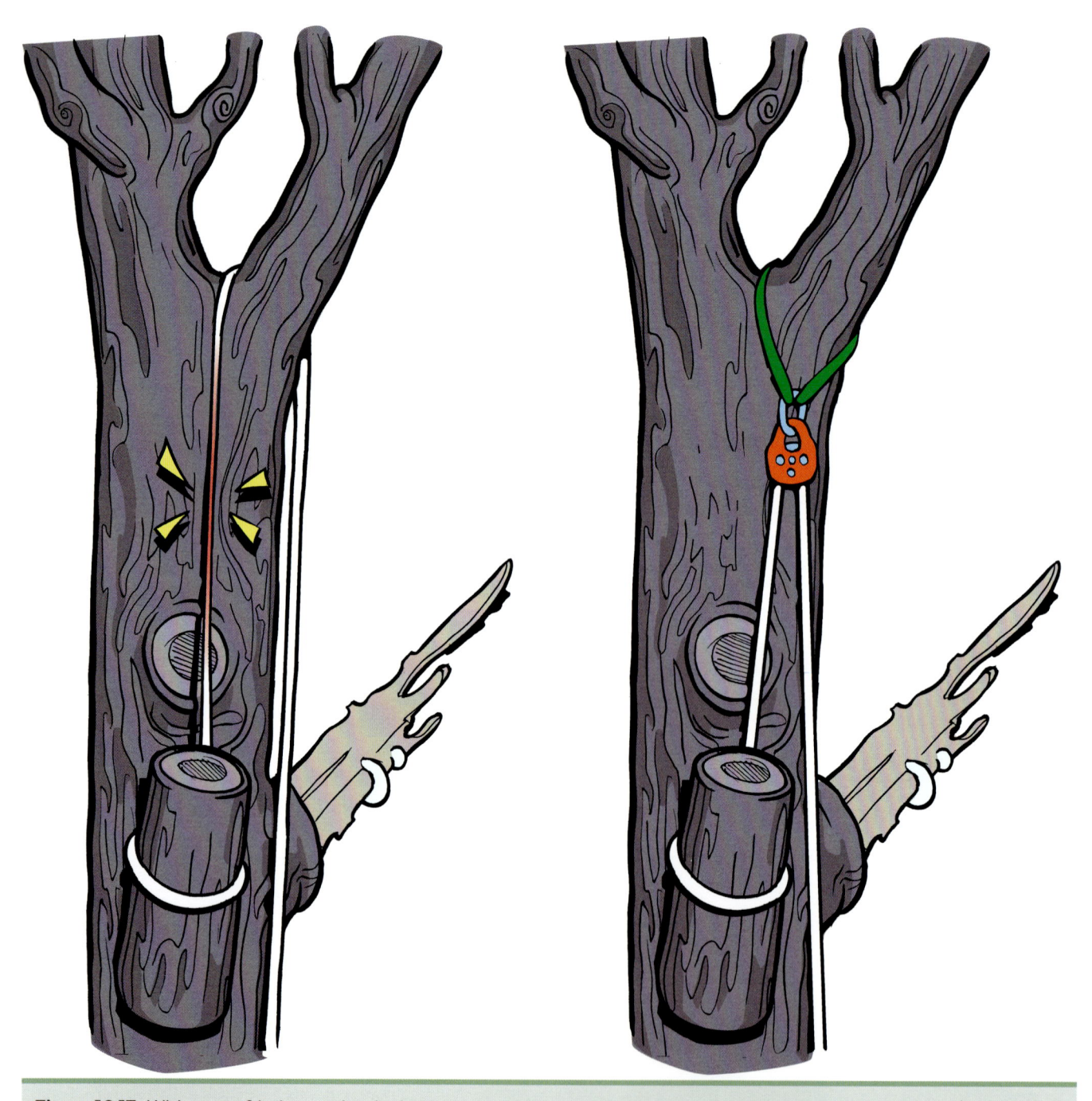

Figure 16.17 With more friction at the rigging point, as with rigging using a natural branch union, the lead (the part of the line from the rigging point to the piece being cut) experiences significantly more of the force in this relatively short length of line than does the rest of the line. Using an arborist block helps distribute the force along a greater length of line.

can be purchased in differently sized sewn loops, or they can be knotted from tubular webbing. The strength of a sling depends on the material, the way the loop is formed (spliced, sewn, or knotted), and the way the loop is used. It should be noted that knots decrease the tensile strength of a rope. Depending on several factors, a splice may maintain the strength of the rope.

Chapter 16: Climbing and Working in Trees

Figure 16.18 A branch is said to be butt-tied if the load line is attached at the butt end.

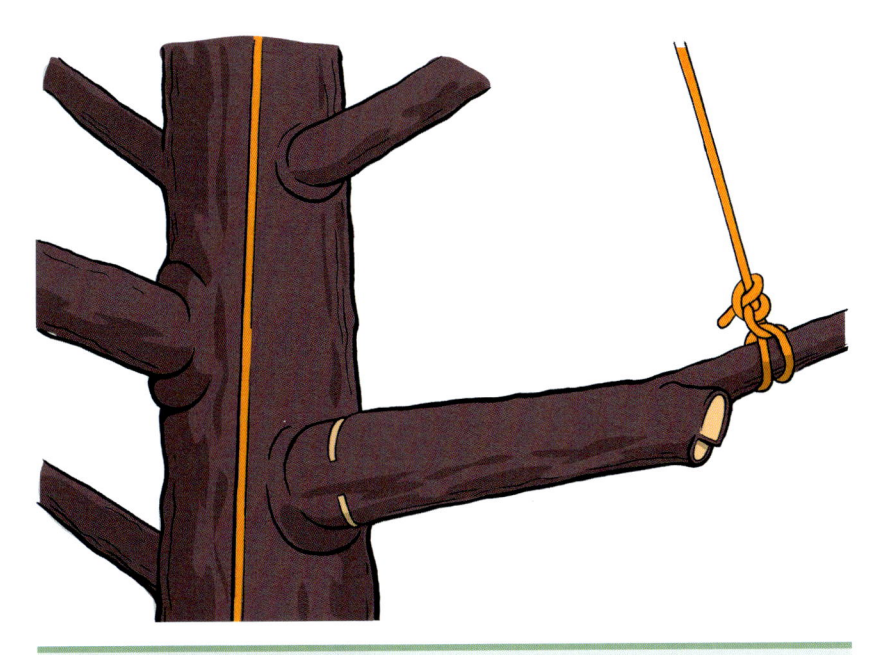

Figure 16.19 If the rigging line is attached at the brush end, the branch is tip-tied.

Fundamental Rigging Techniques

There is always more than one way to lower tree parts from the tree to the ground, but the best method is the one that maximizes productivity while still maintaining safety. One of the first decisions is the choice of the rigging point and whether to run ropes through a natural branch union or through a rigging block. Natural branch unions can be fast and effective, but the running rope can injure the tree and damage the rope. The consistent, predictable friction and versatility of placement of blocks are often great advantages over using natural branch unions. Just as important as how friction is reduced at the rigging point is how it is added to the rigging system when lowering larger pieces. Wrapping the rigging line around a tree trunk will work, but the amount of friction added is inconsistent, and the wear on ropes and trees can be excessive. Friction devices offer alternatives to these drawbacks, and some offer the additional ability to help in raising wood.

Another decision is how and where to tie off branches. Most branches or wood sections are tied off with one of two knots, a running bowline or a clove hitch with two half hitches. A running bowline can be tied at a distance from the desired attachment point and then pulled along the line and cinched at the attachment point. If a clove hitch is used, it must be followed up with two half hitches to prevent the hitch from rolling out. With a clove hitch, bends in the knot are more favorable (less strength loss) than with the running bowline.

If the rigging line is attached at the butt end of the piece to be removed, it is said to be **butt-tied**. The piece will normally drop tip down. The climber must be positioned to avoid contact with the rigging line or avoid being hit by the butt of the branch.

Tip-tying attaches the rigging line toward the tip of the branch being removed. If the branch is tip-tied and cut, the butt end will drop away from

Figure 16.20 Rather than tip-tying or butt-tying, a separate rope tool can be used to balance a branch, helping to reduce swing and dynamic loading. A tagline may be added to any branch being removed to help control the swing.

the cut, and the swing of the branch will depend on the placement of the rigging point. The climber must be positioned to avoid being struck by the swinging branch. A branch might also be tip-tied and lifted to avoid hitting an obstacle below.

At times, it is essential to minimize the swing or drop of a branch. Rather than tip-tying or butt-tying, a separate rope tool can be used to **balance** the branch, keeping either the butt or tip from dropping. Balancing can also help reduce swing and dynamic loading.

A **tagline** is a rope tied to the piece to be removed and controlled by a ground worker, but it is neither run through a rigging point nor used for lowering. A tagline may be used with any of the other techniques.

As mentioned previously, it is always preferable to establish a rigging point above the work. There are situations, however, where a rigging line cannot be anchored above the work, such as in removals. After the last of the branches is removed, the climber no longer has an overhead rigging point to use. Without any overhead options, all rigging points will be below the piece of wood being cut. Rigging from below is a common technique in which a piece is tied with the rigging line above the point where it will be cut, and the line is run through a block or branch union below the cut. This can be one of the most demanding techniques for a rope (as well as the hardware and the tree) because of shock-loading. Because there are no high tie-in points, it can also be dangerous for the climber. Specific training for this technique is essential.

Most of the time, there are many methods for removing a given branch or tree on any site. Arborists call on their experience or the experience of others to choose the safest, easiest, and most efficient technique in each situation. One of the best rules of rigging is, when there is any doubt, take a smaller piece.

Cutting Techniques

Once the branch is rigged for removal, the climber must decide on the appropriate method of cutting the branch. The **drop cut** (three-point cut) consists of an undercut and a top cut on the branch, then a final cut to remove the stub after the branch has been removed. When using a chain saw, arborists often form the top cut directly above the undercut to avoid getting the bar stuck in the **kerf** of the cut as the branch breaks free.

A cut that is handy for controlling relatively small sections of wood that may not require roping is the **snap cut** (bypass cut). This is made by cutting slightly more than halfway through a section from one side, then cutting from the opposite side, about

Figure 16.21 Rigging from below can be one of the most demanding techniques for a rope (as well as the hardware and the tree) because of the potential for shock-loading.

Figure 16.22 Make a snap cut by cutting slightly more than halfway through a section from the side, then cutting from the opposite side, about 2.5 cm (1 in) or more offset from the first cut. The distance apart will need to be greater for larger branches. The two cuts will bypass, but the fibers should hold. The saw can be shut off and put away and the remaining piece broken off manually.

2.5 cm (1 in) or more offset from the first cut. The distance apart will need to be larger for larger branches. The two cuts will bypass, but the fibers should hold. The remaining piece is broken off manually. A snap cut can be made on the top and bottom sides of a branch stub from a large branch that has been removed.

The **hinge cut** is a variation of standard tree-felling techniques and is referred to as a topping cut when taking a top out of a tree, or a directional notch when the worker wants the branch to swing to one side or another. It employs the use of a notch and back cut to form a hinge that controls the direction the branch falls or swings.

Rigging Strategy

The strategy for piecing out a tree depends on the circumstances. The climber must plan the order of removal to avoid being left with a branch that is too difficult or dangerous to remove. A general rule of thumb is to clear a pathway for the branches, removing brush first. It is often better to leave a few branches for damping the rigging forces (dissipating the energy), especially when taking out the top.

As with all aspects of removal, the key to safety is to use equipment that can handle the load and to ensure that the tree can handle the load. It is important to understand that the choices made in rigging setup can affect the loads experienced. Workers must avoid trying to take out a section that is too heavy or large. Every crew member must always consider what could happen if some component of the rigging failed and should have a plan to account for this. The climber and all ground workers must be clear of danger in the event of something breaking unexpectedly.

CHAPTER 16 WORKBOOK

1. List at least three pieces of equipment considered to be part of a climber's life-support equipment.

   a.

   b.

   c.

2. _____ used for climbing must be self-closing and self-double locking (triple-action).

3. True/False—Hitch cord is typically a smaller-diameter cordage used to tie the friction hitch.

4. _____ are small, light-duty pulleys, often used to tend slack and assist advancement of the friction hitch.

5. Climbers may choose to use a _____ - _____ device when tying in. This can reduce the wear on the rope and damage to the tree and can, in some cases, facilitate climbing.

6. True/False—Climbing spikes are used for ascending trees, but only for removal or during a rescue.

7. List three times that climbing gear should be inspected.

   a.

   b.

   c.

8. List two braided rope types that are popular for tree climbing.

   a.

   b.

9. _____ - _____ ropes are not recommended for natural branch union rigging, where the friction of the cover with the tree causes an imbalance in the load taken by the core and cover braids.

Chapter 16: Climbing and Working in Trees

10. True/False—If the load in a rope is equal to its tensile strength, the rope will definitely fail on its first use (one cycle to failure).

11. Arborists generally design rigging systems such that the _____ _____ is the weakest link in the system and all loads are within the working-load limit of the rigging line.

12. It is important to know and abide by _____ _____ _____ for electrical conductors.

13. List the two types of climbing systems.

    a.

    b.

14. True/False—In SRS, there is a 2:1 mechanical advantage; however, this corresponds to only 50 percent gain in upward movement.

15. Carabiners must always be loaded along their _____ _____ and never across the gate.

16. _____ _____ is an efficient ascent method that incorporates ascenders.

17. A _____ is a change in the direction of the climbing line to create a safer or more efficient rope angle and to reduce the chances of an uncontrolled swing.

18. True/False—When a climber is injured or unresponsive in a tree, the first step is to ascend and get to the injured worker as quickly as possible.

19. True/False—Natural branch unions can be fast and effective for use as a rigging point, but the minimal, consistent friction and versatility of placement of an arborist block is often a great advantage.

20. True/False—The forces in rigging are affected by the weight of the piece, the distance of fall, the type and amount of rope in the system, and the angles involved.

21. When the piece must be removed without dropping either the butt or the tip, it can be tied so it is _____ , then lowered to the ground.

22. True/False—It is always preferable to establish a rigging point above the work, if possible.

23. True/False—Rigging from below can be one of the most demanding techniques for a rope (as well as the hardware and the tree) because of shock-loading.

24. The _____ _____ is a variation of standard tree-felling techniques that employs the use of a notch and a back cut to form a hinge and steer the branch.

25. Label the following knots.

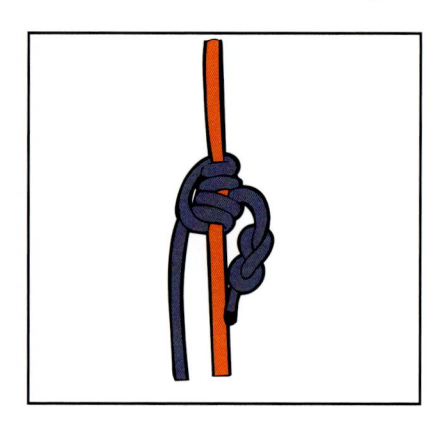

a. _____ b. _____ c. _____

Matching

____ tagline   A.   used to install a climbing line

____ scabbard   B.   usually high and central in the tree is best

____ tie-in point   C.   knot used to secure a rope to an object or another rope

____ kernmantle   D.   rope used to control the swing of a branch

____ quick link   E.   may be used to attach a block

____ rope sling   F.   rope with a cover and a core

____ throwline   G.   must be tightened with a wrench

____ hitch   H.   sheath for a handsaw

Challenge Questions

1. Describe the inspection process that every climber should follow before climbing a tree. Include inspection of gear and inspection of the tree.

2. Compare and contrast the methods of ascending a tree.

3. Describe the steps involved in rigging large branches for removal. How can specialized equipment be used to make the job easier and safer?

Sample Test Questions

1. A "rope inside a rope" is better known as a

    a. hollow-braid rope
    b. 12-strand rope
    c. double-braid rope
    d. 3-strand rope

2. A separate, short length of cordage used to tie a climbing hitch is a

    a. tagline
    b. split-tail
    c. throwline
    d. climbing line

3. Carabiners used for climbing must be

    a. self-double locking
    b. loaded along their minor axis
    c. tightened with a wrench
    d. constructed of steel

4. The first steps of the emergency response process are to

    a. assess the situation and call for emergency help
    b. shut off the electricity
    c. reach the injured worker and begin first aid
    d. reach the injured worker and secure them for descent

5. A cut that consists of an undercut followed by a top cut and then a final cut is called a

    a. drop cut
    b. jump cut
    c. hinge cut
    d. topping cut

Recommended Resources

(See Recommended Resources in back of book for detailed information.)

Tree Climbers' Guide (Lilly and Julius 2021)

The Art and Science of Practical Rigging [DVD and book] (Donzelli and Lilly 2001)

Arborists' Knots for Climbing and Rigging [DVD and workbook] (ISA 2006)

The Tree Climber's Companion (Jepson 2000)

CHAPTER 1 ANSWERS TO WORKBOOK QUESTIONS

Workbook Questions

1. Sites of rapid cell division in the shoot tips, root tips, and cambium are called **meristems**.

2. Meristems located at the end of the roots and shoots are called primary, or **apical**, meristems.

3. The tendency for terminal buds to inhibit the growth of lateral buds is called **apical dominance**.

4. The "food factories" of trees are the **leaves**.

5. The process of **photosynthesis** combines carbon dioxide and water in a reaction driven by light to produce sugars. **Oxygen** is also a product of this reaction.

6. The green color of leaves is created by the presence of the pigment **chlorophyll**, which is necessary for photosynthesis to take place.

7. **Transpiration** is the loss of water vapor from the leaves.

8. The opening and closing of **stomata** on the undersides of leaves allow for gas exchange.

9. Water and dissolved essential minerals are transported within the tree in the **xylem**. The **phloem** conducts carbohydrates.

10. The **cambium** is a layer of meristematic cells located between the phloem and the xylem.

11. The **branch collar** is formed when trunk tissue grows around branch tissues. As the branch and trunk tissues expand against each other in the branch union, the **branch bark ridge** is formed.

12. **Bark** protects the branches and trunk of a tree from mechanical injury and desiccation.

13. Name four functions of the root system.

    a. **absorption**

    b. **anchorage**

    c. **storage**

    d. **conduction**

14. The sugar products of photosynthesis are sometimes referred to as **photosynthates (or carbohydrates)**.

Answers to Workbook Questions

15. The orientation of growth in response to an external stimulus is called **tropism**. Two examples are **phototropism** and **geotropism**.

16. CODIT stands for **Compartmentalization of Decay/Damage/Dysfunction in Trees**.

17. Trees with upright growth and a strong, central leader are said to exhibit **excurrent** growth. More rounded trees, which are often broader than they are tall, have **decurrent** growth habits.

18. Roots and fungi form **mycorrhizae**, which are a symbiotic relationship, aiding in the uptake of water and minerals.

19. The process by which chemical energy, stored as sugar and starch, is released is called **respiration**.

20. Trees that lose their leaves in the autumn are called **deciduous**. Trees that maintain their leaves for more than one year are called **evergreen**.

Label the Following Diagrams

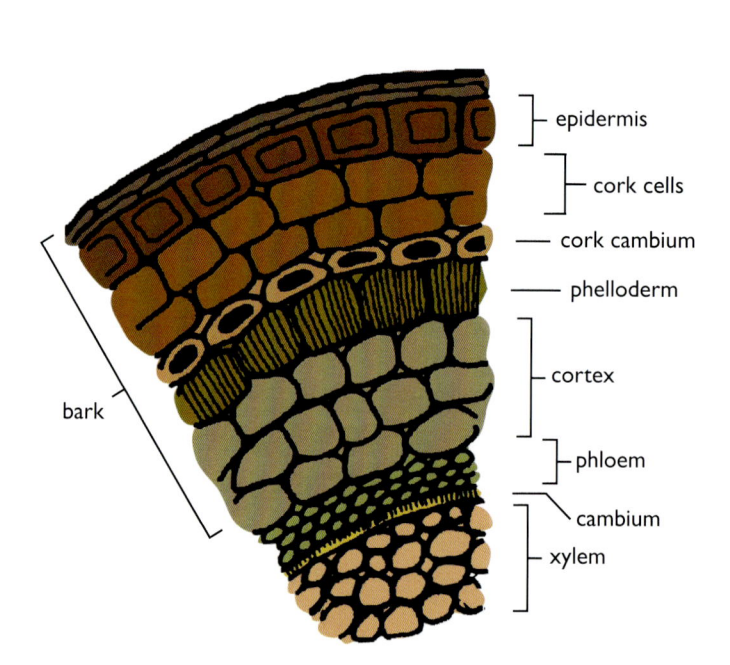

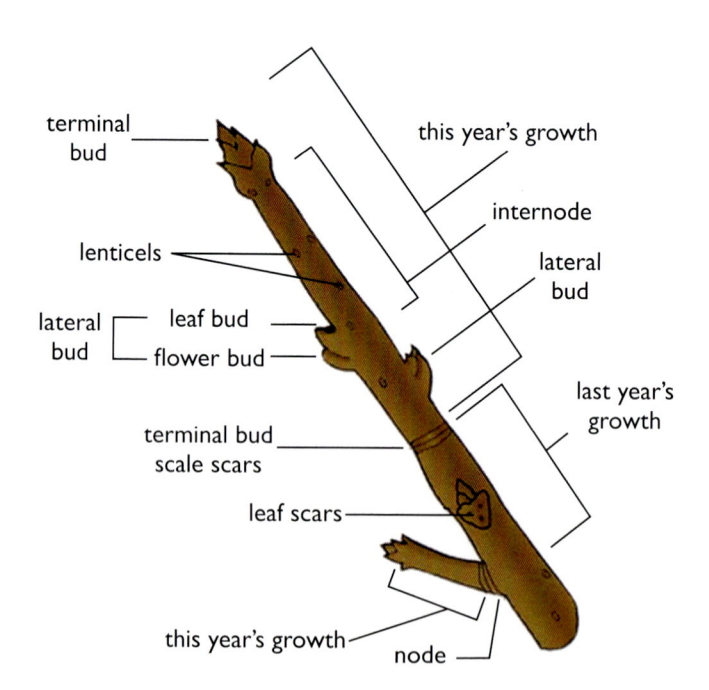

Left diagram labels: bark, epidermis, cork cells, cork cambium, phelloderm, cortex, phloem, cambium, xylem

Right diagram labels: terminal bud, this year's growth, internode, lenticels, lateral bud, leaf bud, flower bud, lateral bud, last year's growth, terminal bud scale scars, leaf scars, this year's growth, node

Matching

G auxin   A. uses more energy than it produces

J chlorophyll   B. mostly located in the upper 12 inches (30 cm) of soil

E cuticle   C. "stalk" of a leaf

C petiole   D. cells that cross the phloem and xylem for radial transport

H internode   E. waxy covering of a leaf

__F__	lenticel	F.	small openings in stems for gas exchange
__D__	ray	G.	plant growth regulator
__B__	absorbing roots	H.	between the nodes of a stem
__I__	source	I.	mature, green leaves—sugar producers
__A__	sink	J.	green pigment

Sample Test Questions

1. When cutting through a tree with a chain saw or drilling into a tree, you would pass through which structures (in order)?

    a. bark, cambium, phloem, xylem
    b. bark, phloem, cambium, xylem
    c. bark, cambium, xylem, phloem
    d. bark, xylem, phloem, cambium

2. If the terminal bud is removed in pruning,

    a. growth may be stimulated in lateral buds
    b. flowering is stimulated to enhance fruit production
    c. the branch will die back
    d. all of the above

3. The growth rings of many trees

    a. are visible because of anatomical differences between earlywood and latewood
    b. are distinguished by the rays that separate them
    c. contain water-conducting xylem, regardless of age
    d. are visible in angiosperms but not in gymnosperms

4. Which layer of cells is responsible for outward trunk growth and increased girth of a tree?

    a. cambium
    b. pith
    c. epidermis
    d. cortex

5. Mycorrhizae are

    a. collar-rot fungi
    b. elongated underground stems producing sucker sprouts
    c. a symbiotic relationship between fungi and roots
    d. cells in which photosynthesis takes place

CHAPTER 2 ANSWERS TO WORKBOOK QUESTIONS

Workbook Questions

1. The classification of living organisms, including plants, is called **taxonomy**.

2. List the levels of classification. The first letter of each term is given.

   Kingdom

   Phylum

   Class

   Order

   Family

   Genus

   Specific epithet

3. **Angiosperms** are vascular plants whose seeds are covered (by an ovary). **Gymnosperms** are vascular plants with "naked seeds."

4. Eudicotyledon (eudicot) refers to plants that have two seed leaves at germination. Grasses, banana, and palms belong to another group called **monocotyledons (monocots)** and have only one seed leaf.

5. The naming of plants is called **nomenclature**.

6. Name five plant characteristics used to identify trees.

   a. form or growth habit

   b. bark texture

   c. leaves

   d. flowers

   e. fruit

   f. seeds

   g. buds

   h. pith

   i. leaf scars

   j. scent

Answers to Workbook Questions

7. Draw a twig with the following leaf arrangements.

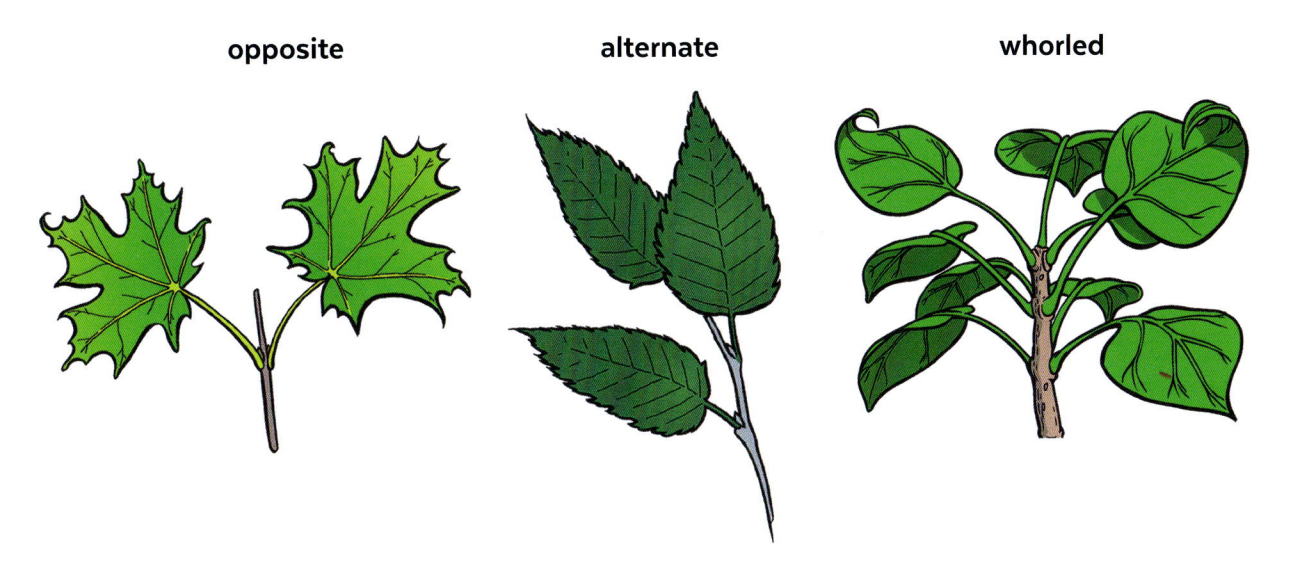

      opposite　　　　　　　**alternate**　　　　　　　**whorled**

8. Name a tree with palmately compound leaves: **example: buckeye or horsechestnut**. Name a tree with pinnately compound leaves: **examples: ash, walnut, Kentucky coffeetree, ailanthus, honeylocust**.

9. Draw a simple leaf with a lobed leaf margin.

10. Draw a compound leaf with serrate margins on the leaflets.

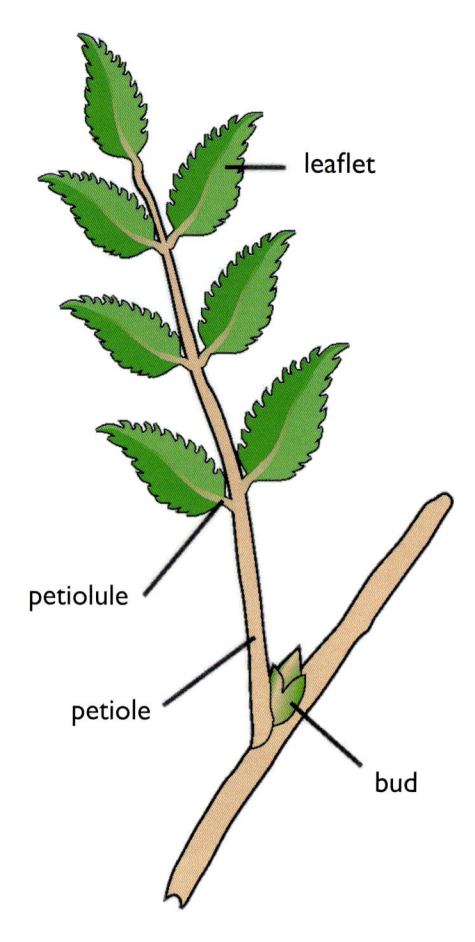

(Illustration of a compound leaf labeled with leaflet, petiolule, petiole, and bud.)

11. A compound leaf with multiple leaflets will have **one** bud(s).

12. Give an example of a tree species that has more than one common name: **_Carpinus caroliniana_ (one of many examples)**.

13. In the botanical name *Acer saccharum*, *Acer* identifies the **genus**, and *saccharum* identifies the **specific epithet**.

14. Species may be subdivided into **subspecies**, **varieties**, **forma**, and/or **cultivars** that have distinct differences from the general species.

15. A **cultivar** is a cultivated variety.

Sample Test Questions

1. Douglas-fir (*Pseudotsuga menziesii*) differs from balsam fir (*Abies balsamea*) in that

   a. they are not in the same genus
   b. they are not in the same family
   c. Douglas-fir is actually a type of hemlock
   d. balsam fir is not a conifer

2. When two leaves and/or buds are located at the same node on a twig, the arrangement is called

   a. opposite
   b. alternate
   c. whorled
   d. bicompound

3. Select the botanical name that is written correctly.

   a. *Quercus Rubra*
   b. *Quercus rubra*
   c. *quercus Rubra*
   d. *quercus rubra*

4. Which genus of trees usually does *not* have an opposite leaf arrangement?

   a. *Acer* (maples)
   b. *Fraxinus* (ashes)
   c. *Quercus* (oaks)
   d. *Aesculus* (horsechestnuts/buckeyes)

5. Which conifers have needles in bundles?

   a. hemlocks
   b. firs
   c. pines
   d. spruces

CHAPTER 3 ANSWERS TO WORKBOOK QUESTIONS

Workbook Questions

1. The majority of the fine, absorbing roots of a tree are in the **A** horizon.

2. Soils are generally described in terms of their **physical**, **chemical**, and **biological** properties.

3. Soil **compaction** is defined as an increase in bulk density and a decrease in total pore space.

4. **True**/False—Negatively charged clay particles hold cations near their surface.

Answers to Workbook Questions

5. **True**/False—A large percentage of tree decline situations can be attributed to soil stress, especially with urban soils.

6. **True**/False—Soil can hold water so tightly in micropores that the ability of tree roots to absorb the water is restricted.

7. Soil texture refers to the relative coarseness or fineness of a soil. Rank the following from the finest texture (1) to the coarsest (3).

    __2__ silt   __1__ clay   __3__ sand

8. On the pH scale, less than 7 is **acid**, 7 is **neutral**, and more than 7 is **alkaline**.

9. **True**/False—Over time, mulching can improve soil structure, reduce compaction, and add organic matter to the soil.

10. A pH of 5 is **100** times more acidic than a pH of 7.

11. The process in which ions of essential elements wash down through the soil profile and are lost is called **leaching**.

12. Many essential elements are dissolved in soil water in the form of positively charged particles called **cations**.

13. Soils with excess levels of soluble salts are called **saline** soils. **Sodic** soils are soils in which the cation sodium (Na+) occupies an unusually high percentage of the CEC.

14. The buffering capacity is the resistance of a soil to changes in pH. Clay soils and soils high in organic matter usually have a **high** buffering capacity.

15. The **rhizosphere** is the zone of intense biological activity near the actively elongating roots.

16. The diversity of organisms living, moving, and interacting in the soil is often referred to as the **soil food web**.

17. Water that drains from the macropores is called **gravitational** water. Following drainage, the soil is said to be at **field capacity**.

18. True/**False**—Most soil organisms cause disease or decay in tree roots.

19. **True**/False—Many tree roots exist in a symbiotic relationship with fungi that assist the tree in water and mineral absorption.

20. Compaction problems around established trees can be improved using an **air-excavation device** that breaks up soil with high-pressure compressed air.

Matching

__F__ sand   A.   "fungus roots"

__J__ buffering capacity   B.   measure of acidity or alkalinity

__H__	field capacity	C.	fine-textured soil particles
__I__	rhizosphere	D.	water that drains from the macropores
__G__	macropores	E.	ability of a soil to attract and hold cations
__A__	mycorrhizae	F.	coarse-textured soil particles
__E__	CEC	G.	tend to be air filled
__C__	clay	H.	soil after gravitational water has drained
__B__	pH	I.	soil zone immediately surrounding roots
__K__	micropores	J.	resistance to change in pH
__D__	gravitational water	K.	tend to be water filled

Sample Test Questions

1. The primary reason that most fine, absorbing roots are typically found near the soil surface is that

    a. roots need both air and water
    b. roots need UV light to drive respiration
    c. the pH of the soil is generally higher near the surface
    d. phosphorus and potassium are more available

2. Microorganisms tend to congregate in the rhizosphere, in part, because

    a. sugar exudates from root tips are a source of food
    b. mycorrhizae fix nitrogen and make it available
    c. root hairs tend to collect cations essential to microorganism growth
    d. bacteria preferentially feed on the meristem tissue at the root tips

3. If a planting hole in a clay soil site is backfilled with sandy soil,

    a. drainage will be improved, helping the tree to establish
    b. nutrients will be more available to the newly established roots within the planting hole
    c. water will drain very slowly out of the planting hole
    d. the improved texture of the backfill will reduce the chances of girdling roots forming later

4. When soil is compacted,

    a. micropores combine to form macropores
    b. soil particles are broken up, giving the soil a finer texture
    c. a high water content will reduce the damaging effects
    d. total pore space and the percentage of macropores are reduced

5. A characteristic of sandy soils in arid regions is that they

   a. **tend to become alkaline, and salts build up due to the lack of heavy rainfall**
   b. tend to become acidic because basic ions leach out
   c. are fine in texture due to the high sand content
   d. have a high water-holding capacity because rainfall is scarce

CHAPTER 4 ANSWERS TO WORKBOOK QUESTIONS

Workbook Questions

1. **True**/False—Infrequent, deep soakings of established trees and shrubs are preferable to frequent, shallow watering.

2. The **infiltration** rate is the rate at which water soaks into the soil.

3. Clay soils generally have a greater **water-holding capacity** than sandy soils, but water percolates through clay soils more slowly.

4. A **water budget** is an estimate of how much water is available to a tree (supply) and how much water is required (demand).

5. Four irrigation factors that affect a water budget are **duration**, **frequency**, **distribution**, and **timing**.

6. **True**/False—In most landscapes, it is preferable to minimize irrigation methods that apply water directly to the foliage of plants.

7. List three plant adaptations for surviving low water availability.

   a. **deep root systems**
   b. **shedding leaves**
   c. **leaf adaptations (small, thick cuticle, leaf hairs)**

8. Name two advantages and two disadvantages of drip irrigation.

   Advantages
   a. **reduces water waste**
   b. **reduces surface compaction**

   Disadvantages
   a. **must be moved outward as root system expands**
   b. **may become clogged**

9. **Minimum irrigation** is designed to maintain plants during periods of reduced rainfall, supplying only enough water to maintain a desired plant quality.

Answers to Workbook Questions

10. In regions where rainfall is insufficient to support landscapes, **recycled/reclaimed** water is becoming an important source for irrigation.

11. A measure of the rate of water use by plants and evaporation from soil is known as the **evapotranspiration** rate.

12. List five benefits of using mulches around trees.

    a. moderate temperature extremes
    b. reduce soil compaction
    c. reduce competition from weeds, grasses
    d. keep mowers away
    e. reduce soil moisture loss
    f. increase organic matter (organic mulch)
    g. aesthetically pleasing

13. Soil probes, **tensiometers**, and electronic moisture sensors are tools used to monitor soil wetness or dryness.

14. **True**/False—If a site is not already developed, adjusting the grade is the first choice to avoid drainage problems.

15. Name three tree health problems that may be associated with flooding.

    a. lack of oxygen in the soil suffocates roots
    b. changes the chemical composition of essential elements
    c. can lead to fermentation in root cells and buildup of toxic compounds
    d. can create mineral toxicities
    e. can harm the living components of the soil ecosystem

Sample Test Questions

1. When irrigating trees,

    a. infrequent, deep soakings are preferable to frequent, shallow watering
    b. the most beneficial and efficient time to water is midafternoon at peak sunlight
    c. the foliage should be kept wet at night to reduce transpiration
    d. keeping the soil moist at the root flare reduces girdling root formation

2. Sandy soils

    a. have a greater water-holding capacity than clay soils
    b. have higher infiltration rates than clay soils
    c. do not ever reach field capacity, because drainage is good
    d. have lower percolation rates than clay soils

3. A soil is at field capacity when

    a. it is completely saturated
    b. the permanent wilting point has been reached
    c. gravitational water has drained away
    d. there is no water available to the roots

4. Which of the following is true about irrigating with recycled (reclaimed) water?

    a. Recycled water rarely contains essential minerals.
    b. Recycled water typically lowers the soil pH.
    c. Recycled water is typically high in salts.
    d. Recycled water is typically chemical free.

5. Which of the following is a problem associated with flooding?

    a. lack of oxygen in the soil leading to root suffocation
    b. changes the chemical composition of essential elements
    c. fermentation in root cells and buildup of toxic compounds
    d. all of the above

CHAPTER 5 ANSWERS TO WORKBOOK QUESTIONS

Workbook Questions

1. Trees take up essential elements, dissolved in **water**, through the roots.

2. **Macronutrients** are elements required by trees in relatively large quantities.

3. The macronutrient **nitrogen** is a constituent of chlorophyll and, if deficient, can cause reduced growth and chlorosis.

4. **Organic** fertilizers release nutrients over an extended period of time and can be a source of carbon.

5. The **fertilizer analysis**, listed on the container, gives the relative percentage of nitrogen, phosphorus, and potassium.

6. A 23 kg (50 lb) bag of 20-10-5 fertilizer contains **4.6** kg (or **10** lb) of actual nitrogen.

7. If fertilizer "burn" or leaching are potential problems, it may be desirable to use a **slow-release** fertilizer.

8. The most important factor for good uptake of fertilizer elements is adequate **water**.

386 Copyright © 2022 International Society of Arboriculture

Answers to Workbook Questions

9. **True**/False—Surface application of fertilizer is relatively inexpensive and makes the fertilizer available in the upper few centimeters (inches) of soil.

10. **True**/False—Fertilization can increase susceptibility to certain pests.

11. True/**False**—An advantage to foliar application of fertilizer is that it usually lasts up to five years.

12. Name two limitations to trunk implants and microinjections.

    a. wounding
    b. generally not suitable for macronutrients
    c. limitation for repeat applications

13. **Leaching** is the washing out of chemicals down through the soil.

14. Fertilization recommendations should be based on **soil** and **foliar** analyses and assessment of the tree and site.

15. Studies show that nitrogen fertilization can trigger a tree's energy allocation toward growth, sometimes at the expense of **defense (and other plant processes)**.

Matching

D	micronutrients	A	element needed in the largest quantity
F	10-0-4	B.	provides pH, CEC, nutrient information
E	leaching	C.	nitrogen, phosphorus, potassium
C	macronutrients	D.	iron, manganese, boron
A	nitrogen	E.	washing down through the soil
B	soil analysis	F.	fertilizer analysis

Sample Test Questions

1. A 36 kg (80 lb) bag of 10-6-4 fertilizer contains how many kilograms (pounds) of actual nitrogen?

    a. 1.8 kg (4 lb)
    b. 2.7 kg (6 lb)
    c. 3.6 kg (8 lb)
    d. 4.5 kg (10 lb)

2. A complete fertilizer contains

    a. all 16 essential elements
    b. nitrogen, phosphorus, and potassium
    c. organic and inorganic nitrogen
    d. equal amounts of N, P, and K

3. A tree may not respond immediately to fertilizer application if

    a. a slow-release fertilizer was applied
    b. there is inadequate soil moisture
    c. the tree is not actively growing
    d. all of the above

4. A soil test may not identify a nutrient deficiency problem in a plant because

    a. the tests are not reliable
    b. the nutrient content can change after collecting
    c. the soil may contain adequate nutrients, but something may be inhibiting uptake
    d. no one knows which levels of nutrients in soils are adequate

5. Which of the following statements about soil additives is true?

    a. Addition of organic matter is always beneficial to plant growth and health.
    b. Mycorrhizal inoculants generally show beneficial effects when added to most established landscape soils.
    c. Biochar has not been shown to have any beneficial effects for plants.
    d. Some composts and other sources of organic matter can contain heavy metals.

CHAPTER 6 ANSWERS TO WORKBOOK QUESTIONS

Workbook Questions

1. Name five tree species that would *not* be appropriate for planting under utility wires.

    any species that grows higher than the wires at maturity would be a correct answer

2. **Hardiness** is the ability of a tree to withstand low temperatures and winter stresses in a given site.

3. **True**/False—Although a tree may be considered hardy in a given area, it may decline or die if the roots are unprotected.

4. Name three site characteristics that must be considered in tree selection.

   Potential answers include the following:
   a. growing space
   b. light conditions
   c. soil conditions
   d. climate
   e. functional requirements

5. Upright, pyramidal, and weeping are three examples of tree **growth habit (form)** that are important in selection.

6. If a particular disease is known to be a problem, a tree species or cultivar should be selected that has **resistance** to that disease.

7. Name three plant characteristics that may make a tree aesthetically desirable.

   Potential answers include the following:
   a. flowers
   b. attractiveness to birds
   c. fall color
   d. exfoliating bark
   e. growth habit

8. **Acclimation** is the gradual process by which a tree adapts to changes in its environment.

9. True/**False**—A tree listed as adaptable to wet soil conditions will always thrive if planted in those conditions.

10. Name five characteristics to consider when selecting a tree in the nursery.

    Potential answers include the following:
    a. light-colored, healthy roots
    b. solid root mass
    c. good twig extension growth in previous years
    d. no major scars or injuries
    e. no insect or disease problems
    f. good branch structure

Sample Test Questions

1. The primary climatic factor that determines hardiness zones is

   a. north-south location
   b. temperature, rainfall, and winds
   c. east-west location
   d. low temperature extremes

2. Trees to be planted under utility lines should be

   a. tolerant of heavy top pruning
   b. low growing to remain below the lines
   c. vase shaped or overarching to clear conductors
   d. all of the above

3. Some trees acclimate to shade conditions by

   a. developing larger, thinner leaves
   b. developing smaller, thicker leaves
   c. producing stomata mostly on the upper leaf surfaces
   d. developing variegated foliage

4. Fastigiate trees have a growth form that is

   a. upright
   b. weeping
   c. overarching
   d. vase shaped

5. Which of the following is a true statement?

   a. Floodplain species will always grow well in wet soils.
   b. Trees adapted to grow in full shade will perform well in full sun.
   c. Some tree species are adapted to hot, dry, or bright light conditions with small, thick foliage and sunken stomata.
   d. Most evergreen conifers are very shade tolerant and tend to scorch in full sunlight.

CHAPTER 7 ANSWERS TO WORKBOOK QUESTIONS

Workbook Questions

1. Trees are generally available from the nursery in one of four forms:

   a. bare root
   b. containerized
   c. container-grown
   d. balled and burlapped (B&B)

2. Bare-root trees are normally planted when **<u>dormant</u>**, before buds begin to grow.

Answers to Workbook Questions

3. **Girdling** roots can become a problem because they can constrict the vascular system in the trunk or in other roots.

4. Planting holes should be dug **two** to **three** times the width of the root ball at the surface, with the sides sloping down to the diameter at the base of the root ball.

5. Trees that are dug in the nursery are often wrapped with **burlap** to help keep the root ball intact and to reduce exposure of the roots to air.

6. The planting hole should never be **deeper** than the root ball.

7. **True**/False—Planting a tree too deeply can stress, drown, or suffocate roots; it may also enable soilborne pathogens to enter the trunk.

8. **True**/False—Research has shown that soil amendments used in the planting hole generally do not promote establishment and growth.

9. **True**/False—Digging a tree for transplanting can remove as much as 90 percent of the absorbing roots.

10. In temperate climates, the best times to transplant most trees are **spring** and **autumn**.

11. **True**/False—When a tree is dug for transplanting, the size of the root ball is usually based on tree caliper.

12. True/**False**—If trees have wire baskets to help maintain the integrity of the root ball, these baskets should never be removed, nor should the top portion be cut off at planting.

13. **True**/False—Adding fill to the bottom of the hole can cause the root ball to settle in the days and weeks after planting, resulting in a final depth that is deeper than originally planted.

14. Predigging around a tree to create a more densely rooted ball is called **root pruning**.

15. **True**/False—Staking of newly planted trees is not always necessary.

16. Name three adverse effects of staking or guying trees.

    a. **produce less trunk taper**
    b. **develop smaller root systems**
    c. **more subject to breaking or tipping after stakes are removed**

17. **True**/False—The material used to attach the tree to the stake should be broad, smooth, and flexible.

18. Warm soil temperatures and adequate soil **moisture** are the optimal conditions for new root growth.

19. True/**False**—In general, palms are more difficult to transplant than most other kinds of trees.

20. When transplanting palms, the fronds are often tied up to protect the solitary **terminal bud**.

21. Transplant shock is mainly due to **water** stress from the greatly reduced root system.

22. If fertilizer is applied at planting, it should be a **slow-release** type to avoid excess salt buildup in the root zone.

23. **True**/False—There is no advantage to pruning one-third of the tree crown at the time of planting (i.e., compensatory pruning).

24. The most important maintenance factor in the survival of a newly planted tree is proper **watering**.

25. **True**/False—Tree roots may suffocate if the tree receives too much water after planting.

Sample Test Questions

1. Staking or guying when planting a tree is
   a. done only for bare-root trees
   b. not necessary for trees greater than 15 cm (6 in) in diameter
   c. not always required or necessary
   d. for promoting a larger and stronger root system and better trunk taper

2. When mulching a newly planted tree, care should be taken to ensure that
   a. the mulch extends all the way to the base of the tree trunk
   b. the depth is not excessive to the point where soil aeration is negatively impacted
   c. the mulch is sourced only from quality, inorganic sources
   d. an underlying layer of plastic is laid first to prevent weed growth

3. Planting a row of palms at various depths to create a row of uniform height
   a. will not be harmful as long as the soil is sandy and well drained
   b. can lead to death of the deeply planted trees
   c. is acceptable due to the deep rooting of palms
   d. is recommended to achieve deep rooting in sandy soils

4. When planting a container-grown tree,
   a. slice or shave off the outer roots if they are circling or matted
   b. place soft fill in the bottom of the planting hole to encourage taproot growth
   c. backfill the hole with a soilless growth medium to encourage root growth
   d. plant the tree slightly deeper in the planting hole than ground level

5. The most important reason to prune a tree when transplanting is to

   a. compensate for root loss
   b. invigorate the tree
   c. reduce growth at the tips
   d. **remove damaged branches and improve structure**

CHAPTER 8 ANSWERS TO WORKBOOK QUESTIONS

Workbook Questions

1. **True**/False—Poor pruning practices can shorten the life of a tree and can increase the likelihood of tree or structural failure.

2. Name five objectives for pruning trees.

   a. reducing risk
   b. providing clearance
   c. improving or restoring structure
   d. managing size
   e. improving aesthetics

3. The majority of pruning by most arborists follows a **natural** pruning system.

4. If two branches (stems) develop at the tip of the same stem, they may form **codominant stems (codominant branches)**.

5. **Included bark** is bark that becomes enclosed inside the attachment as the two branches grow and develop.

6. When training young trees, a dominant leader should be selected and competing leaders should be removed or **subordinated**.

7. When practical, temporary lower branches should be left on a young tree to help develop trunk **taper**.

8. Removing too much foliage can have adverse effects on the tree, resulting in production of **watersprouts** on interior branches, which is often a tree's reaction to overpruning.

9. Label the branch bark ridge and the branch collar on this drawing. Show where the undercut, top cut, and final cut should be made in removing the branch.

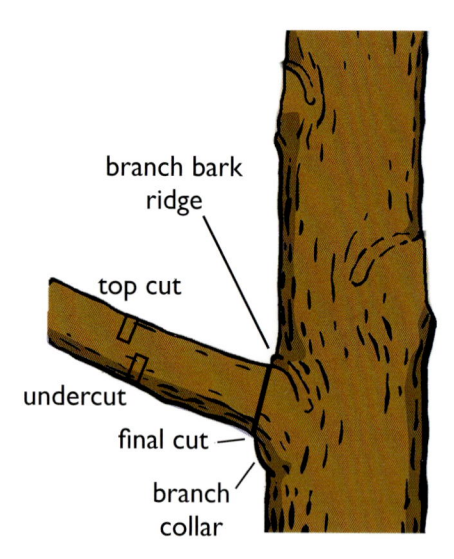

[Diagram labeled: branch bark ridge, top cut, undercut, final cut, branch collar]

10. The swollen area at the base of a branch where it arises from the trunk is called the **branch collar**.

11. **True**/False—In the absence of included bark, the size of a branch in relation to the trunk is more important for branch attachment strength than is the attachment angle.

12. A **branch removal** cut removes the smaller of two branches at a union with the parent stem.

13. A **reduction** cut removes the larger of two or more branches, stems, or codominant stems to a live lateral branch or stem.

14. True/**False**—Tree wound dressings accelerate wound closure and prevent decay.

15. When pruning palms, if older, live fronds must be removed for clearance, avoid removing fronds that initiate above **horizontal (90°)**.

16. Name three negative effects of topping a tree.

    **a. reduction of photosynthetic capacity
    b. creating many entry points for insects and disease
    c. development of weakly attached branches
    d. creating a disfigured crown/loss of natural form**

17. **True**/False—Arborists should be aware that removing or disturbing active bird nests, especially of protected species, is illegal in many locations.

18. True/**False**—Trees that tend to "bleed" should never be pruned in the early spring, because doing so is likely to cause a major decline in vigor.

19. **True**/False—As a general rule, mature trees are less tolerant of heavy pruning than juvenile trees.

20. **Plant growth regulators** are substances, usually effective in small quantities, that enhance or alter the growth and development of a plant.

Sample Test Questions

1. When pruning young trees, it is important to train for a dominant leader and well-spaced scaffold branches to

    a. minimize the need for future pruning
    b. develop a structurally sound crown
    c. minimize codominant branching on the trunk
    d. all of the above

2. To prune trees that flower on the previous year's growth and to maximize flowering, you should prune

    a. anytime during the dormant season
    b. shortly after flowering
    c. in late summer after seed formation
    d. in the fall, just after leaf drop

3. When pruning a branch from a tree, the final cut should be

    a. flush with the parent stem
    b. at a 45° angle to the parent stem
    c. parallel to the branch bark ridge
    d. just outside the branch collar

4. When it comes to pruning, as a rule, mature trees are

    a. more tolerant of extreme pruning than young trees
    b. capable of tolerating heading better than young trees
    c. not as tolerant of severe pruning as young trees
    d. less likely to produce watersprouts than young trees

5. If the height of a tree must be reduced,

    a. branches should be cut to a lateral one-third or more the diameter of the branch removed
    b. all cuts should be made at internodes to avoid cutting through buds
    c. the tree should be root pruned to compensate for foliage loss
    d. pruning should only take place during the dormant season

CHAPTER 9 ANSWERS TO WORKBOOK QUESTIONS

Workbook Questions

1. Common-grade cable is relatively malleable (bendable) and easy to work with. **Extra-high-strength** cable is much stronger but less flexible than common-grade cables.

2. One advantage of **wire rope** is that it's both strong and flexible. A potential drawback is the limitation of choices for attachment.

3. As a general guideline, cables to support codominant stems should be installed **two-thirds** the distance from the weak branch union to the top of the tree, as long as the wood is solid and large enough to install the hardware.

4. Branches may be brought closer together while installing the cable so that when released, the cable will be just **taut**.

5. **True**/False—Attachment hardware should be installed with the cable's pull in direct line with the attachment.

6. **True**/False—When more than one cable is installed on the same branch, the hardware should be spaced at least as far apart as the diameter of the branch.

7. True/**False**—When installing multiple cables, use common anchors whenever possible.

8. Extra-high-strength cable should be attached to hardware using **dead-end grips**.

9. True/**False**—The installation of steel cables, if done properly, will not wound the tree.

10. **True**/False—Rope support systems may reduce the potential for shock loading the system, which can occur if two stems move in opposite directions with great force.

11. **True**/False—An advantage of most rope support systems is that they do not require drilling the tree to install.

12. True/**False**—If a single rod is being used to support a branch union that is not split, it should be installed just below the branch union.

13. **Guying** is the installation of a cable between a tree and an external anchor to provide supplemental support and reduce tree movement.

14. Rigid structures mounted or built on the ground to support a branch or trunk are called **props**.

15. Name three circumstances in which lightning protection for trees might be recommended.

   a. historic trees
   b. valuable trees
   c. large trees where people may seek refuge in a storm, such as on a golf course

Sample Test Questions

1. An advantage of the amon-eye system over the use of an eye bolt is

   a. no washers are needed on the terminations
   b. the length of the rod can be adjusted
   c. they are considered stronger than eye bolts
   d. the eye bolt is not drop forged

2. The purpose of a lightning protection system is to

   a. reduce the voltage of the strike
   b. prevent the tree from being struck
   c. conduct the electrical charge into the soil away from the tree
   d. all of the above

3. If two bracing rods are installed to support a weak union, they should be placed

   a. no more than 10 cm (4 in) apart
   b. in vertical alignment, one above the other
   c. staggered and no closer together than the diameter of the trunk
   d. below the union and perpendicular to one another

4. To install a cable directly across a branch union, install it perpendicular to

   a. an imaginary line that bisects the union
   b. the main stem or larger branch
   c. the branch being supported or the smaller branch
   d. both stems arising from the union

5. An advantage of rope support systems is

   a. the cables are more durable than steel cables
   b. they don't require drilling for installation
   c. they last a very long time
   d. they don't require follow-up inspections

CHAPTER 10 ANSWERS TO WORKBOOK QUESTIONS

Workbook Questions

1. True/**False**—Information about a tree's history and symptoms gained from a client can always be considered accurate.

2. If a tree is not well suited for the site in which it has been planted, it may become **stressed**, predisposing it to other problems.

3. A common mistake in diagnosis is to carefully examine the aboveground portion of the tree while ignoring the **roots**.

4. **True**/False—If a tree declines or dies within the first year following installation, a likely cause is excess or insufficient water.

5. Pollution damage, girdling roots, and mineral deficiencies are examples of **abiotic** disorders.

6. Name five causes of physical or mechanical injuries to trees.

    a. **lawn mower/string trimmer**
    b. **vandalism**
    c. **construction**
    d. **rodents**
    e. **guy wires**

7. Insect damage to trees is usually the result of feeding or **ovipositing (egg laying)**.

8. Name five insect pests of trees with chewing mouthparts. Name five with piercing-sucking mouthparts.

Chewing	Piercing-Sucking
a. **beetles**	a. **aphids, adelgids**
b. **caterpillars**	b. **scales**
c. **weevils**	c. **leafhoppers**
d. **leafminers**	d. **mealybugs**
e. **borers**	e. **true bugs**
f. **webworms**	f. **psyllids**

9. Insects that carry plant pathogens and introduce them into hosts to result in disease are known as **vectors**.

10. **True**/False—Mites are not insects.

398

Answers to Workbook Questions

11. Microscopic wormlike organisms that sometimes feed on or in trees, causing disease, are called **nematodes**.

12. Name the four factors required for a tree disease.

    a. **susceptible host**
    b. **pathogenic organism**
    c. **suitable environment**
    d. **proper timing**

13. True/**False**—Vascular diseases of trees are rarely fatal.

14. **True**/False—Diseases that affect only the foliage of a deciduous tree may not be a serious problem unless defoliation occurs in several consecutive years.

15. True/**False**—Most fungi cause plant disease.

16. **True**/False—The pathogens that cause plant diseases are primarily fungi.

17. Fire blight is an example of a disease caused by a **bacterium**.

18. **Allelopathy** is the chemical inhibition of growth and development of one plant by another.

19. **True**/False—Pollution damage is often difficult to diagnose because the symptoms may mimic other problems such as insect injury and mineral deficiencies.

20. Curling and cupping of the foliage, and parallel venation, are common symptoms of **herbicide** damage.

Matching

D	witch's broom	A.	abnormal, enlarged plant structure, often insect or mite induced
B	vector	B.	carrier of pathogens
C	canker	C.	localized dead tissue, often shrunken and discolored
A	gall	D.	abnormal growth of multiple shoots
H	stunting	E.	may predispose a plant to other problems
E	stress	F.	causal agent of disease
F	pathogen	G.	natural chemical inhibition of growth
I	dieback	H.	abnormally reduced growth
G	allelopathy	I.	progressive death of twigs and branches from the tips back

Sample Test Questions

1. A condition characterized by a cluster of dwarfed shoots on affected twigs is called

    a. witch's broom
    b. anthracnose
    c. chlorosis
    d. Verticillium wilt

2. Twig dieback from periodical cicadas is primarily a result of

    a. ovipositing (egg laying)
    b. adults feeding on the foliage
    c. larvae feeding on the roots
    d. feeding-induced galls on the twigs and foliage

3. Plant damage associated with a sap-feeding insect pest might appear as

    a. leaves that have been skeletonized
    b. distorted leaves or shoots
    c. leaf mines or blotches
    d. webs or tents in the tree

4. Scale damage to plants is the result of

    a. fungal spore growth depleting xylem reserves
    b. a type of sucking insect causing a loss of vigor
    c. vascular damage from fungal invasion
    d. a physiological disorder due neither to insects nor to disease

5. Damage caused by nonliving factors tends to be

    a. uniform and may affect more than one species
    b. uniform but generally not affecting the new growth
    c. random and concentrated on the new growth
    d. random with irregular borders

CHAPTER 11 ANSWERS TO WORKBOOK QUESTIONS

Workbook Questions

1. True/**False**—Plant Health Care and Integrated Pest Management are essentially the same thing.

2. Carbohydrates, produced through photosynthesis, are allocated to these primary functions:

   a. **growth**
   b. **maintenance**
   c. **reproduction**
   d. **storage**
   e. **defense**

3. **Stress** factors are often directly related to soil quality or other environmental conditions (for example, pH, drought, compaction, or excess water), and many can be attributed to human activity such as land development and construction.

4. The **cellulose** and **lignin** in tree cells are indigestible to many insects and other animals and even to some pathogens.

5. Trees produce **allelochemicals** such as tannins and other phenols that have toxic or deterrent effects on certain insects.

6. **True**/False—Rapidly growing trees may be less resistant to certain insects and diseases.

7. **Monitoring** is the process of observing, identifying, recording, and analyzing what happens with plants in the landscape.

8. The process of gathering information, assessing the severity and implications of the problem, determining client expectations, and deciding on a course of action is called the **appropriate response process**.

9. **Integrated Pest Management** is a systematic approach to insect and disease management that incorporates a combination of techniques including resistant plants as well as cultural, biological, and chemical control tactics.

10. **True**/False—A simple degree-day model uses an established threshold temperature and the daily average temperature to predict pest development stages.

11. When possible, arborists should select trees that are **resistant** to known insects or diseases.

12. **Key pests** are organisms that are frequently encountered in landscapes, predictably cause injury to landscape plants, and may include particularly noxious pests in the area.

13. Extensive plantings of the same species, known as **monocultures**, can have catastrophic consequences if a fatal insect pest or disease is introduced.

14. Plant Health Care practitioners often choose from three pest management goals: **prevention**, **eradication**, and **suppression**.

15. **True**/False—Chemical pesticides often kill the targeted pest within minutes or hours of application, whereas biological control can take days or weeks to suppress a pest population.

16. **Systemic** pesticides are taken up by the plant and translocated throughout the branches and into the leaves.

17. **Pest resurgence** occurs when the pest population rapidly rebounds in the absence of natural enemies, which are slower to repopulate than the pest.

18. True/**False**—The use of multiple pesticides with different active ingredients or modes of action in a rotation system will increase the incidence of pesticide resistance.

19. **True**/False—Insecticidal soaps disrupt the cell membranes of soft-bodied insects and are effective on some scales, aphids, mealybugs, and spider mites.

20. True/**False**—Horticultural oil applications are always safe to use on trees in leaf because they have no phytotoxic properties.

21. **Insect growth regulators** are synthetic compounds that act like insect hormones.

22. **Microbial pesticides** are derived from certain bacterial pathogens of insects.

23. Products that contain proteins of ***Bacillus thuringiensis*** (Bt) are examples of microbial pesticides that utilize insect pathogens or lethal microbial byproducts derived from extracts of bacterial pathogens of insects.

24. The biological control strategy is based on the concept that many insect pests live in a natural, dynamic balance with **predators**, **parasites**, and **pathogens** that control pest populations.

25. **True**/False—Plant Health Care practitioners should identify short- and/or long-term stress factors and remediate them using appropriate management techniques.

Sample Test Questions

1. Plant Health Care is a comprehensive program to manage

   a. insects and diseases of plants
   b. tree health without the use of pesticides
   c. the appearance, structure, and health of plants
   d. pests, pathogens, and abiotic disorders of trees

2. The mortality spiral describes the

   a. process of infection and spread of disease in a tree
   b. cumulative effects of stress causing decline of a plant over time
   c. process in which pesticides eradicate both pests and beneficial insects
   d. allocation of resources among growth, storage, and defense

3. The process of gathering information, assessing the severity and implications of the problem, determining client expectations, and deciding on a course of action is called

   a. the appropriate response process
   b. Integrated Pest Management
   c. the cultural control mechanism
   d. Plant Health Care

4. A systemic pesticide is one that

   a. kills all living organisms
   b. kills insects on direct contact
   c. is translocated throughout the plant
   d. has no harmful effect on the environment

5. Releasing predators or parasites of an insect pest is an example of a

   a. cultural control
   b. mechanical control
   c. biological control
   d. chemical control

CHAPTER 12 ANSWERS TO WORKBOOK QUESTIONS

Workbook Questions

1. List the two components of risk.

   a. likelihood
   b. consequences

2. In tree risk assessment, likelihood is a combination of **likelihood of impact** and **likelihood of failure**.

3. List the three levels of assessment.

   a. **Level 1**: **Limited Visual Assessment**
   b. **Level 2**: **Basic Assessment**
   c. **Level 3**: **Advanced Assessment**

4. **Targets** are people, property, or activities that could be injured, damaged, or disrupted by a tree failure. The **target zone** is the area that the tree or branch is likely to hit if it fails.

5. Targets can be categorized by the amount of time that they are within the target zone—their **occupancy rate**.

6. Without a stated **time frame**, the rating for likelihood of failure is meaningless.

7. List at least three factors to consider when assessing the likelihood of impact.

   a. occupancy rate
   b. direction of fall
   c. location in the target zone
   d. protection factors

8. List at least five factors to consider when assessing the likelihood of failure.

   a. site conditions/changes
   b. species failure profile
   c. loads (wind/weather)
   d. structural defects and conditions
   e. tree health
   f. response growth

9. List at least three factors to consider when assessing the consequences of failure.

   a. target value/importance
   b. tree or tree part size
   c. distance of fall
   d. protection factors

10. List at least five conditions that could increase the likelihood of failure.

    a. wood decay
    b. cracks
    c. dead branches
    d. broken branches
    e. weakly attached branches
    f. codominant stems with included bark
    g. damaged or cut roots
    h. leans/unusual tree architecture

Answers to Workbook Questions

11. **White rot** fungi primarily decay the lignin within and between cell walls in the wood, reducing the wood's stiffness but leaving some flexibility.

12. **Brown rot** gets its name because, after the cellulose is decayed, the remaining lignin is dark or brown in color.

13. **True**/False—A tree may appear to be solid and structurally sound or may have a thick, green crown yet can have significant decay inside.

14. List three definite indicators of decay and three potential indicators of decay in a tree.

    Definite indicators
    a. **cavities**
    b. **fruiting bodies on the wood**
    c. **carpenter ants nesting inside**

    Potential indicators
    a. **cracks/seams/bulges**
    b. **old pruning wounds**
    c. **cavity nesting birds or bees**

15. **Response growth** is new wood produced in response to damage or loads.

16. **True**/False—As trees grow and develop, they adapt to the various loads that they experience (gravity and wind) by developing wood where it is needed to support the loads.

17. True/**False**—Woundwood is less dense than and chemically different from other wood, and it resists decay better than normal wood.

18. Two advanced assessment devices/techniques of assessing internal decay are **resistance-recording drill** and **sonic tomography**.

19. List three target-based and three tree-based options for mitigating tree risk.

    Target-based options
    a. **restrict access to the target**
    b. **move the target**
    c. **reroute traffic**

    Tree-based options
    a. **pruning**
    b. **installing structural support**
    c. **removal**

20. **True**/False—Each combination of a failure mode and a target represents a separate risk to analyze.

21. True/**False**—The overall tree risk is the risk of whole tree failure.

22. **Acceptable risk** is the degree of risk that is within the owner's/manager's or controlling authority's tolerance, or that which is below a defined threshold.

23. The risk remaining after mitigation is the **residual risk**.

24. **True**/False—Once the highest risk factor has been mitigated, the tree risk rating goes to the next highest risk factor.

25. True/**False**—The tree risk assessor bears the responsibility for tree risk management.

Sample Test Questions

1. Which of the following is *not* a type of Level 1 assessment?

   a. climbing the tree
   b. walk-by
   c. drive-by
   d. aerial patrol

2. Which type of decay primarily affects the lignin within and between cell walls in the wood, reducing the tree's strength under compression?

   a. brown rot
   b. white rot
   c. soft rot
   d. sapwood rot

3. Following construction, forest trees on the edge of remaining stands are prone to failure due to

   a. losing the protection of the trees that used to surround them
   b. less trunk stability and poor taper
   c. increased exposure to the weather elements
   d. all of the above

4. Trees that lean because of ground failure or root injury

   a. have a high potential to fail
   b. are less of a risk than those that lean due to phototropism
   c. are not a threat unless located at the edge of a wooded area
   d. are a risk only if they begin to grow in compensation for the lean

5. Which of the following is the responsibility of a tree risk assessor?

   a. determining acceptable risk
   b. presenting mitigation options
   c. prioritizing work
   d. choosing among mitigation options

CHAPTER 13 ANSWERS TO WORKBOOK QUESTIONS

Workbook Questions

1. Name five ways that trees can be adversely affected by construction.

   a. **root injury**
   b. **soil compaction**
   c. **injury to trunk or branches**
   d. **grade change (smothering, altering hydrology)**
   e. **excavation/severing root system**
   f. **injury from chemicals**
   g. **increased exposure to wind and sun**

2. **True**/False—Evaluating suitability of individual trees or groups of trees for preservation is an important task for the arborist.

3. True/**False**—The goal of an arborist involved in a development project is to save every tree on the site.

4. **True**/False—The largest, most mature trees are not always the best candidates for preservation.

5. The **critical root zone** is the area around a tree where the minimum amount of roots that are biologically essential to the structural stability and health of the tree are located.

6. The **tree protection zone** is an area defined during site development, where construction activities and access are limited to protect the tree(s) and soil from damage, and to sustain tree health and stability.

7. **True**/False—Less injury is caused by tunneling directly under a tree than by cutting directly across the root system of a tree when excavating for utility lines.

8. True/**False**—Preferably, the critical root zone will be much larger than the tree protection zone.

9. **True**/False—The purpose of the barriers, limitations, and specified work zones should be clearly communicated to each person on the jobsite.

10. True/**False**—It is easier for an arborist to treat trees that have been damaged by construction than to prevent the damage.

11. Written **specifications** should detail exactly what can and cannot be done to and around the trees.

12. Use of an **air-excavation device** has proven effective for soil aeration and radial trenching, causing much less root injury than mechanical excavation equipment.

13. True/**False**—Small-diameter wells built around the trunks of trees are usually adequate to protect the tree and ensure survival.

14. **True**/False—If roots must be severed, they should be cut cleanly with sharp tools and prevented from drying out.

15. In **radial trenching**, trenches are made in a radial pattern throughout the root zone and should extend at least as far as the drip line.

Sample Test Questions

1. Measures to reduce compaction on building sites are not always an option, because
   a. there is no way to effectively reduce compaction of soils with high clay content
   b. if the site has a high water table, compaction reduction efforts will be ineffective
   c. projects must comply with specific engineering standards regarding soil compaction
   d. soil structure and aggregate types prevent changes to bulk density

2. Arborists should be involved early in the construction planning process because
   a. tree preservation measures should be incorporated into the project specifications
   b. once construction has begun, it may be too late to save the trees
   c. there is often little arborists can do to treat construction damage
   d. all of the above

3. A measure that can be taken to minimize compaction on a construction site is
   a. watering the site thoroughly before equipment is brought in
   b. permanently raising the soil grade to protect tree roots
   c. spreading a temporary, thick layer of mulch over the site
   d. root pruning the trees in advance

4. A common strategy for tree preservation that can retain more trees and promote sustainability is
   a. retaining groups of trees with a shared root space and a protected perimeter
   b. preserving all of the largest, most mature trees on the site
   c. preserving only the youngest trees on the site
   d. selecting for preservation only the species known to tolerate soil compaction

5. What flexible guideline for a multiple of tree diameter is commonly used when determining where to establish a tree protection zone?
   a. 3 to 5
   b. 5 to 10
   c. 6 to 18
   d. 18 to 24

CHAPTER 14 ANSWERS TO WORKBOOK QUESTIONS

Workbook Questions

1. Urban forestry is the management of naturally occurring and planted trees and associated plants in urban areas. Whereas arboriculture focuses on the **trees**, urban forestry focuses on the **forest (population of trees)**.

2. List five professionals or groups that an urban forester should learn to communicate with.

    a. urban planners
    b. civil engineers
    c. public works officials
    d. government agencies
    e. the public (and many others)

3. **True**/False—The leaves and branches of trees catch and slow rainwater to reduce soil erosion from runoff.

4. **Carbon sequestration** occurs when trees absorb carbon from CO_2 in the atmosphere and store it in the form of wood and other carbon-based tissues.

5. List three economic benefits of trees.

    Any of the following:

    a. increase residential and business property values
    b. increase tax base for community
    c. attract visitors, businesses, and new residents
    d. increase rental of apartments and offices and reduce vacancy rates
    e. encourage shoppers to linger
    f. protect infrastructure materials that are degraded by heat, such as pavements

6. List three environmental benefits of trees.

    Any of the following:

    a. improve air quality
    b. sequester carbon, mitigate climate change
    c. conserve energy by protecting surroundings from sun and wind
    d. cool the air through transpiration
    e. reduce stormwater runoff and soil erosion
    f. provide habitat for wildlife

Answers to Workbook Questions

7. List three social and health benefits of trees.

    Any of the following:

    a. reduce stress and mental fatigue
    b. enhance mental health
    c. enhance recuperation rates
    d. reduce psychological precursors to crime
    e. enhance community pride
    f. heal and restore communities
    g. increase recreational opportunities
    h. serve as cultural or historic symbol

8. **Sustainability** is the ability to maintain ecological, social, and economic benefits over time.

9. List four types of data that are typically collected in a tree inventory.

    Any of the following:

    a. tree species
    b. diameter
    c. location
    d. tree condition
    e. maintenance history
    f. maintenance recommendations
    g. notes

10. A **risk management** policy statement should set out the policies for identifying, assessing, reporting, and mitigating risks.

11. **Tree ordinances (bylaws)** are legal regulations drafted and instituted to protect trees within a given jurisdiction.

12. List three component plans that are commonly part of an urban forest management plan.

    Any of the following:

    a. planting plan
    b. preservation plan
    c. maintenance and operations plan
    d. tree removal and replacement plan
    e. risk management plan
    f. storm-response/emergency plan
    g. public outreach and education plan
    h. community engagement plan

13. A **tree preservation order** is a legal regulation, established by the local authority, that protects a tree or multiple trees.

14. **True**/False—Even where standards do not carry direct legislative authority, they may be recognized in a court of law.

15. Urban foresters should establish detailed **specifications** for all tree work, including planting, pruning, fertilizing, pest control and monitoring, installation of support or protection systems, construction near trees, and removals.

16. The **management plan** is a document laying out how a municipality will balance the maintenance of its large population of trees within the common urban pressures and financial restraints of a municipality.

17. The urban forest provides **habitat** and food for a wide range of wildlife.

18. **True**/False—Deferring maintenance can lead to higher costs and risks and is likely to reduce the benefits trees provide.

19. True/**False**—Widespread planting of a single species is recommended to bring a uniform appearance to the urban forest.

20. **True**/False—Community engagement and communication are key parts of successful urban forestry programs and their urban forest management plans.

Sample Test Questions

1. A social benefit of trees and natural areas that has been identified through research is

   a. stress reduction from settings with trees
   b. faster healing of patients in hospitals
   c. crime reduction in communities
   d. all of the above

2. A commonly used set of methods for appraising trees was developed by the

   a. Council of Tree and Landscape Appraisers
   b. Society of Consulting Tree Workers
   c. Society of Urban Foresters
   d. Consortium of Landscape Professionals

3. A typical tree ordinance or bylaw will define the jurisdiction's authority and

   a. describe the conditions and requirements of the ordinance
   b. establish penalties for noncompliance
   c. specify the responsibility for enforcement
   d. all of the above

4. Detailed plans, requirements, and statements of particular procedures and/or standards used to define and guide are called

   a. laws
   b. best management practices
   c. specifications
   d. ordinances

5. A problem associated with overplanting of a single species or a few species is

   a. all of the trees maturing and dying within a short period of time
   b. increased biodiversity of associated pest populations
   c. unsustainable management due to uniform and consistent maintenance needs
   d. the risk of catastrophic loss due to an insect or disease outbreak

CHAPTER 15 ANSWERS TO WORKBOOK QUESTIONS

Workbook Questions

1. As used in many standards and regulations, the term **shall** denotes a mandatory requirement, and the term **should** denotes an advisory recommendation.

2. Name four pieces of personal protective equipment (PPE) that are generally required for all tree workers.

   a. head protection
   b. eye protection
   c. hearing protection, when applicable
   d. leg protection, with chain saws

3. **True**/False—All communications wires and cables shall be considered to be energized with potentially fatal voltages and shall never be touched directly or indirectly.

4. **True**/False—Industry-based standards are not, by themselves, legally enforceable, but regulatory agencies often refer to industry standards when citing or fining an employer for unsafe work practices.

5. **True**/False—The ANSI Z133 is the standard used in the United States, and arborists in other countries have adopted the ANSI Z133 in part or have similar standards.

6. True/**False**—Head protection need only be worn while there are climbers in the trees.

7. True/**False**—Eye protection is not required for tree work.

8. **True**/False—Hearing protection may be in the form of earplugs or earmuff-type devices.

Answers to Workbook Questions

9. **True**/False—Workers must not wear gauntlet-type gloves while chipping brush.

10. The **job briefing** summarizes what has to be done and who will be doing each task, the potential hazards and how to prevent or minimize them, and what special PPE may be required.

11. The voice **command**-and-**response** system ensures that warning signals are heard, acknowledged, and acted on.

12. All workers should receive some education and training in **emergency response** procedures, including CPR, first aid, and aerial rescue.

13. **True**/False—Any tree workers who work in proximity to electrical conductors must receive training in electrical hazards.

14. **True**/False—"Electrical conductor" is defined as any overhead or underground electrical device, including wires and cables, power lines, and other such facilities.

15. True/**False**—Rubber footwear and gloves provide absolute protection from electrical hazards.

16. Always engage the **chain brake** before starting a chain saw.

17. True/**False**—A chain saw engine does not need to be stopped for refueling.

18. **True**/False—Kickback can occur when the upper tip of the chain saw guide bar contacts an object.

19. True/**False**—A well-trained climber, in good condition, should be able to dodge the kickback of a chain saw.

20. **True**/False—An open-face notch is preferred for felling trees because the wider notch allows the hinge to work until the tree is almost on the ground.

21. True/**False**—If using a conventional notch to fell a tree, the back cut does not need to be stepped up higher than the hinge.

22. The **hinge** is critical in controlling the direction of fall of a tree.

23. Sometimes, if the tree is leaning in the direction of fall or has internal faults, it can split upward from the back cut. This is called a **barber chair**, and it can be very dangerous.

24. Workers should feed brush into chippers from the **side** and not allow any part of the body to cross the plane of the infeed chute.

25. **True**/False—Safety is the responsibility of all employees from the owner to the ground worker.

Matching

__F__	shall	A.	leg protection for chain saw use
__G__	approved	B.	advisory recommendation
__C__	CPR	C.	cardiopulmonary resuscitation
__D__	direct contact	D.	body touches energized conductor
__B__	should	E.	mobile elevating work platform
__H__	indirect contact	F.	mandatory requirement
__E__	MEWP	G.	meets applicable standards
__A__	chaps	H.	touching an object in contact with an energized conductor

Sample Test Questions

1. According to many standards, the term "shall" denotes
   a. an advisory recommendation
   b. a mandatory requirement
   c. a safety suggestion by ISA
   d. a legal requirement

2. The area within the work zone where the crew expects cut branches or logs to be dropped or lowered from above is called the
   a. landing pad
   b. danger zone
   c. drop zone
   d. barrier zone

3. Head protection is required for tree workers
   a. whenever performing tree care operations
   b. when specified by the supervisor
   c. whenever there are climbers working aloft
   d. only if chain saws or chippers are in use

4. The most common situation that can cause chain saw kickback is
   a. failure to maintain adequate chain tension
   b. a worn sprocket or guide bar
   c. when the upper quadrant of the guide bar tip contacts an object
   d. uneven sharpening of the cutter teeth

5. A leading cause of fatalities among climbers in palms is

   a. insect bites
   b. small mammal attack
   c. chain saw cuts
   d. **sloughing of fronds**

CHAPTER 16 ANSWERS TO WORKBOOK QUESTIONS

Workbook Questions

1. List at least three pieces of equipment considered to be part of a climber's life-support equipment.

   a. climbing line
   b. climbing harness
   c. work-positioning lanyard
   (also, split-tails, connecting links, mechanical friction devices, ascenders, and various other pieces of equipment)

2. **Carabiners** used for climbing must be self-closing and self-double locking (triple-action).

3. **True**/False—Hitch cord is typically a smaller-diameter cordage used to tie the friction hitch.

4. **Micropulleys** are small, light-duty pulleys, often used to tend slack and assist advancement of the friction hitch.

5. Climbers may choose to use a **friction-saving** device when tying in. This can reduce the wear on the rope and damage to the tree and can, in some cases, facilitate climbing.

6. **True**/False—Climbing spikes are used for ascending trees, but only for removal or during a rescue.

7. List three times that climbing gear should be inspected.

   a. daily inspection
   b. periodic inspection
   c. post-incident inspection

Answers to Workbook Questions

8. List two braided rope types that are popular for tree climbing.

   a. 16-strand

   b. 24-strand

9. **Double-braid** ropes are not recommended for natural branch union rigging, where the friction of the cover with the tree causes an imbalance in the load taken by the core and cover braids.

10. True/**False**—If the load in a rope is equal to its tensile strength, the rope will definitely fail on its first use (one cycle to failure).

11. Arborists generally design rigging systems such that the **rigging line** is the weakest link in the system and all loads are within the working-load limit of the rigging line.

12. It is important to know and abide by **minimum approach distances (MAD)** for electrical conductors.

13. List the two types of climbing systems.

    a. moving rope system

    b. stationary rope system

14. True/**False**—In SRS, there is a 2:1 mechanical advantage; however, this corresponds to only 50 percent gain in upward movement.

15. Carabiners must always be loaded along their **major axis** and never across the gate.

16. **Rope walking** is an efficient ascent method that incorporates ascenders.

17. A **redirect** is a change in the direction of the climbing line to create a safer or more efficient rope angle and to reduce the chances of an uncontrolled swing.

18. True/**False**—When a climber is injured or unresponsive in a tree, the first step is to ascend and get to the injured worker as quickly as possible.

19. **True**/False—Natural branch unions can be fast and effective for use as a rigging point, but the minimal, consistent friction and versatility of placement of an arborist block is often a great advantage.

20. **True**/False—The forces in rigging are affected by the weight of the piece, the distance of fall, the type and amount of rope in the system, and the angles involved.

21. When the piece must be removed without dropping either the butt or the tip, it can be tied so it is **balanced**, then lowered to the ground.

22. **True**/False—It is always preferable to establish a rigging point above the work, if possible.

23. **True**/False—Rigging from below can be one of the most demanding techniques for a rope (as well as the hardware and the tree) because of shock-loading.

Answers to Workbook Questions

24. The **hinge cut** is a variation of standard tree-felling techniques that employs the use of a notch and a back cut to form a hinge and steer the branch.

25. Label the following knots.

a. Blake's

b. Sheet bend

c. Running bowline

Matching

__D__ tagline   A. used to install a climbing line

__H__ scabbard   B. usually high and central in the tree is best

__B__ tie-in point   C. knot used to secure a rope to an object or another rope

__F__ kernmantle   D. rope used to control the swing of a branch

__G__ quick link   E. may be used to attach a block

__E__ rope sling   F. rope with a cover and a core

__A__ throwline   G. must be tightened with a wrench

__C__ hitch   H. sheath for a handsaw

Sample Test Questions

1. A "rope inside a rope" is better known as a

   a. hollow-braid rope
   b. 12-strand rope
   c. double-braid rope
   d. 3-strand rope

Answers to Workbook Questions

2. A separate, short length of cordage used to tie a climbing hitch is a
   a. tagline
   b. split-tail
   c. throwline
   d. climbing line

3. Carabiners used for climbing must be
   a. self-double locking
   b. loaded along their minor axis
   c. tightened with a wrench
   d. constructed of steel

4. The first steps of the emergency response process are to
   a. assess the situation and call for emergency help
   b. shut off the electricity
   c. reach the injured worker and begin first aid
   d. reach the injured worker and secure them for descent

5. A cut that consists of an undercut followed by a top cut and then a final cut is called a
   a. drop cut
   b. jump cut
   c. hinge cut
   d. topping cut

GLOSSARY OF TERMS

3-strand rope—rope construction in which three strands are twisted together in a spiral pattern.

7-strand, common-grade cable—steel-cable construction in which seven strands are twisted together in a spiral pattern. Used to limit movement or add supplemental structural support to trees.

12-strand rope—for arborist ropes, a braided rope consisting of 12 strands, typically without a core. There are two types of 12-strand construction: a tight braid that is not easily spliceable, used for climbing and rigging lines, and a loose, easily spliceable braid, commonly used for slings.

16-strand rope—for arborist ropes, a braided rope construction that has a 16-strand, load-bearing cover and a filler core that is not significant in load carrying.

24-strand rope—for arborist ropes, a braided rope that has a 24-strand cover and a core.

A

abiotic agents—nonliving causes of plant disorders.

abscission zone—area at the base of a petiole, small branch, or flower where cellular breakdown leads to leaf, flower, or fruit drop.

absorbing roots—fine roots with functional root hairs that are responsible for the uptake of water and minerals.

acceptable risk—degree of risk that is within the tolerance or threshold of the owner, manager, or controlling authority.

access route—defined entrance and exit route for a property during construction, tree work, or landscape operations.

acclimation—(acclimatization, in British English) physiological adaptation process of plants and other living organisms to a climate or environment different from their prior growing conditions.

action threshold—(1) pest population or plant damage level that requires action to prevent irreversible or unacceptable physiological and/or aesthetic harm. (2) point at which the level of incompatible plant species, density, height, location, or condition threatens the stated management objectives and requires implementation of a control method(s).

acute—(1) disorder or disease that occurs suddenly or over a short period of time. (2) leaves with straight sides tapering to a pointed apex.

adaptability—genetic ability of plants and other living organisms to adjust or acclimate to different environments.

advanced assessment—assessment performed to provide detailed information about specific tree parts, defects, targets, or site conditions. Specialized equipment, data collection and analysis, and/or expertise are usually required.

adventitious branch—branch arising from a stem or parent branch and having no connection to apical meristems.

adventitious bud—bud arising peripherally from a place other than a leaf axil or shoot tip, usually as a result of hormonal triggers.

aerial inspection—inspection of parts of a tree not visible from the ground, including the trunk, stems, and branches; typically done by climbing or from an aerial lift. Aerial inspections may include evaluation of internal decay.

aerial lift—any one of the following types of vehicle-mounted apparatus used to elevate personnel to work positions aloft: (1) extensible boom platform, (2) aerial ladder, (3) articulating boom platform, (4) vertical tower, or (5) a combination of any of the preceding. In the United Kingdom, such a device is called a mobile elevating work platform (MEWP).

aerial rescue—method of bringing an injured worker down from a tree or aerial lift device.

aerial roots—aboveground roots. Usually adventitious in nature and sometimes having unique adaptive functions.

aerobic—having sufficient oxygen; for example, in soil.

aggregate—(1) close cluster or mix of small particles of soil and/or organic matter of varying sizes that are bonded together. (2) sand, gravel, or small rocks in soil and/or used under paved surfaces. (3) clusters of flowers or fruits that appear as a single unit. (4) individual tree crowns that form a canopy.

air-excavation device—device that directs a jet of highly compressed air to excavate soil. Used to avoid or minimize damage to tree roots or underground structures such as pipes and wires. May also reduce hazards associated with excavation near pipes or wires.

air terminal—uppermost part of a tree lightning protection system, located near the top of a tree or large leader, intended to provide a lightning strike termination or attachment point; may be either a manufactured terminal or the end of the conductor.

allelochemicals—substances produced naturally by plants as part of a defense against pests and other plants. May adversely affect the growth and development of other plants.

allelopathy—the influence, usually detrimental, of one plant on another, by the release of chemical substances.

alternate—pertaining to bud or leaf arrangement, one leaf or bud at each node, situated at alternating positions along the stem. In this arrangement, the leaves are not directly across from each other.

alternate host—one of a number of separate plants of certain obligate pathogens (e.g., rust fungi) or insects (e.g., adelgids) on which successive life stages develop.

amon-eye nut—drop-forged nut used to fashion a through-hardware anchor.

anaerobic—without—or with a restricted supply of—air. Process that occurs in the absence of oxygen.

anchor hardware—hardware installed to affix and/or terminate a cable or guy to a tree, the ground, or to another device.

angiosperm—plant with seeds borne in an ovary. Consists of two large groups: monocotyledons (grasses, palms, and related plants) and dicotyledons (most woody trees, shrubs, herbaceous plants, and related plants).

anion—ion that carries a negative charge.

ANSI A300—in the United States, industry-developed, national consensus standards of practice for tree care.

ANSI Z133—in the United States, industry-developed, national consensus safety standards of practice for tree care.

anthocyanin—red or purple pigment responsible for those colors in some parts of plants.

antigibberellin—plant growth regulator that inhibits the action of the plant hormone gibberellin, which, among other things, regulates cell elongation.

apical bud—bud at the tip of a twig or shoot.

apical dominance—condition in which the terminal bud inhibits the growth and development of the lateral buds on the same stem formed during the same season.

apical meristem—growing point in buds and at the tips of shoots and roots.

appraisal—(1) act or process of developing an opinion of value, cost, or some other specified assignment result. (2) a report stating an opinion of appraised value. (3) particularly outside the United States, an evaluation of nonmonetary landscape or plant characteristics.

appropriate response process (ARP)—process of systematically acquiring and using information about the plant, the stressor, and the client to determine which course of action, if any, is recommended.

approved—in the context of guidelines, standards, and specifications, that which is acceptable to federal, state, provincial, or local enforcement authorities or is an accepted industry practice.

arboriculture—practice and study of the care of trees and other woody plants in the landscape.

arborist block—heavy-duty pulley with an integrated connection point (bushing for attaching a rope sling), a rotating sheave for the rope, and extended cheekplates. Used in tree rigging operations.

ascender—mechanical device that enables a climber to ascend a rope. Attached to the rope, it will grip in one direction (down) and slide in the other (up).

atmospheric deposition—the movement of particles, gases, and nutrients from the air to earth by settling (dry particles) or with precipitation (wet).

augmentation—in Plant Health Care, the release of beneficial organisms to suppress pest populations.

auxins—plant hormones that promote or regulate the growth and development of plants. Produced at sites where cells are dividing, primarily in the shoot tips. Auxin-like compounds may be synthetically produced.

available water—water remaining in the soil after gravitational water has drained and before the permanent wilting point has been reached.

axial transport—movement of water, minerals, or photosynthates longitudinally within a tree.

axillary bud—bud in the axil of a leaf. Lateral bud.

B

back cut—cut made in a tree limb or trunk on the side opposite of the intended direction of fall, to complete felling or branch removal.

backfill—(1) soil or amended soil used to fill the hole when planting a tree. (2) soil, common fill, aggregates, or contaminants in various combinations put back after an excavation. May not be suitable for tree root growth and function.

bactericides—pesticides that are used to kill or inhibit bacteria in plants or soil.

balance—in rigging, a technique for lowering a tree limb without allowing either end to drop.

balled and burlapped (B&B)—tree or other plant dug and removed from the ground for transplanting, with the roots and soil wrapped in burlap or a burlap-like fabric.

barber chair—dangerous condition created when a tree or branch splits upward vertically from the back cut, slab up.

bare root—(1) tree or other plant removed from the ground for replanting without soil around the roots. (2) the harvesting or transplanting of a tree or other plant without soil around the roots.

barrier zone—chemical and anatomical barrier formed by the cambium in response to wounding. Inhibits the spread of decay into xylem tissue formed after the time of wounding. Wall 4 in the CODIT model.

basal anchor—in SRS climbing, the means of securing the climbing line at the base of a tree.

basal rot—decay of the lower trunk, trunk flare, or buttress roots. Also called butt rot.

basic assessment—detailed visual inspection of a tree and surrounding site that may include the use of simple tools. It requires that a tree risk assessor inspect completely around the tree trunk, looking at the visible aboveground roots, trunk, branches, and site.

bend—type of knot used to join two rope ends together.

bend radius—radius of an object around which a line passes.

best management practices (BMPs)—best-available, industry-recognized courses of action, in consideration of the benefits and limitations, based on scientific research and current knowledge and standards.

biodiversity—biological diversity in an environment as indicated by the variety of different species of plants and animals.

biological control—(1) method of managing plant pests or weeds through the use of natural predators, parasites, or pathogens. (2) biological methods—management of vegetation by establishment and conservation of compatible, stable plant communities using plant competition, allelopathy, animals, insects, or pathogens. Cover-type conversion is a type of biological control.

biorational control product—product or pesticide formulated from naturally occurring plant extracts, microbes, or microbial byproducts that poses very low risk to nontarget organisms and has limited environmental persistence.

biotic agent—living organism capable of causing disease.

bipinnate—double pinnate.

blight—any disease or disorder, regardless of the causal agent, that rapidly kills flowers, leaves, or young stems that are then typically retained (i.e., not shed) by the plant.

blotch—irregularly shaped necrotic area on leaf, stem, or fruit.

body-thrust—method of ascending a tree using a climbing rope.

botanical pesticide—pesticide derived from plants.

bracket—fruiting body of a decay fungus.

branch bark ridge—raised strip of bark at the top of a branch union, where the growth and expansion of the trunk or parent stem and adjoining branch push the bark into a ridge.

branch collar—swollen area where a branch joins the trunk or another branch that is created by the overlapping vascular tissues from both the branch and the trunk.

branch protection zone—chemically and physically modified tissue within the trunk or parent branch at the base of a smaller, subordinate branch that slows the spread of discoloration and decay from the subordinate stem into the trunk or parent branch.

branch removal cut—pruning cut that removes the smaller of two branches at a union or a parent stem; pruning cut that removes a branch at its point of origin.

branch union—point where a branch originates from the trunk or another branch.

broad-leaved tree—tree that has flat, blade-type leaves and produces seeds inside of fruits.

brown rot—fungal wood rot characterized by the breakdown of cellulose.

bubblers—localized, low-pressure irrigation devices that apply water into basins around trees; often used in groves.

bud—undeveloped flower or shoot containing a meristematic growing point. Small lateral or terminal protuberance on the stem of a plant that may develop into a flower or shoot.

buffering capacity—ability of a soil to maintain (i.e., resist change in) its pH.

bulk density—mass of soil per unit volume; used as a measure of soil compaction. Often written as grams/cubic centimeter.

buttress roots—roots at the trunk base that help support the tree and equalize mechanical stress.

butt-tie—tying off a limb at the butt (larger) end for rigging, allowing the branch to be lowered tip-end first.

C

cable stop—in tree support systems, metal fitting that can be affixed to the ends of steel cable strands to terminate a cable installation.

cambium—thin layer(s) of meristematic cells that give rise (outward) to the phloem and (inward) to the xylem, which results in secondary growth (increasing diameter) of stems and roots.

canker—discrete, localized, usually necrotic area on stems, roots, and branches. Often sunken and discolored. Most canker diseases require laboratory isolation and microscopic examination to be positively identified.

canopy anchor—in SRS climbing, the means of securing the climbing line in the canopy of a tree.

canopy cover assessment—a determination of the proportion of land covered by the vertical projection of tree crowns.

capillary water—water held in the capillary pores of the soil; much of this water can move in any direction and is readily available to plant roots.

carabiner—aluminum or steel connecting device used in climbing and static rigging that is opened and closed by a spring-loaded gate.

carbohydrate—chemical compound, combining carbon, hydrogen, and oxygen in a proportion of C:2H:O (CH2O), that is produced by plants as a result of photosynthesis (sugars) or derived from assimilates (starches, cellulose, hemicellulose, lignin).

carbon sequestration—capturing and storage of carbon. Most often used in reference to the capturing and retention of atmospheric carbon dioxide through biological, chemical, or physical processes.

cardiopulmonary resuscitation (CPR)—emergency procedure in which chest compressions are used to maintain circulation when the heart has stopped beating.

carotenoid—yellow, orange, or red pigment often responsible for those colors in some parts of trees and other plants.

cation—positively charged ion. In soils, the most abundant cations are calcium (Ca), magnesium (Mg), potassium (K), sodium (Na), and aluminum (Al).

cation exchange capacity (CEC)—ability of a soil to adsorb and hold cations. Affected by soil pH and particle size. A measure of soil fertility and clay composition.

cavity—open or closed hollow within a tree stem, branch, or root, usually associated with decay.

cell turgor—distension in a plant cell caused by its fluid contents.

cellulose—long-chain, insoluble glucose polymer found in the cell walls of the majority of plants.

chelates—chemical compounds that keep plant nutrients—usually iron (Fe)—soluble and available for plant absorption over a broad range of pH.

chemical control—(1) control of pests using conventional, manufactured pesticides. (2) management of incompatible vegetation through the use of herbicides or growth regulators.

chlorophyll—green pigment of plants found in chloroplasts. Captures the energy of the sun and is essential in photosynthesis.

chloroplast—specialized organelle found in some cells. Site of photosynthesis.

chlorosis—whitish or yellowish leaf discoloration caused by lack of chlorophyll. Often caused by nutrient deficiency.

chronic—disorder or disease occurring or recurring over a very long period of time, typically multiple growing seasons.

city forester—individual specializing in the fields of arboriculture and urban forestry and having responsibility for the management of all or part of planted and naturally occurring green spaces on public land in communities.

class—taxonomic group below the division level but above the order level.

clay—(1) soil particles with a typical grain size less than 0.004 mm. (2) a soil predominantly composed of such particles.

clearance pruning—pruning to reduce interference with people, activities, infrastructure, buildings, traffic, lines of sight, desired views, or the health and growth of other plants.

climate change—change in global or regional climate patterns.

climbing harness (saddle)—work-positioning harness designed for climbing trees.

climbing hitch—hitch used for securing a tree climber to the climbing line, permitting controlled ascent, descent, and work positioning. Examples of climbing hitches include, but are not limited to, the tautline hitch, Blake's hitch, and the Prusik hitch.

climbing line—rope that meets specifications for use in tree climbing.

climbing spikes (spurs)—sharp devices strapped to a climber's lower legs to assist in climbing poles or trees that are being removed. Also called gaffs, irons, hooks, or climbers.

CODIT—acronym for compartmentalization of decay in trees. Sometimes interpreted as compartmentalization of damage or dysfunction in trees.

codominant stems—forked branches of nearly the same diameter, arising from a common union and lacking a branch collar; may have included bark.

command-and-response system—system of vocal communication in tree care operations used to convey critical information and ensure understanding by another worker, often between a worker aloft and a ground worker.

compartmentalization—natural defense process in trees by which chemical and physical boundaries are created that act to limit the spread of disease and decay organisms.

complete fertilizer—fertilizer containing the three primary elements: nitrogen (N), phosphorus (P), and potassium (K).

compost—(1) n. organic matter that has been intentionally subjected to decay processes and is more or less decomposed. (2) v. to subject organic matter to decay and decomposition processes.

composting—subjecting organic matter to decay and decomposition processes.

compound leaf—leaf with two or more leaflets.

compression wood—reaction wood in gymnosperms, and some angiosperms, that develops on the underside of branches or leaning trunks and is important in load bearing.

conductor—(1) in an electric utility system, metal wires, cables, and bus-bar used for carrying electric current. Conductors may be solid or stranded (i.e., built up by an assembly of smaller solid conductors). (2) any object, material, or medium (e.g., guy wires, communication cables, tools, equipment, vehicles, humans, animals) capable of conducting electricity if energized, intentionally or unintentionally.

conifer—cone-bearing tree or other plant that has its seeds in a structure called a cone.

conk—bracket or shelf-shaped fruiting body or nonfruiting body (sterile conk) of a basidiomycete decay fungus found both on living or dead trees or coarse woody debris. Annual conks can be fleshy or tough (but not woody). Perennial conks are hard and woody. Often associated with mycorrhizae, or wood or bark decay.

connecting link—component of a rigging or climbing system that connects other components.

consequences—effects or outcome of an event. In tree risk assessment, consequences include personal injury, property damage, or disruption of activities due to the event.

consequences of failure—personal injury, property damage, or disruption of activities due to the failure of a tree or tree part.

contact pesticides—materials that cause pest injury or death on contact.

container-grown—tree or other plant that has been grown and marketed in a container.

containerized—field-grown plant placed into a container for a time and then sold as a potted plant. Term does not include a plant initially grown in a container.

controlled-release nitrogen (CRN)—fertilizer that releases nitrogen gradually into the soil.

conventional notch—directional felling cut into the side of a tree, facing the intended direction of fall and consisting of a horizontal face cut and an angle cut above it, creating a notch of approximately 45°.

cordage—cords or ropes; in the context of arboriculture, usually refers to those used in tree climbing and rigging.

cork cambium—lateral meristem from which the corky, protective outer layer of bark is formed. Also known as phellogen.

cost approach—approach to tree appraisal that produces a cost estimate for repairing, replacing, or restoring the tree and/or site.

crack—separation in wood fibers; narrow breaks or fissures in stems or branches. If severe, may result in tree or branch failure.

critical root zone (CRZ)—area around a tree where the minimum amount of roots that are biologically essential to the structural stability and health of the tree are located. There are no universally accepted methods to calculate the CRZ.

cultivar—cultivated variety of a species that cannot be reproduced without human assistance. Usually propagated asexually (cloned).

cultivation—the growing and raising of plants.

cultural control—(1) method of controlling plant pests by providing a growing environment favorable to the host plant and/or unfavorable to the pest. (2) management of vegetation through alternative use of the right-of-way that precludes growth of incompatible vegetation through establishment of crops, pastures, prairies, parks, successful cover-type conversion, or other managed landscapes.

cuticle—waxy layer outside the epidermis of a leaf that slows water loss and helps protect the leaf from insects and diseases.

cycles to failure—number of times a rope or other piece of equipment can be used with a given load before mechanical failure.

cytokinins—plant hormones involved in cell division, leaf expansion, and other physiological processes. Compounds with cytokinin-like activity may be synthetically produced.

D

dead-end grips—manufactured wire wrap designed to form a termination in the end of a 1 × 7 left-hand lay cable.

decay—(1) n. the process of decomposition. (2) v. the process of degradation by microorganisms.

deciduous—tree or other plant that sheds all of its foliage annually.

decomposition—breakdown or separation of a substance into simpler substances.

decurrent—rounded or spreading growth habit of the tree crown.

definite indicator—indicator that decay is definitely present.

defoliation—loss of leaves from a tree or other plant by biological, chemical, or mechanical means (as opposed to natural shedding).

degree day—measure of heat accumulation over time that allows comparison of the daily average temperature and a given temperature base. Used to plan, monitor, or schedule plant or pest management. Also called growing degree day.

design factor—factor by which the rated or minimum breaking strength of a rope or piece of equipment is divided to determine its working-load limit. For example, a rated strength of 10,000 lb (4,500 kg) and a design factor of 2.0 would result in a safe working load of 5,000 lb (2,250 kg).

dieback—condition in which the branches in a tree die from the tips toward the main stem.

differentiation—process in the development of cells in which they become specialized for various functions.

diffuse porous—pattern of wood development in which the vessels and vessel sizes are distributed evenly throughout the growth ring.

direct contact—any part of the body touching an energized conductor.

directional pruning—selective removal of branches to guide and/or discourage growth in a particular direction.

disorder—abnormal condition that impairs the performance of one or more vital functions. Often associated with noninfectious agents or abiotic factors.

dormant—having normal physical functions suspended or slowed down for a period of time.

double braid rope—rope construction consisting of a braided core within a braided rope, both of which carry part of the load.

drill-hole fertilization—applying fertilizer by drilling holes in the soil within the root zone.

drip irrigation—method of minimizing evaporation and runoff by applying water slowly through small emitters.

drip line—imaginary line defined by the branch spread of a single plant or group of plants, projected onto the ground.

drop cut—branch-removal technique consisting of an undercut and then a top cut, usually made farther out on the branch, or with a chain saw, directly over the undercut.

drop zone—predetermined area beneath workers aloft where cut branches or wood sections will be dropped or lowered from a tree and where the potential exists for struck-by injuries.

drum lace—method of tying the root ball of a balled-and-burlapped tree for moving.

dynamic load—forces created by a moving load. Load that changes with time and motion.

E

earlywood—portion of an annual ring (growth ring) that forms after a period of dormancy, characterized by large-diameter cells and thin walls (in ring-porous species). Also called springwood.

ecosystem—system of interacting organisms and their physical environment.

electrical conductor—any overhead or underground electrical device capable of carrying an electric current, including communications wires and cables, power lines, and other such fixtures or apparatus.

emergency response—predetermined set of procedures by which emergency situations are assessed and handled.

entire—term describing a leaf margin without teeth.

epicormic shoot—shoot arising from a dormant bud or from newly formed adventitious tissue.

eradication—total removal of a species from a particular area. May refer to pathogens or insect pests or to unwanted plants.

eriophyid mites—mites in the family Eriophyidae. Typically smaller than other mites, requiring higher magnification to see, and often inducing development of leaf galls.

espalier—(1) n. pruning system that develops a plant in a plane, such as along a wall or a fence. (2) n. a plant trained in that manner. (3) v. to train plants in that manner.

essential elements—minerals essential to the growth and development of trees. These minerals are essential because plants cannot complete their life cycle without them.

eudicotyledon (eudicot)—plant with tricolpate (three-grooved) pollen grains and two cotyledons in its embryo. Eudicotyledons constitute the larger of the two great divisions of flowering plants and typically have broad, stalked leaves with net-like veins.

evapotranspiration (ET)—loss of water by evaporation from the soil surface and transpiration by plants.

evergreen—tree or other plant that sheds all of its foliage progressively over a period of years rather than annually.

excurrent—pattern of tree branching characterized by a dominant leader and an upright or pyramidal, cone-shaped crown.

exfoliating—peeling off of bark in shreds or layers.

extra-high-strength (EHS) cable—in tree support systems, type of 7-strand steel cable, often used to cable trees. Stronger but less flexible than common-grade cable. Must be terminated with dead-end tree grips or terminal fasteners.

eye bolt—drop-forged, closed-eye bolt, used to anchor cables to a tree in a through-fastened system.

eye splice—(1) in cabling, a closed termination loop, hand formed in common-grade cable by wrapping the successive strands back upon the standing part to attach the cable to anchor hardware. (2) a rope or cable splice that forms a closed eye or loop.

F

failure mode—location or manner in which failure could occur or has occurred; for example, stem failure, root failure, or soil failure.

failure potential—in tree risk assessment, the professional assessment of the likelihood of a tree or tree part to fail within a defined time frame.

family—taxonomic group under the order level and above the genus level.

fastigiate—having an upright growth habit, with upward sloping branches.

feller—the person felling a tree.

fertilizer analysis—composition of a fertilizer, expressed as a percentage by weight of total nitrogen (N), available phosphoric acid (P_2O_5), soluble potash (K_2O), and other nutrients.

fertilizer burn—injury to a plant resulting from excess fertilizer salts in the surrounding soil.

fertilizer ratio—ratio of total nitrogen (N), available phosphoric acid (P_2O_5), and soluble potash (K_2O), expressed as percentages of total fertilizer weight; for example, the ratio of a 30-10-10 fertilizer is 3:1:1.

fiber—(1) elongated, tapering, thick-walled cell that provides strength to wood. (2) smallest component of a rope.

field capacity—maximum soil moisture content following the drainage of water due to the force of gravity.

first aid—emergency care or treatment to stabilize a person with injury or illness before medical help is available.

flexure wood—response growth triggered by the continued flexing of a tree stem or branch.

foliage—leaves of a plant.

foliar analysis—laboratory analysis of the mineral content of foliage.

foliar application—applying a fertilizer, pesticide, or other substance on foliage.

footlock—to climb up a suspended rope by pulling with the hands and arms and pushing with the feet. The loose end of the rope is wrapped under the middle and over the top of one foot and is locked in place with pressure from the other foot.

form (pl. forma)—group of plants within a species having distinct variations that occur sporadically and naturally.

frass—fecal material and/or wood dust or shavings produced by insects.

friction—specific type of force that resists the relative motion between two objects in contact. The direction is always opposite the motion.

friction device—device used to take wraps in a load line to provide friction for controlled lowering or climbing.

friction hitch—any of numerous knots used in tree climbing or rigging that may alternately slide along and then grip a rope.

friction-saving device—type of artificial tie-in point used to reduce damage to the tree and climbing line.

frond—large, divided leaf structure found in palms and ferns.

fruiting body—reproductive structure of a fungus that usually develops in diseased tissues. The presence of certain species may indicate decay in a tree. Many fruiting bodies are small and can only be seen with a hand lens, and require microscopic examination to determine the species of fungus present.

functional goals—in landscape design, the set of goals pertaining to the future needs and practical purpose of the site.

fungicides—chemical compounds that are toxic to fungi.

G

gall—abnormal plant structure that develops in the cells, tissues, or organs of a plant only when it is colonized by certain parasitic organisms such as bacteria, fungi, nematodes, mites, or insects.

genus (pl. genera)—taxonomic group, composed of species having similar fundamental traits. Botanical classification under the family level and above the species level.

geotropism—plant growth produced as a response to the force of gravity, either positive as in the direction of gravity (roots) or negative as in opposite the direction of gravity (shoots).

girdling—restriction, compression, or destruction of the vascular system within a root, stem, or branch that inhibits movement of sap in the phloem.

girdling root—root that encircles all or part of the tree trunk or the tree's other roots, constricting the vascular tissue and inhibiting secondary growth and the movement of water and photosynthates.

gravitational water—water that drains from soil macropores due to the force of gravity.

ground rod—metal rod used in grounding a lightning protection system.

ground terminal—in a tree lightning protection system, a conductive plate or rod used to ground.

growth rings—rings of xylem that are visible in a cross section of the stem, branches, and roots of some trees. In temperate zones, the rings typically represent one year of growth and are sometimes referred to as annual rings.

guard cells—pair of specialized cells that regulate the opening and closing of a stomate due to a change in water pressure within cells.

gummosis—exudation of sap, gum, or resin often in response to disease or insect damage.

guying—installation of a steel cable or synthetic-rope cabling system and an external anchor to provide supplemental support.

gymnosperm—plants with exposed seeds, usually within cones. The classes Ginkgopsida and Coniferopsida are members of the group.

H

habit (growth habit)—characteristic form or manner of growth.

habitat—environment suitable for sustaining a population of a given organism.

hardened off—(1) plant tissue that is acclimated to the cold or a new environment. (2) a process that acclimates balled-and-burlapped trees to water stress when dug with foliage.

hardiness—genetically determined ability of a plant to survive winter growing conditions; often referring specifically to low temperatures.

hardscape—constructed elements of a landscape, such as walls, pathways, and seats made of wood, stone, and/or other materials.

heading cut—pruning cut that removes a branch or stem between nodes (leaving a stub), to a bud, or to a live branch that is less than one-third the diameter of the stem being removed.

heartwood—central wood in a branch or stem characterized by being composed of dead cells, more resistant to decay, generally darker, and harder than the outer sapwood. Trees may or may not have heartwood.

heartwood rot—any of several types of fungal decay of tree heartwood, often beginning with infected wounds in the living portions of wood tissue. Also called heart rot.

herbicides—pesticides used to kill, slow, or suppress plant growth by interfering with botanical pathways.

herbivore—an animal that feeds primarily on plants.

hinge—strip of uncut wood fibers created between the face cut or notch and the back cut that helps control direction in tree felling or limb removal. Holding wood.

hinge cut—sequence of cuts used to control the direction of a limb being removed.

hitch—(1) type of knot made when a rope is secured around an object or its own standing part. (2) a mechanical device for connecting a towing vehicle to a trailed or towed vehicle or implement.

hitch cord—short length of cordage used to tie a friction hitch in climbing or rigging.

honeydew—sugary substance excreted by certain insects, including aphids and some scales, when feeding on plants.

horticultural oils—highly refined petroleum oils applied to plants to control certain insects and other pests by disrupting their respiration.

hybrid—plant resulting from a cross between two different genera, species, or highly inbred lines within a single species.

hydrology—study of the properties, distribution, and effects of water on the earth's surface, underground, and in the atmosphere.

hydrozone—landscape area for grouping plants in a landscape according to their water and/or irrigation requirements.

I

identification key—dichotomous guide used to help identify plants and other organisms.

implant—(1) device, capsule, or pellet inserted into the tree's xylem system to treat or prevent diseases, disorders, or pest problems. Requires a relatively large diameter and deep hole in the trunk. (2) a microchip device implanted into a tree and containing information about the tree that can be retrieved and updated by a compatible chip reader.

included bark—bark that becomes embedded in the union between branch and trunk or between codominant stems. Lacks wood connections, resulting in a weak structure.

income approach—approach to tree appraisal used to appraise income-producing plants or property.

indirect contact—when any part of the body touches any conductive object, including tools, tree branches, trucks, equipment, or other objects, that is in contact with an energized electrical conductor. Such contact can also be made as the result of communication wires and cables, fences, or guy wires being accidentally energized.

infectious—capable of being spread from plants to result in infection of other plants or organisms.

infiltration—(1) downward entry of water into the soil. (2) entry of fine particles into drainage or aeration systems; can lead to system clogging and failure. (3) downward entry of materials from one soil or fill layer to another, as when a gravel road surface mixes with underlying soil.

infiltration rate—speed at which water penetrates the soil.

inflorescence—cluster of flowers.

injection—injection of a liquid substance into a plant or soil.

inorganic—compound or substance not containing carbon or not containing organic material.

inorganic mulch—mulch, such as stone, lava rock, or pulverized rubber, not derived from an organic source.

insect growth regulators—substances, man-made or naturally occurring in insects, that affect growth and development of insects.

insecticidal soaps—soap-based pesticides approved for application to plants to kill insects and certain mites by disrupting the cell membranes or the insect's respiratory tracheas.

insecticides—substances toxic to insects.

Integrated Pest Management (IPM)—systematic method of managing pests that uses cultural, biological, physical, and chemical methods to manage pests to acceptable levels.

internal cycling—recycling of essential elements (sometimes called nutrients) within a plant for use in other plant parts.

internodal—between the nodes on a stem.

internode—region of the stem between two successive nodes.

introduced species—organisms not native or endemic to a region; typically introduced by human activity in the modern era.

invasive species—an alien species whose introduction causes or is likely to cause economic or environmental harm or harm to human health.

ion—atom or a group of atoms with a positive or negative charge.

i-Tree—suite of software products and management tools, developed by the United States Department of Agriculture Forest Service, that allows the user to inventory the urban forest and analyze its costs, benefits, and management needs.

J

job briefing—the communication of at least the following subjects for arboricultural operations: hazards associated with the job, work procedures involved, special precautions, electrical hazards, job assignments, and personal protective equipment.

K

kerf—space created by a saw cut (the width of the chain or blade).

kernmantle rope—rope construction with a cover and a core in which the core yarns are not braided but consist of twisted fibers.

kickback—sudden, sometimes violent and uncontrolled backward or upward movement of a chain saw.

kickback quadrant—upper quadrant of the tip of a chain saw bar.

kingdom—taxonomic group separating plants from animals.

knot—(1) any of various fastenings formed by looping and tying a rope (or cord) upon itself or to another rope or to another object. (2) imbedded remnant of a branch in a tree or cut timber, often harder than surrounding wood.

L

lag eye—lag-threaded, drop-forged, closed-eye cable anchor used for dead-end systems.

lag hook (J-hook)—lag-threaded, J-shaped cable anchor.

lag-threaded—cable anchor or bracing rod with a coarse thread pattern; typically screwed into a predrilled hole that is smaller in diameter than the anchor or rod.

landing zone—predetermined area where cut branches or wood sections will be dropped or lowered from a tree.

lanyard—short rope or strap, often equipped with carabiners, snaps, and/or eye splices.

lateral—secondary or subordinate branch or root.

lateral root—root that arises by cell division in the pericycle of the parent root and then penetrates the cortex and epidermis.

latewood—portion of an annual ring (growth ring) that forms during summer, characterized by small-diameter cells with thick walls. Summer wood.

leach—(1) tendency for elements or compounds to wash down through and/or out of the soil. (2) tendency for elements or compounds to wash into the soil.

leader—primary terminal shoot or trunk of a tree. Large, usually upright stem. A stem that dominates a portion of the crown by suppressing lateral branches.

leaf apex—tip of the leaf blade.

leaf axil—edge of a leaf petiole where it meets the stem.

leaf base—bottom part of the leaf blade.

leaf margin—outer edge of the leaf blade.

lean—predominant angle of the trunk from vertical.

leg protection—personal protective equipment intended to reduce the risk of injury to the legs during chain saw operations.

lenticel—small opening in the bark that permits the exchange of gases.

level of assessment—categorization of the breadth and depth of analysis used in an assessment.

life-support equipment—components of climbing gear that are involved with keeping the climber aloft.

lignin—organic substance that impregnates secondary cell walls to thicken and strengthen the cell and, at times, to reduce susceptibility to decay and pest damage.

likelihood—chance of an event occurring. In the context of tree failures, the term may be used to specify (1) the chance of a tree failure occurring, (2) the chance of impacting a specified target, or (3) the combination of the likelihood of a tree failing and the likelihood of impacting a specified target.

likelihood of failure—chance of a tree or tree part failure occurring within the specified time frame.

likelihood of impact—chance of a tree or tree part failure impacting a target if failure were to occur.

limb walking—technique of moving laterally along limbs while keeping the climbing line taut.

limited visual assessment—visual assessment from a specified perspective such as foot, vehicle, or aerial (airborne) patrol of an individual tree or a population of trees near specified targets to identify specified conditions or obvious defects of concern.

lion tailing (lion's tailing)—poor pruning practice in which an excessive number of branches are removed from the inside and lower part of specific limbs or a tree crown, leaving mostly terminal foliage. Results in poor branch taper, poor wind load distribution, and a higher risk of branch failure.

liquid injection fertilization—applying liquid formulations of fertilizer by injection into the root zone of a tree or by application to soil surface or to foliage.

load—(1) a general term used to indicate the magnitude of a force, bending moment, torque, pressure, or such, applied to a substance or material. (2) cargo; weight to be borne or conveyed.

load line—rope used to lower a tree branch or segment that has been cut. Lowering line.

loam—soil texture classification based on a certain ratio of sand, silt, and clay. Considered ideal for plant growth.

M

machine-threaded—fine-thread pattern on hardware, used with a nut and washer and installed through a pre-drilled hole that is larger in diameter than the hardware being installed.

macronutrient—essential element that is required by plants in relatively large quantities, such as nitrogen (N), phosphorus (P), potassium (K), and sulfur (S).

macropore—relatively large space between soil particles that is usually air filled and allows for water movement and root penetration.

mechanical friction device—device used to provide friction for controlled climbing, either in place of, or together with, a friction hitch.

meristem—undifferentiated tissue in which active cell division takes place. Found in the root tips, buds, cambium, cork cambium, and latent buds.

microbial pesticides—pesticides that contain insect pathogens or lethal microbial byproducts that are derived from extracts of bacterial pathogens of insects.

microclimate—the climate of a small or restricted area.

micronutrient—essential element that is required by plants in relatively small quantities, such as iron (Fe), manganese (Mn), zinc (Zn), copper (Cu), and boron (B).

micropore—space between soil particles that is relatively small and likely to be water filled.

micropulley—small, light-duty pulley used in climbing operations. Often used as a knot tender.

mineralization—process in which an organic substance is converted to or trapped in an inorganic substance.

minimum approach distance (MAD)—the closest distance an employee may approach or bring any conductive object near an energized or a utility system grounded object; or the closest distance the employee may be to an energized or utility system grounded object.

minimum irrigation—practice of minimizing irrigation needs through the use of drought-tolerant plants and watering only when necessary.

miticides—chemical compounds that are toxic to mites.

mitigation—the act of making a condition or consequence less severe. In tree risk management, the process for reducing risk.

mobile elevating work platform (MEWP)—general term used primarily in the United Kingdom for mobile aerial platforms that are widely used on work sites in place of ladders and tower scaffolds.

monitoring—keeping a close watch; performing regular checks or inspections.

monocotyledon (monocot)—plant with an embryo that has one single seed leaf (cotyledon). Examples are grasses and palms.

monoculture—cultivation or planting of a single species on agricultural land, in a forest setting, or within an urban landscape.

morphology—study of the form and structure of plants and other living organisms.

mortality spiral—combination of events or conditions that accelerate decline and, if left untreated, may cause the eventual death of a tree.

moving rope system (MRS)—climbing system in which the rope adjustment device advances along a moving climbing line. The doubled rope technique (DdRT) is an example of a moving rope system.

municipal arborist—individual specializing in the fields of arboriculture and urban forestry and having responsibility for the management of all or part of planted and naturally occurring green spaces on public land in communities.

mycorrhizae—symbiotic association between certain fungi and absorbing roots of plants.

N

native species—plants endemic (indigenous) to a region within the modern era; naturally occurring and not introduced by humans.

naturalized species—nonnative species that has become established in a region and propagates without human assistance.

necrosis—localized or general death of cells or parts of a living organism.

nematode—small, often microscopic, unsegmented roundworm. Many are beneficial organisms, but some feed on plant tissues to cause disease or spread viruses.

nitrogen fixation—process by which molecular nitrogen in the air is converted into ammonium, nitrites, or nitrates in the soil, making it available for uptake by plants.

node—point on a stem from which leaves, branches, and aerial roots are attached.

nomenclature—scientific naming system for living organisms. Scientific names are Latin (or Latinized forms of other languages) and written in italics, the genus first (always starting with capital letter), followed by the specific epithet (always starting with lowercase letter), and together making the species name (e.g., *Quercus alba*).

noninfectious—disorders that are not caused by a pathogen and cannot be passed from one host to another.

nutrient cycling—movement of mineral elements (sometimes called nutrients) within an ecosystem as organic matter decomposes, releasing bound nutrients back to plants.

nutrient deficiency—condition in which the supply or availability of an essential element causes cessation of critical plant processes, resulting in visible, physical symptoms such as chlorotic leaves or necrotic margins, and which, if left untreated, will reduce tree health and may eventually lead to premature death.

nutrient limitation—condition in which the supply or availability of an essential element reduces the growth rate but does not cause plant dysfunction or premature death.

O

occupancy rate—amount of time targets are within a target zone.
oozing—seeping or exudation from a tree cavity or other opening.
open-face notch—directional felling cut into the side of the tree, facing the intended direction of fall and consisting of two cuts that create a notch greater than 70°.
opportunistic—pathogens or other pests that do not usually attack or infect healthy plants but tend to attack stressed plants.
opposite—pertaining to leaf or branch arrangement, leaves or branches situated two at each node, across from each other on the stem.
order—taxonomic group below the class level but above the family level.
organic—(1) containing carbon. (2) of animal or vegetable origin, especially when referring to a fertilizer or pesticide.
organic layer—layer of organic matter at the soil surface.
organic matter—material derived from the growth (and death) of living organisms. The organic components of soil.
organic mulch—mulch derived from plant material such as wood chips, bark chips, or pine needles.
osmosis—diffusion of water through a semipermeable membrane from a region of higher water potential (lower salt concentration) to a region of lower water potential (higher salt concentration).

P

palmate—type of leaf with veins or leaflets radiating in a fanlike pattern.

palm skirt—dead or dying palm fronds gathered down the stem of a palm, which may be removed for aesthetic or safety reasons.
parasite—organism living in or on another living organism (host) from which it derives nourishment to the detriment of the host.
parenchyma cells—thin-walled, living cells capable of dividing and essential in photosynthesis, radial transport, energy storage, and production of defense compounds.
parent material—soil bedrock or base material from which a soil profile develops.
pathogen—causal agent of disease. Usually refers to microorganisms.
percolation—movement of water through the soil.
percolation rate—the rate that water moves through the soil.
permanent branches—in pruning, branches that will be left in place for the life of the tree, often forming the initial scaffold framework of a tree.
permanent wilting point—point at which a plant cannot pull any more water from the soil and suffers permanent damage.
permit—written order granting permission to conduct a specified task or action.
personal protective equipment (PPE)—personal safety gear such as helmet, safety glasses, hearing protection, gloves, and leg protection, including chaps.
pest—organism (including, but not limited to, weeds, insects, bacteria, or fungi) that is damaging, noxious, or a nuisance.
pesticide—any chemical used to control or kill unwanted organisms such as weeds, insects, or fungi.
pesticide resistance—ability to withstand certain pesticides; survival of just a few genetically resistant pests whose reproduction can lead to populations that are resistant.
pest resistance—(1) in plants, the ability to resist pest infestation or infection. (2) in pests, the genetically acquired ability of an organism to survive a pesticide application at doses that once killed most individuals of the same species.

pest resurgence—increase in the population of a pest following a reduction in the population of natural predators or parasites of that pest. Usually the result of a broad-spectrum pesticide application or an unfavorable environmental condition.

petiole—stalk or support axis of a leaf between the stem and the blade.

pH—unit of measure that describes the alkalinity or acidity of a solution. Negative log of the hydrogen ion concentration. Measured on a scale from 0 to 14. Greater than 7 is alkaline, less than 7 is acid, and 7 is neutral (pure water).

phenology—relationship between the climate and biological events, such as flowering or leafing out in plants.

phloem—plant vascular tissue that transports photosynthates and growth regulators bidirectionally (up and down). Situated on the inside of the bark, just outside the cambium.

photosynthate—general term for the sugars and other carbohydrates produced during photosynthesis.

photosynthesis—process in green plants (and in algae and some bacteria) by which light energy is used to form glucose (chemical energy) from water and carbon dioxide.

phototropism—influence of light on the direction of plant growth. Tendency of plants to grow toward light.

phylum (pl. phyla)—primary taxonomic group within a kingdom.

phytotoxic—term to describe a compound that is poisonous to plants.

pinnate—type of compound leaf, with leaflets along each side of a common axis.

pith—central core of a stem. Often a lighter color than surrounding tissue.

plant disease triangle—conceptual model showing three factors required for plant disease: a susceptible host, a pathogen or an abiotic agent, and a conducive environment.

plant growth regulator—compound effective in small quantities that affects the growth and/or development of plants. May be naturally produced (hormone) or synthetic.

Plant Health Care (PHC)—comprehensive program to manage the health, structure, and appearance of plants in the landscape.

plant hormone—substance produced by a plant that, in low concentrations, affects physiological processes such as growth and development, often at a distance from the substance's point of origin.

planting specifications—detailed plans and statements of particular procedures, requirements, tools, materials, and standards for planting.

pollarding—semiformal pruning system that maintains crown size by initial heading of branches on young trees or young portions of older trees, followed by removal of sprouts to their point of origin at appropriate intervals, without disturbing the resulting pollard heads.

pore space—air- or water-filled spaces between soil particles.

portable watering device—portable devices or systems designed to deliver water to plants slowly over a period of time.

potential indicator—indicator that decay might be present.

predator—any organism that preys on another organism.

prescription fertilization—basing fertilization recommendations on plant needs as determined by conducting soil and/or foliar nutrient analysis, setting plant health goals, and selecting a fertilizer to achieve the goals.

prevention—proactive process intended to guard against an adverse impact by avoiding or reducing the risk of its occurrence.

primary growth—root and stem growth in length. Occurs in apical and lateral meristems.

primary suspension point—the position in the tree where the climbing line is anchored or crosses over and that experiences the highest loads during the climb; this term is used mostly with stationary rope systems (SRS).

propagation—process of increasing plant numbers, both sexually and asexually.

propping—the installation of rigid structures between the ground and a branch or trunk to provide support.

protection factors—structures, trees, branches, or other factors that would prevent or reduce harm to targets in the event of a tree failure.

pruning objectives—the defined purpose(s) for pruning (e.g., provide clearance, reduce risk).

pruning system—technique or procedure that is applied to develop the desired long-term form of the plant.

Q

quick link—metal connector hardware with a screw-type fastener, used to attach ropes or other climbing hardware.

R

radial transport—lateral movement of substances, perpendicular to the longitudinal axis of the tree or stem.

radial trenching—technique for relieving soil compaction and improving rooting conditions around a tree by excavating trenches in a spoke-like pattern radially from the trunk. Air excavation minimizes root damage.

ray—parenchyma tissues that extend radially across the xylem and phloem of a tree and function in transport, storage, structural strength, and defense.

reaction wood—wood formed in leaning or crooked stems or on lower or upper sides of branches as a means of counteracting the effects of gravity.

reaction zone—natural boundary formed chemically within a tree to separate damaged wood from existing healthy wood. Important in the process of compartmentalization.

reactive force—force generated in response and opposite to another force. Often demonstrated when operating a chain saw.

redirect—(1) v. to change the path of a climbing or rigging line to modify the forces on, or the direction of, the line. (2) n. a system installed to change the path of a climbing or rigging line.

reduction—pruning to decrease height and/or spread of a branch or crown.

reduction cut—pruning cut that removes the larger of two or more branches or stems, or one or more codominant stems, to a live lateral branch, typically at least one-third the diameter of the stem or branch being removed.

rescue pulley—light-duty pulley used in light rigging operations.

residual risk—risk remaining after mitigation.

resilience—the ability to respond and recover from disturbances such as shocks and stresses.

resistance-recording drill—device consisting of a specialized micro-drill bit that drills into trees and graphs resistance to penetration; used to detect internal differences in the wood, such as decay.

resource allocation—(1) in plant physiology, distribution and use of photosynthates for various plant functions and processes. (2) in management, distribution of materials or other assets to accomplish objectives.

respiration—in plants, process by which carbohydrates are converted into energy by using oxygen.

response growth—new wood produced in response to loads to compensate for higher strain experienced (adaptive growth); includes reaction wood (compression and tension), flexure wood, and woundwood.

restoration—(1) pruning to improve the structure, form, and appearance of trees that have been vandalized, damaged, or improperly trimmed. (2) management and planting to restore altered or damaged ecosystems or landscapes. (3) replacement or reproduction of damaged or destroyed plants.

retreat path—predetermined escape route away from a tree that is to be felled; should be a 45° angle back and away from the direction of the falling tree and clear of obstruction.

rhizosphere—soil area immediately adjacent to, and affected by, plant roots. Typically has a high level of microbial activity.

rigging—(1) in tree pruning or removal, method of using ropes and hardware to control or direct the descent of cut material or to handle heavy loads. (2) with cranes, loaders, or other equipment, method of using ropes and hardware to lift heavy loads.

rigging line—rope, usually the load line, used in rigging operations.

rigging point—place in the tree (natural branch union or an installed arborist block) that the load line passes through to control limb removal in rigging operations.

ring porous—pattern of wood development in which the large-diameter vessels are concentrated in the earlywood.

risk—the combination of the likelihood of an event and the severity of the potential consequences. In the context of trees, risk is the likelihood of a conflict or tree failure occurring and affecting a target, and the severity of the associated consequences.

risk management plan—plan that establishes policies, procedures, and practices to identify, evaluate, mitigate, monitor, and communicate risk.

root ball—soil containing all or a portion of the roots that are moved with a plant when it is planted or transplanted.

root collar excavation (RCX)—process of removing soil to expose and assess the root collar (root crown) of a tree.

root crown—area where the main roots join the plant stem, usually at or near ground level. Root collar.

root exudates—sugar and other substances that are released into the soil as the root caps and external layers are sloughed off.

root initiation zone—region at the base of a palm stem where lateral roots emerge.

root mat—dense network of roots. In palms, near the base of the stem.

root pruning—severing roots selectively. (1) in transplanting, the process of cutting roots to increase the density of root development within the root ball. (2) in tree conservation and preservation, the process of cutting roots to prevent tearing and splintering of remaining roots. (3) in tree disease management, severing tree roots to prevent disease transmission through root grafts.

root:shoot ratio—relative proportion of root mass to crown mass.

rope angle—in tree climbing, the angle at the tie-in point or redirect that the climbing line makes away from vertical, straight down.

rope snap—connecting device used by tree climbers, primarily for connecting the work-positioning lanyard to the harness.

rope support system—cabling system that utilizes elastic materials (usually rope of various constructions) for tree support systems.

rope walking—technique for ascending a rope using a combination of ascenders, including at least one knee or foot ascender to grip the line, allowing the climber to "walk" up the rope, maintaining a vertical body position.

S

safety standards—established or widely recognized authority of acceptable practices, techniques, and equipment to ensure safety.

sales comparison approach—approach to appraisal that identifies sale transactions (sales history, comparable sales, listing prices, etc.) and analyzes them to develop an opinion of market value.

saline soil—soil with a high concentration of soluble salts. Can cause poor plant growth.

salinity—amount or percentage of salt in the soil.

sand—soil particles with a size between 0.06 mm and 2.0 mm in diameter.

sapwood—outer wood (xylem) that has living cells that are active in longitudinal transport of water and solutes.

sapwood rot—decay located in the sapwood. Bark and/or cambium may be damaged or dead. Usually, signs of this classification of rot are fruiting bodies along the bark's surface.

scabbard—protective sheath for a pruning saw or other tool.

scaffold branches—permanent branches that form the scaffold or structure of a tree.

scorch—browning and shriveling of foliage, especially at the leaf margin.

secondary growth—increase in root and stem girth or diameter. Occurs at lateral or secondary meristems in some vascular plants such as dicots.

secondary pest outbreak—increase in a secondary pest population following a reduction in the population of natural predators or parasites due to broad-spectrum pesticide application.

self-double locking—pertaining to a carabiner, requiring three distinct motions to open the gate and to auto-close and auto-lock when the gate is released.

serrate—sawtooth margin of a leaf, with the teeth pointed forward.

shake—separation of wood at the growth rings (ring shake) or rays (radial shake), often along the barrier zone that forms in the compartmentalization process (CODIT).

shall—as used in the ANSI standards (United States), denotes mandatory requirement.

shearing—cutting leaves, shoots, and branches to a desired plane, shape, or form, using tools designed for that purpose, as with topiary (hedge).

shear-plane crack—crack at the neutral plane between tension and compression stresses.

ship auger—type of drill bit with an open spiral form. Used to drill holes in trees for cable or bracing installation.

shock-loading—force exerted by a falling or moving object on the structure supporting it, which is greater than the weight of the object (also called dynamic load).

should—as used in the ANSI standards (United States), denotes an advisory recommendation.

sign—physical evidence of a causal agent (e.g., insect eggs, borer hole, frass, mycelium, fruiting body).

silt—soil particles with a grain size between 0.004 mm and 0.062 mm (coarser than clay particles but finer than sand).

simple leaf—single-bladed leaf. Not composed of leaflets.

sink—plant part that uses or stores more energy than it produces.

sinker roots—downward-growing roots that provide anchorage and take up water and minerals. Especially useful during periods of drought.

site analysis—(1) consideration or evaluation of the conditions, restrictions, and environment of a planting site. (2) consideration or evaluation of a construction or development site requiring a tree conservation or preservation plan. (3) consideration during a tree risk assessment.

size diversity—measure of the number and variety of different sizes of trees found in a given area.

skeletonized—leaf-feeding damage caused by insects (skeletonizers), characterized by the loss of tissue between the leaf veins.

sling—device used in rigging to attach, connect, or secure equipment or pieces being rigged.

slow-release fertilizer—fertilizer containing plant nutrients in a form that delays availability for plant uptake and use after application or that extends availability to the plant.

snap cut—cutting technique in which offset, bypassing cuts are made so that a section can be broken off easily.

sodic soil—soil with relatively low levels of soluble salts and a concentration of sodium high enough to adversely affect soil structure (symptoms include waterlogging, erosion, soil surface crusting, and poor plant growth).

soft rot—decay of plant tissues characterized by the breakdown of tissues within the cell walls.

soil analysis—analysis of soil to determine pH, mineral composition, texture, structure, salinity, organic matter, and other characteristics.

soil biological properties—characteristics of soil determined by the living inhabitants of the soil, such as fungi, bacteria, nematodes, earthworms, insects, and other animals.

soil chemical properties—soil characteristics determined by the molecular content that affects pH, cation exchange capacity, buffering capacity, and other factors.

soil compaction—compression of the soil, often as a result of vehicle or heavy-equipment traffic, that breaks down soil aggregates and reduces soil volume and total pore space, especially macropore space.

soil food web—complex network of interconnected food chains within the soil ecosystem.

soil grade—(1) n. surface level of the ground. (2) n. the slope, or percentage of change, in the surface level of the ground; important for drainage and for the safe operation of equipment. (3) v. to change or groom the surface level or contours of the ground.

soil horizon—layer or zone of the soil profile with physical, chemical, and biological characteristics that differ from adjacent layers.

soil moisture reservoir—the volume of water available depending on the volume of soil occupied by plant roots and the water-holding capacity of the soil.

soil physical properties—characteristics of soil, such as texture, structure, and bulk density, that are determined by the size and arrangement of the soil's mineral particles and other content.

soil profile—vertical section through the soil and all of the soil horizons.

soil structure—arrangement of soil particles into aggregates.

soil texture—relative fineness or coarseness of a soil due to particle size (sand, silt, and clay).

sonic tomography (acoustic tomography)—use of multiple sensors placed around a trunk or limb to record sound waves traveling through the wood, with measurements resulting in a picture of internal density characteristics. Typically used to measure the extent of decay.

sooty mold—fungus that appears as a black coating on the surface of leaves, branches, fruits, and other surfaces (including those of buildings), resulting from deposits of sugary excrement from aphids and scale insects.

source—in physiology, plant part that produces carbohydrates. Most green parts are sources because the presence of chlorophyll is indicative of photosynthesis, including mature leaves and green bark.

species—taxonomic group of organisms composed of individuals of the same genus that can reproduce among themselves and have similar offspring.

species diversity—measure of the number and variety of different species found in a given area.

species failure profile—common mode(s) of structural failure or resistance to failure for a tree species.

specifications—detailed plans, requirements, and statements of particular procedures and/or standards used to define and guide work.

specific epithet—classification name that follows the genus name in scientific nomenclature.

spliced eye—a loop spliced into the end of cordage.

split-tail—separate, short length of rope used to tie the climbing hitch in a climbing system.

spot—discrete, localized, and usually small necrotic area of a leaf or needle, stem, flower, or fruit.

spray irrigation—method of applying water to plants through a network of spray emitters (sprinkler irrigation).

stakeholder—(1) person or group that has an interest in, or is affected by, an activity or a decision. (2) person that either creates policy, is affected by those policies, or perceives to be affected by those policies.

staking—supporting a tree with stakes and ties. Usually used in reference to newly planted trees.

standard—established or widely recognized authority of acceptable performance.

static load—constant load exerted by a mass due to its weight.

stationary rope system (SRS)—climbing system in which the rope adjustment device moves along a stationary climbing line.

steel cable system—cabling system that utilizes steel cable to limit movement and provide support of branches.

stippling—speckled or dotted areas in which chlorophyll is absent on foliage.

stomata—small apertures, between two guard cells on the undersides of leaves and other green plant parts, through which gases are exchanged and water loss is regulated.

structural cells—modular system consisting of integrated support structures filled with soil that serves as both a foundation for paved surfaces and a hospitable environment for tree root growth.

structural defect—feature, condition, or deformity of a tree that indicates a weak structure or instability that could contribute to tree failure.

structural pruning—pruning to influence the orientation, spacing, growth rate, strength of attachment, and ultimate size of branches and stems.

structural soil—pavement substrate that can be compacted to meet engineering specifications yet remain penetrable by tree roots in the urban environment. Cornell University developed and trademarked CU-Structural Soil™, composed of angular crushed stone, clay loam, and hydrogel mixed in a weight ratio of 100:25:0.03.

stunting—growth reduction of plants or plant parts.

subordinate—(1) v. prune to reduce the size and ensuing growth of a branch in relation to other branches or leaders. (2) adj. dominated by other trees, branches, or parts; suppressed.

subspecies—group of plants within a species having distinct differences that occur naturally, usually within a specific geographic region.

substrate—layer of material below the soil surface; materials used to provide plant support, regulate moisture, and (in many cases) provide mineral nutrients to container plants. Growing media.

subsurface application—placement of fertilizer or other material below the soil surface.

succession—in urban forestry, the planned replacement of trees over time.

suppression—(1) (pest management) management practices intended to reduce the pest population and associated plant injury to a tolerable level. (2) (fire) the fighting of forest, brush, rangeland, wildland, or bush (Australia and New Zealand) fires to reduce the area of burning or to protect other resources or assets, particularly inhabited structures.

surface application—placement of fertilizer or other material on the ground surface. Also called broadcast application.

susceptibility—potential for infection and/or injury from exposure to a biotic or an abiotic factor.

suspended pavement—paving option that transfers the load to the subsoil rather than to the top of the soil surface.

sustainability—ability to maintain ecological, social, and economic benefits over time.

swage—hollow metal fitting used to terminate a wire rope (aircraft cable).

symbiosis—n. association of two different types of living organisms that is not detrimental to either organism and is beneficial to one, and usually both, of the organisms.

symbiotic—adj. association of two different types of living organisms that is not detrimental to either organism and is beneficial to one, and usually both, of the organisms.

symptom—plant reaction to a disease or disorder (e.g., spot, wilt, dieback).

systemic—(1) when referring to a substance: moving throughout an organism after absorption. (2) when referring to a condition, disease, disorder, or pest: affecting the entire organism.

systemic pesticide—pesticide that moves throughout a tree after it has been injected or absorbed (often by roots or foliage or by injected cambium).

T

tagline—rope used to control the swing of a limb being removed or to control the direction or fall of a tree or limb being removed.

taproot—central, vertical root growing directly below the main stem or trunk that may or may not persist into plant maturity.

target—(1) people, property, or activities that could be injured, damaged, or disrupted by a tree failure. (2) intended application site or pest.

target zone—area where a tree or tree part is likely to land if it were to fail.

taxonomy—science that studies the description, denomination, and classification of living organisms based on their similarities and differences.

temperate—region lying between the tropics and permafrost zones where temperature ranges and conditions permit plant growth.

temporary branches—branches (generally lower branches) that are left in place or subordinated but will be removed later.

tensile strength—force at which a new piece of equipment or rope in testing fails in tension under a static load.

tensiometer—instrument used to measure soil moisture.

tension wood—form of reaction wood in angiosperms that forms on the upper side of branches or the trunks of leaning trees.

terminal bud—bud at the tip of a twig or shoot. Apical bud.

thimble—(1) an oblong, galvanized, or stainless-steel fitting with flared margins and an open-ended base. (2) a device used to increase the bend radius of and reduce wear on a rope when attached to hardware.

threaded rod—machine-threaded steel rod used for through-brace installations.

three-cut method—branch-removal technique consisting of an undercut and then a top cut, usually made farther out on the branch, followed by a third cut to remove the stub.

threshold—(1) in Integrated Pest Management, pest population levels requiring action. (2) in risk assessment and management, level of risk requiring action.

throwline—thin, lightweight cord attached to a throwbag or throwing ball used to set climbing or rigging lines in trees.

tie-in point—position in a tree (in a natural branch union or installed device) through which the climbing line is set to serve as the top rope placement for work positioning; this term is used mostly with moving rope systems (MRS).

time frame—in tree risk assessment, the time period for which an assessment is defined.

tip-tie—tying a line on the tip (brush) end of a branch to be removed, allowing the branch to be lowered butt-end first.

tomogram—image, produced by processing signals (including, but not limited to, sound waves, X-rays, and magnetic fields) and transmitted through an object that depicts the relative density of that object.

topiary—a formal pruning system that uses a combination of pruning, supporting, and training branches to orient a plant into a desired shape. Hedging is a subset of topiary.

topping—reduction of tree size by cutting live branches and leaders to stubs, without regard to long-term tree health or structural integrity.

tracheid—elongated, tapering xylem cell that is dead at maturity and is adapted for the support and transport of water and elements.

transpiration—water vapor loss, primarily through the stomata of leaves.

transplant shock—plant stress following transplant; characterized by reduced growth, wilting, dropping foliage, or death.

transverse crack—crack in a tree that forms across the short axis of the trunk, stem, or root.

tree bylaws—tree-specific laws that define a public agency's authority, describe required conditions or actions, establish penalties for nonconformance, and identify who is responsible for enforcement and oversight (also called ordinances or municipal code, depending on region).

tree guard—device for protecting the trunk of a tree from animal or mechanical injury.

tree inventory—record of trees within a designated area that provides specified identification and condition information to be used for management decisions and actions.

tree island—enclosed planting bed surrounding a tree, often within a paved area or adjacent to a street.

tree officer—in the United Kingdom and other countries, an individual responsible for the care and protection of public trees and green spaces.

tree ordinance—tree-specific law that defines a public agency's authority, describes required conditions or actions, establishes penalties for nonconformance, and identifies who is responsible for enforcement and oversight (also called bylaws or municipal code, depending on region).

tree preservation order (TPO)—in the United Kingdom, Australia, and some other countries, a legal regulation, established by a local authority, that protects a tree or multiple trees.

tree protection plan—detailed plan that outlines specific measures to protect trees during construction.

tree protection zone (TPZ)—area defined during site development, where construction activities and access are limited to protect the tree(s) and soil from damage, and to sustain tree health and stability. There are several methods to establish the TPZ.

tree risk assessment—a systematic process to identify, analyze, and evaluate tree risk.

tree risk management—application of policies, procedures, and practices to identify, evaluate, mitigate, monitor, and communicate tree risk.

tree spade—mechanical equipment to dig, transport, and replant trees with a sufficiently large volume of roots and soil.

tree stress—factor that affects the health or condition of a tree.

tree warden—title given to an individual who cares for trees on public town lands, mostly used in several New England states (United States) and in the United Kingdom, where they are also called tree officers.

tree well—wall constructed around a tree to maintain the original grade between the trunk and the wall, with a raised grade beyond the wall. Used to protect the trunk and root zone from the negative effects of excessive soil accumulation.

tree wrap—material used to wrap the trunks of newly planted or transplanted trees or to protect thin-barked mature trees when they are newly exposed to the sun.

trenching—linear or curvilinear open excavation, often used to install utilities or structural footings. Can cause tree root damage.

triple-action—pertaining to a carabiner, requiring three distinct motions to open the gate.

tropism—tendency of growth or variation of a plant in response to an external stimulus, such as gravity (geotropism) or light (phototropism).

trunk flare—transition zone from trunk to roots where the trunk expands into the buttress or structural roots.

trunk formula technique (TFT)—technique for developing a cost basis that involves extrapolating the purchase cost of a nursery-grown tree up to the size of the subject tree being valued.

tunneling—digging or boring, often with special machinery and shoring or other supports, below the surface of the ground without an open trench. Alternative for installation of underground utilities that avoids cutting of tree roots or damage to hardscape or existing utilities.

turgid—fully hydrated to a normal state of distension.

U

urban forest—the sum of all woody and associated vegetation in and around dense human settlements, from small communities to metropolitan regions, including street, residential, and park trees, greenbelt vegetation, trees on unused public and private land, trees in transportation and utility corridors, and forests on watershed lands.

urban forester—individual trained in or practicing urban forestry.

urban forest management plan—document that describes how urban forestry goals are to be accomplished within a defined time frame; includes tasks, priorities, best management practices, standards, specifications, budgets, and staffing analyses.

urban forestry—management of naturally occurring and planted trees and associated plants in the built environment.

V

variety—naturally occurring subdivision of a species having a distinct difference and breeding true to that difference.

vascular discoloration—darkening of the xylem or phloem of woody plants in response to disease, insect boring, or injury.

vascular plant—plant with xylem and phloem elements for conducting water, nutrients, and photosynthates.

vector—(1) in pathology, biotic or abiotic agent that transmits a pathogen. (2) in mechanics or rigging, quantity that has both a magnitude and a direction (e.g., force).

vertical mulching—an aeration or fertilization technique. Drilling vertical holes in the soil and filling them with materials to improve aeration.

vessel—end-to-end, tubelike, water-conducting cells in the xylem of angiosperms.

volatilization—conversion of a solid or liquid into a gas or vapor.

W

water budget—a calculation, based on precipitation and environmental factors, that establishes the minimum irrigation needed to maintain plant health.

water-holding capacity—ability of a soil to hold moisture.

water-insoluble nitrogen (WIN)—nitrogen fertilizer in a form that is not readily soluble in water.

watersprout—upright, epicormic shoot arising from the trunk or branches of a plant above the root graft or soil line. Incorrectly called a sucker.

white rot—fungal decay of wood in which both cellulose and lignin are broken down.

whorled—leaves, twigs, or branches arranged in a circle around a point on the stem.

wildlife—animals living in a natural, undomesticated state.

wilt—(1) n. loss of turgor and subsequent drooping of leaves and young stems; a symptom.
(2) n. infectious disease caused by a particular agent on a particular host or range of hosts.
(3) v. to lose turgor or to wilt.

wire basket—type of metal basket used to support the root ball of a balled-and-burlapped tree or a tree dug with a tree spade.

wire rope—cable fabricated from individual wires twisted together in a uniform helical arrangement. Wire rope used in arboricultural applications typically contains 7 groups of 19 wires each.

witch's broom—plant disorder characterized by a shortening of the internodes and a proliferation of terminal shoots forming a dense, brushlike mass of twigs.

working-load limit (WLL)—tensile strength divided by design factor. Maximum load that should not be exceeded in a piece of equipment, rope, or rope assembly when performing its normal working function.

work plan—predetermined, orderly means for job completion.

work-positioning lanyard—lanyard used in climbing, often as a secondary means of attachment.

work zone—defined area of a jobsite, marked with caution signs and/or cones, where potential hazards exist and safety measures are in place to avoid accidents.

wound dressing—compound applied to tree wounds or pruning cuts.

woundwood—lignified, differentiated tissues produced on woody plants as a response to wounding.

X

xeriscaping—use of plant materials (usually native plants) and practices that minimize water use. Term for environmentally friendly form of landscaping.

xylem—main water- and mineral-conducting (unidirectional, up only) tissue in trees and other plants. Provides structural support. Arises (inward) from the cambium and becomes wood after lignifying.

RECOMMENDED RESOURCES

The following are all the works and sources cited in each chapter as Recommended Resources.

American National Standards Institute. 2017. *American National Standard for Arboricultural Operations—Safety Requirements* (ANSI Z133-2017). Champaign (IL, USA): International Society of Arboriculture.

American National Standards Institute. 2017. *American National Standard—Tree, Shrub, and Other Woody Plant Management—Standard Practices (Pruning)* (ANSI A300, Part 1). Londonderry (NH, USA): Tree Care Industry Association.

American National Standards Institute. 2017. *American National Standard—Tree, Shrub, and Other Woody Plant Management—Standard Practices (Tree Risk Assessment a. Tree Failure)* (ANSI A300, Part 9). Londonderry (NH, USA): Tree Care Industry Association.

American National Standards Institute. 2018. *American National Standard—Tree, Shrub, and Other Woody Plant Management—Standard Practices (Planting and Transplanting)* (ANSI A300, Part 6). Manchester (NH, USA): Tree Care Industry Association.

American National Standards Institute. 2018. *American National Standard—Tree, Shrub, and Other Woody Plant Management—Standard Practices (Soil Management a. Assessment, b. Modification, c. Fertilization, and d. Drainage)* (ANSI A300, Part 2). Manchester (NH, USA): Tree Care Industry Association.

Arborists' Knots for Climbing and Rigging [DVD and workbook]. 2006. Champaign (IL, USA): International Society of Arboriculture.

Bond J. 2013. *Tree Inventories*. 2nd Ed. Champaign (IL, USA): International Society of Arboriculture. 35 p. (Best Management Practices).

British Standards Institution. 2010. *Tree Work. Recommendations.* (BS 3998:2010). London (UK): BSI Standards Limited. 76 p.

Costello LR, Perry EJ, Matheny NP, Henry JM, Geisel PM. 2003. *Abiotic Disorders of Landscape Plants: A Diagnostic Guide*. Richmond (CA, USA): University of California, Agriculture and Natural Resources. 242 p.

Council of Tree and Landscape Appraisers. 2019. *Guide for Plant Appraisal*. 10th Ed. Atlanta (GA, USA): International Society of Arboriculture. 181 p.

Dirr MA. 2009. *Manual of Woody Landscape Plants*. 6th Ed. Champaign (IL, USA): Stipes Publishing LLC. 1325 p.

Dirr MA. 2011. *Dirr's Encyclopedia of Trees and Shrubs*. Portland (OR, USA): Timber Press. 952 p.

Donzelli PS, Lilly SJ, Arbor Master Training Inc. 2001. *The Art and Science of Practical Rigging* [DVD and book]. Champaign (IL, USA): International Society of Arboriculture. 172 p.

Dujesiefken D, Liese W. 2015. *The CODIT Principle: Implications for Best Practices*. Champaign (IL, USA): International Society of Arboriculture. 162 p.

Dunster JA, Smiley ET, Matheny N, Lilly S. 2017. *Tree Risk Assessment Manual*. 2nd Ed. Champaign (IL, USA): International Society of Arboriculture. 194 p.

Fite K, Smiley ET. 2016. *Managing Trees During Construction*. 2nd Ed. Champaign (IL, USA): International Society of Arboriculture. 37 p. (Best Management Practices).

Gilman EF. 2012. *An Illustrated Guide to Pruning*. 3rd Ed. Clifton Park (NY, USA): Delmar, Cengage Learning. 476 p.

Recommended Resources

Hirons AD, Thomas PA. 2018. *Applied Tree Biology*. Hoboken (NJ, USA): Wiley Blackwell. 432 p.

Introduction to Arboriculture: Abiotic Disorders [online course]. Atlanta (GA, USA): International Society of Arboriculture. https://wwv.isa-arbor.com/store/product/740

Introduction to Arboriculture: Biotic Disorders [online course]. Atlanta (GA, USA): International Society of Arboriculture. https://wwv.isa-arbor.com/store/product/741

Introduction to Arboriculture: Early Care [online course]. Atlanta (GA, USA): International Society of Arboriculture. https://wwv.isa-arbor.com/store/product/746

Introduction to Arboriculture: Fertilization [online course]. Atlanta (GA, USA): International Society of Arboriculture. https://wwv.isa-arbor.com/store/product/745

Introduction to Arboriculture: General Diagnosis [online course]. Atlanta (GA, USA): International Society of Arboriculture. https://wwv.isa-arbor.com/store/product/742

Introduction to Arboriculture: Identification Principles [online course]. Atlanta (GA, USA): International Society of Arboriculture. https://wwv.isa-arbor.com/store/product/771

Introduction to Arboriculture: Lightning Protection Systems [online course]. Atlanta (GA, USA): International Society of Arboriculture. https://wwv.isa-arbor.com/store/product/747

Introduction to Arboriculture: Pest Management Techniques [online course]. Atlanta (GA, USA): International Society of Arboriculture. https://wwv.isa-arbor.com/store/product/750

Introduction to Arboriculture: Plant Health and Pest Control Practices [online course]. Atlanta (GA, USA): International Society of Arboriculture. https://wwv.isa-arbor.com/store/product/751

Introduction to Arboriculture: Plant Health Basics [online course]. Atlanta (GA, USA): International Society of Arboriculture. https://wwv.isa-arbor.com/store/product/761

Introduction to Arboriculture: Planting [online course]. Atlanta (GA, USA): International Society of Arboriculture. https://wwv.isa-arbor.com/store/product/749

Introduction to Arboriculture: Safety [online course]. Atlanta (GA, USA): International Society of Arboriculture. https://wwv.isa-arbor.com/store/product/759

Introduction to Arboriculture: Selection [online course]. Atlanta (GA, USA): International Society of Arboriculture. https://wwv.isa-arbor.com/store/product/744

Introduction to Arboriculture: Soils [online course]. Atlanta (GA, USA): International Society of Arboriculture. https://wwv.isa-arbor.com/store/product/755

Introduction to Arboriculture: Tree Anatomy [online course]. Atlanta (GA, USA): International Society of Arboriculture. https://wwv.isa-arbor.com/store/product/738

Introduction to Arboriculture: Tree Physiology [online course]. Atlanta (GA, USA): International Society of Arboriculture. https://wwv.isa-arbor.com/store/product/739

Introduction to Arboriculture: Trees and Construction [online course]. Atlanta (GA, USA): International Society of Arboriculture. https://wwv.isa-arbor.com/store/product/754

Introduction to Arboriculture: Trees and Water [online course]. Atlanta (GA, USA): International Society of Arboriculture. https://wwv.isa-arbor.com/store/product/756

Introduction to Arboriculture: Tree Support Systems [online course]. Atlanta (GA, USA): International Society of Arboriculture. https://wwv.isa-arbor.com/store/product/748

Jepson J. 2000. *The Tree Climber's Companion*. 2nd Ed. Longville (MN, USA): Beaver Tree Publishing. 104 p.

Johnson WT, Lyon HH. 1991. *Insects That Feed on Trees and Shrubs*. 2nd Ed. Ithaca (NY, USA): Comstock Publishing Associates, Cornell University Press. 560 p.

Lilly SJ, Gilman EF, Smiley ET. 2019. *Pruning*. 3rd Ed. Atlanta (GA, USA): International Society of Arboriculture. 63 p. (Best Management Practices).

Lilly SJ, Julius AK. 2021. *Tree Climbers' Guide*. 4th Ed. Atlanta (GA, USA): International Society of Arboriculture. 272 p.

Lloyd J. 1997. *Plant Health Care for Woody Ornamentals*. Champaign (IL, USA): International Society of Arboriculture and University of Illinois at Urbana-Champaign Cooperative Extension Service. 223 p.

Luley CJ. 2005. *Wood Decay Fungi Common to Urban Living Trees in the Northeast and Central United States*. Naples (NY, USA): Urban Forestry LLC. 60 p.

Luley CJ, Ali AD. 2009. *Pest Management in the Landscape: An Introduction*. Naples (NY, USA): Urban Forestry LLC. 89 p.

Matheny N, Clark JR. 1998. *Trees and Development: A Technical Guide to Preservation of Trees During Land Development*. Champaign (IL, USA): International Society of Arboriculture. 183 p.

Matheny NP, Clark JR. 2008. *Municipal Specialist Certification Study Guide*. Champaign (IL, USA): International Society of Arboriculture. 279 p.

Mattheck C, Breloer H. 1996. *The Body Language of Trees: A Handbook for Failure Analysis*. Norwich (UK): Stationery Office Books (TSO). 240 p.

Miller RW, Hauer RJ, Werner LP. 2015. *Urban Forestry: Planning and Managing Urban Greenspaces*. 3rd Ed. Long Grove (IL, USA): Waveland Press. 560 p.

Plant Finder. St. Louis (MO, USA): Missouri Botanical Garden. [Accessed 2021 Apr 14]. https://www.missouribotanicalgarden.org/plantfinder/plantfindersearch.aspx

Scharenbroch BC, Smiley ET. 2021. *Soil Management for Urban Trees*. 2nd Ed. Atlanta (GA, USA): International Society of Arboriculture. 70 p. (Best Management Practices).

SelecTree: A Tree Selection Guide. Cal Poly, Urban Forest Ecosystems Institute. [Accessed 2021 Apr 14]. https://selectree.calpoly.edu/

Shigo AL. 1986. *A New Tree Biology: Facts, Photos, and Philosophies on Trees and Their Problems and Proper Care*. Durham (NH, USA): Shigo and Trees, Associates. 595 p.

Sinclair WA, Lyon HH, Johnson WT. 2005. *Diseases of Trees and Shrubs*. 2nd Ed. Ithaca (NY, USA): Comstock Publishing Associates, Cornell University Press. 616 p.

Smiley ET, Graham AW Jr, Cullen S. 2015. *Tree Lightning Protection Systems*. 3rd Ed. Champaign (IL, USA): International Society of Arboriculture. 61 p. (Best Management Practices).

Smiley ET, Lilly S. 2014. *Tree Support Systems: Cabling, Bracing, Guying, and Propping*. 3rd Ed. Champaign (IL, USA): International Society of Arboriculture. 50 p. (Best Management Practices).

Smiley ET, Matheny N, Lilly S. 2017. *Tree Risk Assessment*. 2nd Ed. Champaign (IL, USA): International Society of Arboriculture. 86 p. (Best Management Practices).

Smiley ET, Werner L, Lilly SJ, Brantley B. 2020. *Tree and Shrub Fertilization*. 4th Ed. Atlanta (GA, USA): International Society of Arboriculture. 57 p. (Best Management Practices).

Tree Care Industry Association. 2018. *Tailgate Safety: Manual for Job Site Safety Meetings*. 7th Ed. Manchester (NH, USA): Tree Care Industry Association.

Urban J. 2008. *Up By Roots: Healthy Soils and Trees in the Built Environment*. Champaign (IL, USA): International Society of Arboriculture. 479 p.

Watson G. 2014. *Tree Planting*. 2nd Ed. Champaign (IL, USA): International Society of Arboriculture. 40 p. (Best Management Practices).

Watson GW, Himelick EB. 2013. *The Practical Science of Planting Trees*. Champaign (IL, USA): International Society of Arboriculture. 250 p.

Wiseman PE, Raupp MJ. 2016. *Integrated Pest Management*. 2nd Ed. Champaign (IL, USA): International Society of Arboriculture. 36 p. (Best Management Practices).

Z133 Online Course. Atlanta (GA, USA): International Society of Arboriculture. https://wwv.isa-arbor.com/store/product/4449

BIBLIOGRAPHY

The following resources were used in the writing of this study guide.

American National Standards Institute. 2013. *American National Standard—Tree, Shrub, and Other Woody Plant Management—Standard Practices (Supplemental Support Systems)* (ANSI A300, Part 3). Londonderry (NH, USA): Tree Care Industry Association.

American National Standards Institute. 2014. *American National Standard—Tree, Shrub, and Other Woody Plant Management—Standard Practices (Lightning Protection Systems)* (ANSI A300, Part 4). Londonderry (NH, USA): Tree Care Industry Association.

American National Standards Institute. 2016. *American National Standard—Tree, Shrub, and Other Woody Plant Management—Standard Practices (IPM)* (ANSI A300, Part 10). Londonderry (NH, USA): Tree Care Industry Association.

American National Standards Institute. 2017. *American National Standard for Arboricultural Operations—Safety Requirements* (ANSI Z133-2017). Champaign (IL, USA): International Society of Arboriculture.

American National Standards Institute. 2017. *American National Standard—Tree, Shrub, and Other Woody Plant Management—Standard Practices (Pruning)* (ANSI A300, Part 1). Londonderry (NH, USA): Tree Care Industry Association.

American National Standards Institute. 2017. *American National Standard—Tree, Shrub, and Other Woody Plant Management—Standard Practices (Tree Risk Assessment a. Tree Failure)* (ANSI A300, Part 9). Londonderry (NH, USA): Tree Care Industry Association.

American National Standards Institute. 2018. *American National Standard—Tree, Shrub, and Other Woody Plant Management—Standard Practices (Planting and Transplanting)* (ANSI A300, Part 6). Manchester (NH, USA): Tree Care Industry Association.

American National Standards Institute. 2018. *American National Standard—Tree, Shrub, and Other Woody Plant Management—Standard Practices (Soil Management a. Assessment, b. Modification, c. Fertilization, and d. Drainage)* (ANSI A300, Part 2). Manchester (NH, USA): Tree Care Industry Association.

Arborist Ropes [online course]. Atlanta (GA, USA): International Society of Arboriculture. https://www.isa-arbor.com/store/product/762

Arborists' Knots for Climbing and Rigging [DVD and workbook]. 2006. Champaign (IL, USA): International Society of Arboriculture.

ArborMaster Training Series I: Climbing Techniques [DVD and workbook]. 1997. Champaign (IL, USA): International Society of Arboriculture.

ArborMaster Training Series III: Chainsaw Safety, Maintenance, and Cutting Techniques [DVD and workbook]. 1997. Champaign (IL, USA): International Society of Arboriculture.

Ball DJ, Ball-King L. 2011. *Public Safety and Risk Assessment Improving Decision Making*. New York (NY, USA): Earthscan. 204 p.

Ball J, Johnson M. 2016. *First Responder Arborist Field Guide*. Manchester (NH, USA): Tree Care Industry Association. 90 p.

Ball J, Vosberg SJ, Walsh T. 2020. A Review of United States Arboricultural Operation Fatal and Nonfatal Incidents (2001-2017): Implications for Safety Training. *Arboriculture & Urban Forestry*. 46(2):67-83. https://doi.org/10.48044/jauf.2020.006

Bernick S, Smiley ET. 2015. *Tree Injection*. Atlanta (GA, USA): International Society of Arboriculture. 28 p. (Best Management Practices).

Bond J. 2013. *Tree Inventories*. 2nd Ed. Champaign (IL, USA): International Society of Arboriculture. 35 p. (Best Management Practices).

British Standards Institution. 2010. *Tree Work. Recommendations*. (BS 3998:2010). London (UK): BSI Standards Limited. 76 p.

Costello L, Watson G, Smiley ET. 2017. *Root Management*. Atlanta (GA, USA): International Society of Arboriculture. 41 p. (Best Management Practices).

Costello LR, Perry EJ, Matheny NP, Henry JM, Geisel PM. 2003. *Abiotic Disorders of Landscape Plants: A Diagnostic Guide.* Richmond (CA, USA): University of California, Agriculture and Natural Resources. 242 p.

Council of Tree and Landscape Appraisers. 2019. *Guide for Plant Appraisal.* 10th Ed. Atlanta (GA, USA): International Society of Arboriculture. 181 p.

Davidson JA, Raupp MJ. 2014. *Managing Insects and Mites on Woody Plants: An IPM Approach.* 3rd Ed. Manchester (NH, USA): Tree Care Industry Association. 198 p.

Day SD, Wiseman EP, Dickonson SB, Harris JR. 2010. Contemporary Concepts of Root System Architecture of Urban Trees. *Arboriculture & Urban Forestry.* 36(4):149-159.

Dirr MA. 2009. *Manual of Woody Landscape Plants.* 6th Ed. Champaign (IL, USA): Stipes Publishing LLC. 1325 p.

Dirr MA. 2011. *Dirr's Encyclopedia of Trees and Shrubs.* Portland (OR, USA): Timber Press. 952 p.

Donzelli PS, Lilly SJ, Arbor Master Training Inc. 2001. *The Art and Science of Practical Rigging* [DVD and book]. Champaign (IL, USA): International Society of Arboriculture. 172 p.

Dreistadt SH. 2004. *Pests of Landscape Trees and Shrubs: An Integrated Pest Management Guide.* 2nd Ed. Oakland (CA, USA): University of California, Agriculture and Natural Resources. 501 p.

Dujesiefken D, Liese W. 2015. *The CODIT Principle: Implications for Best Practices.* Champaign (IL, USA): International Society of Arboriculture. 162 p.

Dunster JA, Smiley ET, Matheny N, Lilly S. 2017. *Tree Risk Assessment Manual.* 2nd Ed. Champaign (IL, USA): International Society of Arboriculture. 194 p.

Ennos AR. 2016. *Trees: A Complete Guide to Their Biology and Structure.* Ithaca (NY, USA): Cornell University Press. 128 p.

Farrar JL. 1995. *Trees of the Northern United States and Canada.* Ames (IA, USA): Iowa State University Press. 502 p.

Ferrini F, van den Bosch CCK, Fini A. 2017. *Routledge Handbook of Urban Forestry.* London (UK): Routledge. 574 p.

First Aid: Information to Help You During a Medical Emergency. Mayo Clinic. [Accessed 2021 Mar 28]. www.mayoclinic.org/first aid

Fite K, Smiley ET. 2016. *Managing Trees During Construction.* 2nd Ed. Champaign (IL, USA): International Society of Arboriculture. 37 p. (Best Management Practices).

Franzmeier DP, McFee WW, Graveel JG, Kohnke H. 2016. *Soil Science Simplified.* 5th Ed. Long Grove (IL, USA): Waveland Press. 198 p.

Gilman EF. 1997. *Trees for Urban and Suburban Landscapes.* Albany (NY, USA): Delmar Publishers. 662 p.

Gilman EF. 2012. *An Illustrated Guide to Pruning.* 3rd Ed. Clifton Park (NY, USA): Delmar, Cengage Learning. 476 p.

Gilman EF, Kempf B, Matheny N, Clark J. 2013. *Structural Pruning: A Guide for the Green Industry.* Visalia (CA, USA): Urban Tree Foundation. 83 p.

Gustavsson R, Hermy M, Konijnendijk C, Steidle-Schwahn A. 2005. Management of Urban Woodland and Parks—Searching for Creative and Sustainable Concepts. In: Konijnendijk C, Nilsson K, Randrup T, Schipperijn J, editors. *Urban Forests and Trees: A Reference Book.* New York (NY, USA): Springer. p. 369-397. https://doi.org/10.1007/3-540-27684-X_14

Harris RW, Clark JR, Matheny NP. 2004. *Arboriculture: Integrated Management of Landscape Trees, Shrubs, and Vines.* 4th Ed. Upper Saddle River (NJ, USA): Prentice Hall. 578 p.

Hilbert DR, North EA, Hauer RJ, Koeser AK, McLean DC, Northrop RJ, Andreu M, Parbs S. 2020. Predicting Trunk Flare Diameter to Prevent Tree Damage to Infrastructure. *Urban Forestry & Urban Greening.* 49:126645. [Accessed 2021 Apr 14]. https://doi.org/10.1016/j.ufug.2020.126645

Hirons AD, Thomas PA. 2018. *Applied Tree Biology.* Hoboken (NJ, USA): Wiley Blackwell. 432 p.

Hodel DR. 2012. *The Biology and Management of Landscape Palms.* Portersville (CA, USA): The Britton Fund, Western Chapter of the International Society of Arboriculture. 176 p.

Jepson J. 2000. *The Tree Climber's Companion.* 2nd Ed. Longville (MN, USA): Beaver Tree Publishing. 104 p.

Jepson J. 2009. *To Fell a Tree: A Complete Guide to Tree Felling and Woodcutting Methods.* Longville (MN, USA): Beaver Tree Publishing. 166 p.

Johnson G, Giblin C, Murphy R, North E, Rendahl A. 2019. Boulevard Tree Failures During Wind Loading Events. *Arboriculture & Urban Forestry.* 45(6):259-269.

Bibliography

Johnson WT, Lyon HH. 1991. *Insects That Feed on Trees and Shrubs*. 2nd Ed. Ithaca (NY, USA): Comstock Publishing Associates, Cornell University Press. 560 p.

Lilly SJ, Gilman EF, Smiley ET. 2019. *Pruning*. 3rd Ed. Atlanta (GA, USA): International Society of Arboriculture. 63 p. (Best Management Practices).

Lilly SJ, Julius AK. 2021. *Tree Climbers' Guide*. 4th Ed. Atlanta (GA, USA): International Society of Arboriculture. 272 p.

Lim J, Kane B, Bloniarz D. 2020. Arboriculture Safety Standards: Consistent Trends. *Urban Forestry & Urban Greening*. 53(8). [Accessed 2021 Apr 14]. https://doi.org/10.1016/j.ufug.2020.126736

Lloyd J. 1997. *Plant Health Care for Woody Ornamentals*. Champaign (IL, USA): International Society of Arboriculture and University of Illinois at Urbana-Champaign Cooperative Extension Service. 223 p.

Luley CJ. 2005. *Wood Decay Fungi Common to Urban Living Trees in the Northeast and Central United States*. Naples (NY, USA): Urban Forestry LLC. 60 p.

Luley CJ, Ali AD. 2009. *Pest Management in the Landscape: An Introduction*. Naples (NY, USA): Urban Forestry LLC. 89 p.

Magarey RD, Borchert DM, Schlegel J. 2009. Global Plant Hardiness Zones for Phytosanitary Risk Analysis. *Scientia Agricola*. 64:54-59.

Matheny N, Clark JR. 1998. *Trees and Development: A Technical Guide to Preservation of Trees During Land Development*. Champaign (IL, USA): International Society of Arboriculture. 183 p.

Matheny N, Costello LR, Randisi C. 2021. *Designing and Managing Landscapes Irrigated with Recycled Water: A Guide for the San Francisco Bay Area*. Alexandria (VA, USA): WateReuse Association.

Matheny N, Costello LR, Randisi C, Gilpin RM. 2021. *Irrigating San Francisco Bay Area Landscapes Irrigated with Recycled Water*. California: WateReuse Association. https://watereuse.org/wp-content/uploads/2021/07/Bay-Area-RWGuide-June2021.pdf

Matheny NP, Clark JR. 2008. *Municipal Specialist Certification Study Guide*. Champaign (IL, USA): International Society of Arboriculture. 279 p.

Mattheck C, Breloer H. 1996. *The Body Language of Trees: A Handbook for Failure Analysis*. Norwich (UK): Stationery Office Books (TSO). 240 p.

Miller RW, Hauer RJ, Werner LP. 2015. *Urban Forestry: Planning and Managing Urban Greenspaces*. 3rd Ed. Long Grove (IL, USA): Waveland Press. 560 p.

North EA, D'Amato AW, Russell MB, Johnson GR. 2017. The Influence of Sidewalk Replacement on Urban Street Tree Growth. *Urban Forestry & Urban Greening*. 24:116-124.

Pallardy S. 2010. *Physiology of Woody Plants*. 3rd Ed. Cambridge (MA, USA): Academic Press. 1298 p.

Plant Finder. St. Louis (MO, USA): Missouri Botanical Garden. [Accessed 2021 Apr 14]. https://www.missouribotanicalgarden.org/plantfinder/plantfindersearch.aspx

Ritter M. 2011. *A Californian's Guide to the Trees Among Us*. Berkeley (CA, USA): Heyday Books. 192 p.

Scharenbroch BC, Smiley ET. 2021. *Soil Management for Urban Trees*. 2nd Ed. Atlanta (GA, USA): International Society of Arboriculture. 70 p. (Best Management Practices).

Schwab JC. 2009. *Planning the Urban Forest: Ecology, Economy, and Community Development*. Chicago (IL, USA): American Planning Association. 160 p.

Schwarze FWMR. 2008. *Diagnosis and Prognosis of the Development of Wood Decay in Urban Trees*. Rowville (VIC, Australia): ENSPEC. 336 p.

SelecTree: A Tree Selection Guide. Cal Poly, Urban Forest Ecosystems Institute. [Accessed 2021 Apr 14]. https://selectree.calpoly.edu

Shigo AL. 1986. *A New Tree Biology: Facts, Photos, and Philosophies on Trees and Their Problems and Proper Care*. Durham (NH, USA): Shigo and Trees, Associates. 595 p.

Shigo AL, Marx HG. 1977. *Compartmentalization of Decay in Trees*. USDA Agricultural Information Bulletin. 405. Washington (DC, USA): US Department of Agriculture, Forest Service. 73 p.

Simpson MG. 2010. *Plant Systematics*. 2nd Ed. Cambridge (MA, USA): Academic Press. 752 p.

Sinclair WA, Lyon HH, Johnson WT. 2005. *Diseases of Trees and Shrubs*. 2nd Ed. Ithaca (NY, USA): Comstock Publishing Associates, Cornell University Press. 616 p.

Slater D. 2016. *Assessment of Tree Forks*. Stroud (England, UK): Arboricultural Association.

Slater D, Bradley RS, Withers PJ, Ennos AR. 2014. The Anatomy and Grain Pattern in Forks of Hazel (*Corylus avellana* L.) and Other Tree Species. *Trees: Structure and Function*. 28:1437-1448.

Smiley ET, Graham AW Jr, Cullen S. 2015. *Tree Lightning Protection Systems*. 3rd Ed. Champaign (IL, USA): International Society of Arboriculture. 61 p. (Best Management Practices).

Smiley ET, Lilly S. 2014. *Tree Support Systems: Cabling, Bracing, Guying, and Propping*. 3rd Ed. Champaign (IL, USA): International Society of Arboriculture. 50 p. (Best Management Practices).

Smiley ET, Matheny N, Lilly S. 2017. *Tree Risk Assessment*. 2nd Ed. Champaign (IL, USA): International Society of Arboriculture. 86 p. (Best Management Practices).

Smiley ET, Werner L, Lilly SJ, Brantley B. 2020. *Tree and Shrub Fertilization*. 4th Ed. Atlanta (GA, USA): International Society of Arboriculture. 57 p. (Best Management Practices).

Stearn WT. 2002. *Stearn's Dictionary of Plant Names for Gardeners*. 3rd Ed. London (UK): Orion Publishing Company. 368 p.

Tailgate Safety: Manual for Job Site Safety Meetings. 7th Ed. 2018. Manchester (NH, USA): Tree Care Industry Association.

Urban J. 2008. *Up By Roots: Healthy Soils and Trees in the Built Environment*. Champaign (IL, USA): International Society of Arboriculture. 479 p.

Watson G. 2014. *Tree Planting*. 2nd Ed. Champaign (IL, USA): International Society of Arboriculture. 40 p. (Best Management Practices).

Watson GW, Himelick EB. 2013. *The Practical Science of Planting Trees*. Champaign (IL, USA): International Society of Arboriculture. 250 p.

Wiseman PE, Raupp MJ. 2016. *Integrated Pest Management*. 2nd Ed. Champaign (IL, USA): International Society of Arboriculture. 36 p. (Best Management Practices).

Wolf KL, Lam ST, McKeen JK, Richardson GRA, van den Bosch M, Bardekjian AC. 2020. Urban Trees and Human Health: A Scoping Review. *International Journal of Environmental Research and Public Health*. 17(12):4371. https://doi.org/10.3390/ijerph17124371

Z133 Online Course. Atlanta (GA, USA): International Society of Arboriculture. https://wwv.isa-arbor.com/store/product/4449

INDEX

Page numbers in bold refer to figures. Page numbers followed by "(t)" refer to tables.

A

ABC (airway, breathing, and circulation), 362
abiotic agents, 191, 199–205
abscission zone, 10
absorbing roots, 10–11, **10**, **11**
absorption, 14
    nutrient, 88, 93
    water, 67
acceptable risk, 265
access routes, 283
acclimation, 100
acidity *see* pH, soil
acoustic decay detection, 264, **264**
action threshold, 228–229, 230
adaptability, 100, 204
advanced assessment (risk), 251
adventitious branches, 257
adventitious buds, 7–8
aeration, 47, 68, 75, **287**
aerial lift truck, 324
aerial lifts, 330–331, **330**, **331**
aerial rescue, 331, **360**, 362–363, **363**
aerial roots, 18
aerobic soil conditions, 196
aesthetics, 150, **158**, 296
aggregates, 46–47
air excavation, 55, 264, 285, 287–288, **288**
air quality, 296, 299
air terminal, 183
alkalinity *see* pH, soil
allelopathy, 202–203, 224
alternate leaf arrangements, 32, **32**
alternate-host diseases, 210–211, **212**
aluminum, 200

American beech (*Fagus grandifolia*), **5**
American elm (*Ulmus americana*), 30
American National Standards Institute (ANSI), 312
    cables and hardware, 168
    planting specifications, 132
    pruning specifications, 141
    safety, 319, **319**, 320
amputations, 334
anaerobic fermentation, 204
anaerobic soil conditions, 195
anatomy
    bark, **4**, 6, **6**
    cells and tissues, 3–4
    leaves, 9–10, 30–31, **30**
    palms, **18**
    roots, 10–12, **10**, **11**
    shoot, **3**
    stems, 6–9
    tree, 3–12
    twig, **7**
    xylem and phloem, 4–6
anchor hitch, 352
anchors, 129–130, **129**, 168–170, **168**, **169**, 357
angiosperms, 3, 5, 27
anions, 49
ANSI *see* American National Standards Institute
anthocyanins, 10
antigibberellins, 159
aphids, 207
apical buds, 7, **7**
apical control, 16
apical dominance, 7, 16
apical meristems, 6, 19, **19**
apple scab, **210**
appraisal, 302–303
appropriate response process (ARP), 225

arboriculture, 3, 295
arborist blocks, 365, **366**, 367
arid regions, 67
Armillaria root rot, 232
as low as reasonably practicable (ALARP) principle, 267
ascenders, **346**, 347, **357**, 359
ascent, 359–360
Asian longhorned beetle, 206–207
attachment knots, 352–354
attachments (cabling), 170–171
augmentation (biological pest control), 235
automated external defibrillator (AED), 338
auxin, 15, **15**
available water, 52, **52**, **65**, 66
axial transport, 14
axillary buds, 7, **7**
axillary wood, 151–152, **152**

B

Bacillus thuringiensis, 239
back cuts, 328, 329, 369
back injuries, 334
backfill, 116, 121, 125, **277**, 278
bactericides, 236
balancing (rigging), 368, **368**
balled-and-burlapped (B&B) stock, 107, 117–118, **117**, **120**, 121, 123–124, **133**, 205
barber chair, 329, **329**
bare-root stock, 115–116, **115**, **116**, 129
bark beetles, 206–207
bark, **105**
    anatomy, **4**, 6, **6**
    chips, 73
    exfoliating, 103
    included, 9, 144, **145**, 257, **257**
    injury treatment, 286

bark tracing, 286
barrier zone, 17
barriers, 282, 282–283, 323
basal anchor, 357
basal rot, 260, **260**
basic assessment (risk), 250–251, **250**
basin irrigation, 70
baskets, 117–118, **117**, **118**
battery-powered hand tools, 324
beavers, 209
bee stings, 366, **366**
beech (*Fagus*) trees, 6
bend radius, 170
bends, 350
benefits of trees, 296–299
berms, 121, **122**
best management practices (BMPs), 312, **312**
    construction sites, 275
    fertilization, 83
    lightning protection systems, 184
    planting specifications, 115, 123, 132
    pruning, 141, 266
    transplanting, 123
    tree support, 167
    urban forestry, 312, **312**
biochar, 93
biodegradable burlap, 117, **117**
biodegradable containers, 116
biodiversity, 308
biofuel, 308
biohazard bags, 333
biological control, 235, **235**
biology
    soil, 50–51, **50**
    tree, 1–19, 150–152
biomass, 297
biorational control products, 238–239
biotic agents, 191, 205–212
bipinnately compound leaf, 30, **31**
birds, 209, 261, 309
black vine weevils, 206
Blake's hitch, 351
bleeding, 333–334, **333**

blight, 192, 210
blocks, arborist, 365, **366**, 367
blue infrastructure, 303
Bluetooth communications system, 323, 362, **362**
body-thrusting, 356, **356**, 359
bonsai, 142
boots, 321, 325
borers (insects), 206–207
boron, **85**, 89, 200
botanical names, 28
botanical pesticides, 239
bowline, 353
box cabling, 176, **176**
bracing
    see also cabling
    palms, 125, **130**
    support system, 178–179, **178**, **179**
brackets, 258, 261
braided ropes, 348, **349**
branch bark ridge, **8**, 9, **151**, 152
branch collar, **8**, 9, 151, **151**
branch protection zone, 152, **152**, 153
branch unions, **8**, 9, 144–146, 151, 174, 175, 367
branches, 6, 8
    see also pruning
    adventitious, 257
    attachment, 144, 150, 151, 257
    broken or hanging, 257
    and disorder diagnosis, 195–196
    permanent, 146
    scaffold, 107, 146
    temporary, 146
    and tree selection, 107
    types of, 107, 146
branch-removal cut, 152–153, **153**
broad-leaved evergreens, 27
broken branches, 257
bronzing, 194
brown infrastructure, 303
brown rot fungi, 259, **259**
brush chippers, 321, 329–330, **329**
bubblers, 69
bucket auger, **44**
buds, 4, 7–8, **7**, 19, **19**, 30, 125

buffering capacity, 49
bulk density, 47–48, **47**, 66
burn (fertilizer), 88, 90
burns (injury), 336–337
butterfly knot, 355
buttress roots, 18, 262
butt-tying, 367, 367, **367**
bylaws, 310–311

C

cable stops, 170–171, 172(t)–173(t)
cabling
    see also bracing
    cable types, 168
    configurations, **176**
    hardware, 168–171
    installation techniques, 175–177, **176–177**
    steel cable systems, 167–171, **168**
    systems, comparison of, 172(t)–175(t)
calcitic limestone, 89
calcium, 84, **85**, 89
California black oak (*Quercus kelloggii*), **5**
calipers, 123
cambium, 4, **4**, 6
cambium savers, 347
Canadian Standards Association (CSA), 320
cankers, 68, 192
canopy anchor, 357
canopy cover assessment, 304
capillary water, 52, 66
carabiners, 346, 361, 365
carbohydrates, 4, 6, 8, 12, 13, 14, 146, **225**
carbon, 83
carbon dioxide, 10, 12, 13, 52, 300, 302
carbon sequestration, 296, **300**, 302
cardiopulmonary resuscitation (CPR), 331, 332, 337–338, **337**, 363
    with breaths, 338
    hands-only, 337–338

carotenoids, 10
carpenter ants, 261, **261**
cation exchange capacity (CEC), 49, **49**, 57, 86
cations, 49, **49**
cavities, 249, 261, 264
cedar-apple rust, **212**
cell turgor, 13
cells, 3–4
cellulose, 4, 224, 259
central cylinder, **18**
chain saws, 321, 325–326, **325**, **326**, 333–334, **334**, 361–362
chaps, 321, 325
chelated iron sprays, 92
chelates, 89, 90
chemical control, 235–239, **236**, **237**, **238**, **239**
chemical injury, 203, 278
chewing insects, 206–207
chilling injury, 201
chippers, 329–330, **329**
chloride salts, 72
chlorine, 89, 90
chlorophyll, 9, 12, **12**, 84
chloroplasts, 9, 12, **12**
chlorosis, 67, 85, 89, 92, 192, **192**, 195
cicadas, 206
circling roots, 108, 116–117, **121**, 205
city foresters, 295
class (taxonomy), 27, 28(t)
classification *see* taxonomy
clay, 46, **46**, **47**, 49, 52, **65**, 66, 68, **74**, 75, **121**
clearance pruning, 147, **147**, 159
climate, 102(t), 122, 126
climate change, 107, 296, 302
  adaptation, 302
  mitigation, 296, 302
climbers and climbing
  aids, 346–348
  with chain saw, 326
  design and limitations, 349–350
  equipment, 345–348
  knots, 350–355

climbers and climbing (*continued*)
  limb walking, 360, **361**
  preclimb inspection, 355–356, **355**
  redirects, 360–361, **361**
  rope installation and ascent, 359–360
  ropes, 348–349, **349**
  safe work practices, 361–362
  systems, 356–358
  tie-in point, 347, **356**, 358–359, **358**
climbing harness, 345, **346**
climbing hitches, 350–352, 360
climbing lines, 326, 345, 356–357, **356**, 358–359, 360, 360
climbing saddle *see* climbing harness
climbing spurs/spikes, 155, **157**, 347, 359–360, **360**
closed fracture, 335
coarse-textured soil, 46, **46**, 49, 66
coast live oak, **67**
coconut palm, 34
CODIT *see* Compartmentalization of Decay in Trees
codominant stems, 257, **257**
  pruning, 144, **145**, 146
  support systems for, 167, **167**, 175
collar rot, 71, 74
color
  foliage, 10, 198
  pigments, 10
  soil, 45
command-and-response system, 322
commercial revenue, 297
common names, 28
communication, 160
  construction site, 285–286, **286**
  and emergency response, 362, **362**
  for safety, 322–323, **322**, **323**
  urban forestry, 310, **310**
  urban planning, 304
communities, strengthening, 298
community engagement, 309–310, **310**

compaction *see* soil compaction
Compartmentalization of Decay in Trees (CODIT), 17, **17**, 19, 151
compartmentalization
  CODIT, 17, **17**, 19, 151
  defense mechanism, 17, 224, 261
  and tree biology, 151, 152, 153
compensatory pruning, 130
competition, 202
complete fertilizer, 88
compost, 56, 234
composting, 309, **309**
compound leaf, 30–31, **30**, **31**, 32, **35**
compression wood, 261
conductors, 183, **183**
conifers, 5, 27, 32–33, **33**, **34**, 116, 159, 207, 259
conks, 258, 261
connecting links, 346
consequences of failure, 248, **248**, 255
conservation
  energy, **298**
  water, 71–73
conservation (biological pest control), 235
construction
  belowground damage, 276, **276**
  earlier involvement of arborists in, 278–279, **279**
  land development, 275
  practices, 275–276
  preservation during planning and design phase, 278–282
  storage areas, **283**
  treatment of damaged trees, 286–289
  tree protection during, 276–278, 282–286
contact pesticides, 236
container-grown stock, 116–117, 205
containerized stock, 107, **108**, 116–117, 121, **121**
contaminants *see* pollutants
controlled-release nitrogen (CRN) fertilizers, 88–89

conventional notch, 328
cooling, 296, **299**
cooling packs, 335
copper, **85**, 89
cork, **4**, 6
cork cambium, 4
cork oak (*Quercus suber*), 6
cortex, **4**, 18
cost approach (appraisal/valuation), 302
cost-benefit analysis, 305
costs of trees, 296, 299–301
Council of Tree and Landscape Appraisers (CTLA), 303
Covenants, Conditions, and Restrictions (CCRs), 310
cow hitch with half hitch, 354
CPR *see* cardiopulmonary resuscitation
cracks, 257–258
cranes, 125, 323
crime, 298
critical root zones (CRZ), 281, **281**
crown gall, 211
crowns
    see also topping
    construction injury, 276
    density, reducing, 148–149
    elevation, 147
    injury treatment 286
    reduction, 148
    restoration, 149, **149**
crusts (soil), 48
CTLA *see* Council of Tree and Landscape Appraisers
cultivars, 29, 104, 106
cultural control, 233–235, **233**
cultural heritage of trees, 299
cuticle, **9**, 10, 13, 67, **67**
cut-resistant pants/chaps, 325
cutting techniques
    felling, 327–329, **327**, **328**, **329**, 368–369
    pruning, 152–155, **153**, **154**, **155**
    rigging, 368–369, **369**
cycles to failure, 349
cytokinins, 15

D

data analysis, 305
dead branches, removal of, **154**
dead trees, 256, **256**
dead-end grips, 170, **170**
dead-end hardware, 169, 178
death, 66, 117, 150, 192, 195, 198, 225, **226**
decay, 150, 258–263, 356
    definite indicator, 261
    definition, 192
    indicators, 258, 261
    internal, assessment of, 263–264
    potential indicators, 261
    and support systems, 177, 178
    trunk flare, 200
deciduous trees, 10, 27, 31, 116
decline diseases, 196–197, 199
decurrent growth, 16, **16**
deer, 209
defense mechanisms, 17, 224–225
deficiencies, nutrient, 85, **85**, 158, **192**, 200, 234
defoliation, 202, 207, **210**
degree-day model, 230–231
deodar cedar (*Cedrus deodara*), 105
desiccation, 10, 116, **116**, 117
design criteria, 102–103
design factor, 349
dieback, **199**, **289**, 356
    and belowground construction damage, 276, **276**
    causes of, 66, **196**, 199, **199**, 208
    definition of, 192
    root, 125
    symptoms of, 198
differentiation, 3–4
diffuse porous, 5, **5**
direct contact (electricity), 323–324, **324**
directional pruning, 147
diseases and disorders, 150, 156, 191, 209–212
    see also Plant Health Care (PHC)
    abiotic agents, 191, 199–205

diseases and disorders (*continued*)
    biotic agents, 191, 205–212
    decline diseases, 196–197
    diagnosis, 191–197, **191**, **195**, 198
    exotic pests, 212
    laboratory assistance, 213–214, **213**
    pathogens, 209–212
    plant disease triangle, 209, **209**
    plant inspection for, **228**, **229**, 230
    resistance, 103(t)
disorders *see* diseases and disorders
distel hitch, 352
diversity, 232
    age and size, 307
    biodiversity, 308
    food web, 50, **50**
    plant species, 232, 307
division (taxonomy) *see* phylum (taxonomy)
dolomitic limestone, 89
dormancy, 7, **8**, 66, 67, 116, 156
double fisherman's knot, 354
double-braid ropes, 348
doubled rope technique (DdRT) *see* moving rope systems (MRS)
drainage, 102
    construction site, 286
    flooding, 73–76, **74**, **75**, **76**
    improvements, 55–56, **56**
    planting hole, 121, **121**
    surface, 76
    tiles, 75, **75**, **76**
    and tree selection, 102(t), 103
drill-hole method, 91, **91**
drip irrigation, 69–70
drip line, 281
drop cut, 368
drop zone, 322, 327
drought, 65, 157
drought tolerance, 67, 103(t)
drum lace, 124
Dutch auger, **44**
Dutch elm disease, 207, 210, **211**, 232
dying parts, 256–257, **256**

Index

dynamic loads, 365, **365**, 368, **368**
dynamic support systems *see* rope support systems

E

earlywood, 5
eastern tent caterpillar, **206**
economic benefits of trees, 296–297
economic costs of trees, 301
ecosystems, 222, 308
education, 227, 309–310, 319, **319**
electric lines, 251–252
electric shock, 324
electrical conductors, 324–325, **324**, 355
electrical hazards, 323–325, **324**
electrostatic spraying, 237
elm leaf beetles, 206
emergency medical services (EMS), 332, 333–334, 336, 338, 362, 363
emergency planning, 307
emergency response, 307, 331–338, 362–363, **362**, **363**
endline clove hitch with two half hitches, 353, 367
energy conservation, 296, **298**
environmental benefits of trees, 296
environmental costs of trees, 299–300
epicormic shoots, 8
epidermis, **4**, **9**
epinephrine auto-injector, **366**, 367
equipment *see* tools and equipment
eradication, pest, 233
eriophyid mites, 208, **208**
erosion, soil, 56, 73, 74, 296
escape route, 327, **327**
espalier, 142, **143**
essential elements, 12–13, 14, 72, 83–86, **83**
eudicotyledons (eudicots), 4, 27
evaporation, 13
evapotranspiration (ET), 68, **69**, 73
evergreens, 10, 27, 32–33, **33–34**, 100, 122

excavation, 278
excurrent growth, 16, **16**
exfoliating bark, 103
exotic pests, 212
exotic species, 106
extra-high-strength (EHS) cable, 168
exudates, root, 50
eye bolts, 168, **168**, 172(t)–173(t), **177**
eye protection, 321, **321**
eye splice, 170, **170**

F

face mask/face shield, 321
failure modes, 264
failure potential, 256
fall distance, 255
fall protection, 345, **346**
family (taxonomy), 27–28, 28(t)
felling, tree, 327–329, **327**, **328**, **329**
fences, 266, 281, 282–283
fertilization, 83
  best management practices, 83
  construction site, 289
  excessive application, 222, 234
  and Plant Health Care (PHC), 224
  prescription, 86, **87**
  soil additives, 73, 93
  transplants, 127
fertilizer analysis, 88, **89**
fertilizer burn, 88, 90
fertilizer ratio, 88
fertilizers
  application of, 90–93
  micronutrients, 89–90
  secondary macronutrients, 89
  types of, 88–89
  uptake, 86, 90, 91
fibers, 5
field capacity, 52, **52**, **65**, 66
figure-8 knot, 354
fill, 278, 285
fine-textured soil, 46, **46**, 49, 66
fire blight, 211

fire extinguishers, 323
fire safety, 323
firs, 32–33
first aid, 331, 332–338, 362
first-degree burns, 335
flags, 323
flammable liquids, 323
flat inserts, 171, **171**
flexure wood, 261
flooding, 66, 73–76, **74**, 204
floodplains, 74
flower buds, 7, **7**
flowers, 19, 103, **104**, 147, 156, **156**, 158
flush cuts, 154, **155**
foliage, **104**
  awl-like, 33, **34**
  color, 10, 198
  diagnosis, 195, 198
  diseases, 209–210
  scalelike, 33, **34**
  tree selection, 107
foliar analysis, 57, 86–87, **87**, 88
foliar application, 92–93, **93**
foot ascenders, **346**, 347, 359
footlocking, 359
footwear, 321, **321**, 324, 325
forest therapy, 299
forestry *see* urban forestry
form (taxonomy), 29
fractures, 334–335, **335**
frass, 192, **193**
freezing injury, 201, **201**
friction devices, **346**, 346, 365, 367
friction hitches, 350–352, 359
friction, 360, 364
friction-saving devices, 347, 357
fronds, 19, 34, **35**, 125, 157–158, 331
frost cracks, 201
fruiting bodies, 258, 260, **260**, 261, **261**
fruits, 19, **104**, **105**, 142, 147, **210**, 301
fuel spills, 323
full-thickness burns, 335–336
functional goals, 99–100, **100**

fungi
   see also mycorrhizae
   decay, 258–260, **259**, **260**
   diseases, 211
   rust fungi, 211
   and tree wrap, 130
   vascular wilt disease, 68, 210, **210**
fungicides, 125, 236

G

galls, 192, **194**, 208, **208**
gas exchange *see* transpiration
gasoline-powered equipment, 323
genus (taxonomy), 28, 28(t)
geogrid, 284
geotextile fabric, 54, 284
geotropism, 15–16
girdling, **128**, 129
girdling roots, 68, **108**, 117, 204–205, **204**
girth-hitching, 352, 361
global positioning system (GPS), 305
gloves, 321, 322, 326, 333, **333**
GPS (global positioning system), 305
grade *see* soil grade
grading, 277, 278, **277**
grapevine knot, 354
gravitational water, 52, **52**, 66
gray infrastructure, 303
green infrastructure, 303, **303**
gripper gloves, 322
ground terminal, 183
groundwater, 67
growing space, 101, **102**, **199**
growth, 195–196
   decurrent, 16, **16**
   excurrent, 16, **16**
   fast-growing trees, 104
   habit, 103(t), 104
   hormones, 15
   points, **8**, 19, **19**
   primary, 4
   rate, 224–225, **224**

growth, 195–196 (*continued*)
   regulation, 15–17, 159
   regulators, 15, 16
   response growth, 253–254, 261–263, **262**
   rings, 5
   root, 11, **11**, 118, **118**, 126
   seasonal, **14**
   secondary, 4
   and topping, 150
   twig, **7**, **229**
guard cells, **9**, 10, 13
guards, tree, 130, **131**
Guide for Plant Appraisal, 303
gummosis, 192
guying, 128–129, **128**, **129**, 179–180, **180**
gymnosperms, 3, 5, 27

H

habitat, 309, **309**
hail, **201**, 202
half hitch and running bowline, 354
hand protection, 322
hand tools, 124, 205, 264, 285, 324
hanging branches, 257
hard hats, 321
hardening off, 122
hardiness, 103(t), 105
hardiness zone maps, **106**
hardscapes, 297
hardware
   see also tools and equipment
   cabling, 168–171, **168–170**, 175–177, **177**, 361
   guying, 129, 130
   lightning protection, 183, **183**
   rigging, 365–366
harnesses, 345, **346**, 359
head injuries, 334
head protection, 321, **321**
heading cuts, 148, 149, 153
health benefits of trees, 297, 298–299
health costs of trees, 301

health problems
   see also diseases and disorders
   abiotic agents, 199–205
   biotic agents, 205–212
   diagnosis, 191–197, 198–199, 213
   Plant Health Care (PHC), 222–224
   stress, **196**
hearing protection, 321
heartwood, 6
heartwood rot, 260
heat exhaustion, 336
heat injuries, 336
heat islands, **299**
heat stroke, 336
heat zone maps, 105
heating costs, 296
helmets, 321, **321**, **345**
herbicides, 203, **203**, 236
herbivores, 224
high-pressure water injection, 70
hinge cut, 369
hinges, 328, **328**, 369
hitch cords, 346
hitches, 350
holly leafminer, **206**
honeybees, 261
honeydew, 207
horizons (soil), **43**, 44–45, **44**
hormones, 15, 159
hornets, 366
horticultural oils, 239, **239**
hoses, soaker, 71
hub-and-spoke cabling, 176, **176**
humidity, 13
hybrids, 28–29
hydrogels, 73
hydrogen, 83
hydrology, 284
   hydrozones, 71, **71**
   hygroscopic water, 52, 66

I

ice damage, 202
identification, tree, 25–35, **27**, **29**
   and diagnosis of disorders, 193

identification, tree (*continued*)
  keys, 34–35
  nomenclature, 28–29
  palms, 34, **35**
  principles, 29–33
  taxonomy, 27–28, 28(t)
implants, fertilizer, 92
included bark, 9, 144, **145**, 257, **257**
income approach
    (appraisal/valuation), 302
indirect contact (electricity),
    323–324
industry-based safety standards, 320
infectious agents, 191
infiltration, 52, **53**, 66, 68, 70, 73
inflorescence, 19, 147
infrastructure, 297, 303, **303**
injection method, 91, **91**, 92–93, **92**
injuries, 151
  see also Compartmentalization
    of Decay in Trees (CODIT);
    decay
  burns, 336–337
  chain saw, 333–334, **334**
  fractures, 334–335, **335**
  head, neck, and back, 334
  heat, 336–337
  lightning, 182
  personal, 332–338
  and pruning, 156
  tree, 200–205, **200, 201, 202,
    203, 204**, 276–278, 282,
    286–289
inoculants, 93
inorganic fertilizer, 88
inorganic mulches, 127
insect growth regulators, 239
insecticidal soaps, 239, **239**
insecticides *see* pesticides
insects, 150, 196
  and biotic disorders, **205**,
    206–207, **206, 207**
  chewing, 206–207
  decay indicators, 261, **261**
  life cycle of, 206
  stings/bites, 331, 336–337, **366**
  sucking, 207

insects (*continued*)
  susceptibility to, 105–106
inspections, 266
  of climbing equipment, 347–348,
    347, 348
  landscape, 228–231
  lightning protection systems, 184
  preclimb, 355–356, **355**
  support systems, 181
  of trees after construction, 286
Integrated Pest Management (IPM),
    227–228, **227**
internal cycling, 84
International Society of
    Arboriculture, 363
  see also best management
    practices
  ISA Certified Arborist Municipal
    Specialist® certification, 296
  Tree Risk Assessment
    Qualification (TRAQ), 248
internodal cuts, 143, **148**
internodes, **7**, 8, 31
introduced species, 106
introduction (biological pest
    control), 235
invasive pest management, 308
invasive species, 106, 212, 300, 308
ions, 49
iron, 13, **85**, 89, **192**
  chlorosis, 89
  deficiency, 89, 200
irrigation, 65, **65, 69**, 68–71, **72**, 234
  chronic stress, **197**
  construction site, 286
  distribution, 70
  duration, 70
  excessive, 204, **232**
  frequency, 70
  minimum, 71, **71**
  recycled/reclaimed water, 72–73
  schedules, 121
  systems, 69–71
  timing, 70
  for transplants, 126–127, **127**
  and tree selection, 103
  water budget, 70

irrigation bags, **127**
i-Tree (software), 302, 305

J

Japanese beetles, **205**, 206
J-hook, 168, 172(t)–173(t)
J-lag-anchored synthetic rope,
    174(t)–175(t)
job briefing, 322, **322**, 355

K

kerf, 369
kernmantle rope, 348
keys, identification, 34–35
kickback, 326, **326**
kickback quadrant, 326
kingdom (taxonomy), 27, 28(t)
kinked roots, 108
knobs, 143
knots, 350–355, 365–366

L

laceration, bleeding from, 333
ladders, 359
lag eye, 169
lag hook, 169–170, **169**,
    172(t)–173(t), **177**
lag-threaded rods, 178, **178**
land development, 275–276
landing zone, 322
landscape inspection, 228–231
landscape sanitation, 231, 235
lanyards
  see also climbing lines
  safety, 326
  work-positioning, 345, **346**, 360,
    362
lateral branches, 147, 153, **148**
lateral buds, 7, **7**
lateral roots, 11
latewood, 5
lawn, 43, 91
  care, 222

lawn (*continued*)
   irrigation for, **69**
lawn mowers, 73, **180**, 200, **200**, 205
leaching, 85, 90
leaders, 131, 144, **145**, 146
leaf apex, 31, **31**
leaf base, 31, **31**
leaf blade, **9**, 10, **30**
leaf blotch, 192, **194**, 198
leaf drop, 10, 65, 122, 198
leaf margin, 31, **32**
leaf scars, **7**, 8
leaf scorch, 68, 72, 192, **194**, 198, 211
leaf spot, 192, **194**, 198, **210**
leaf stalks, 10, **30**
leafhoppers, 207
leaflets, **30**, **31**
leafminers, 206, **206**
leans, 258, 327
leaves
   see also needles
   anatomy, 9–10, 19, 30–32, **30**, **31–32**
   arrangements, 31–32, **32**
   diagnosis of disorders, 198, 199
   identification, 30–32, **30**, **31**, **32**
   skeletonized, **205**, 206
leg protection, 321, **321**, 325
legislation, 106, 266, 309, 310–312
legumes, 51
lenticels, 6, **7**
Licuala spinosa, **18**
LiDAR (light detection and ranging), 250
life-support equipment, 345
light requirements, 99, 100, 101–102, 103(t)
lightning, 182
   candidates for protection, 182
   damage, 68, 182, **182**, 202, **202**
   protection system, 183–184
lignin, 4, 224, 259, **259**
likelihood matrix, **247**
likelihood of failure, 253
   loads, 254–255, **254**

likelihood of failure (*continued*)
   reducing, 146–147
   site factors, 254
   species, 255
   time frame, 253, **254**
   tree health, 255
   tree structural conditions, 253–254
likelihood of impact, 251, **253**
   occupancy rate, 252–253, **253**
   target zone, 252, **252**
   targets, 251–252, **251**
limb walking, 360, **361**
limited visual assessment (risk), 249–250, **249**
lines of sight, blocking, 300
lion tailing (lion's tailing), 149, **149**
liquid injection fertilization, 91, **91**
live oak (*Quercus virginiana*), 30
loads and loading
   dynamic, 365, **365**, 368, **368**
   and likelihood of failure, 254–255, **254**
   lines, 365
   and response growth, 261–262
   static, 365
   working load limit, 349
loam, 46, **65**
lodged branches, 257
low and slow principle (rigging), 364
lowering devices, 365

M

machine-threaded rods, 178, **178**
macronutrients, 84
   deficiencies, 89
   secondary, 89
macropores, 47, 48, **51**, 52, **65**, 76
magnesium, 84, **85**, 89, 200
maintenance
   costs, 300
   lightning protection systems, 184
   planning, 306–307, **306**
   site, 102(t)
   support systems, 181
   tree, 103(t), 306–307, **307**

mammals, 209, 224, 261
manganese, **85**, 89–90, 200
mechanical control, 235
mechanical friction devices, **346**, 346, 365
mechanical injuries *see* physical injuries
mechanical root pruning, 283
memory devices
   leaf arrangement, 32
   taxonomy, 28
mental health, 298
meristems, 3, **3**, 4, 125, 151
mice, 209
Michoacán hitch, 352
microbial pesticides, 239
microclimate, 101
micronutrients, 48–49, 85, 89–90, 92
micropores, 47, 52, **65**
micropulleys, 346, **346**, 347, 359
midline clove hitch, 353
midrib, **30**
Migratory Birds Convention Act (Canada), 309
mineralization, 51
minimum approach distances (MAD), 331, **331**, 355
minimum irrigation, 71, **71**
mites, 207–208, **208**, 235, 237
miticides, 236
mitigation, risk, 266–267
mobile elevating work platforms (MEWPs) *see* aerial lifts
mold, sooty, 207
molybdenum, **85**, 89
monitoring
   construction site, 289
   plant health, 228–231, **228**, **229**
monocotyledons (monocots), 18–19, 27, 34
monoculture, 232
morphology, and identification, 29–33
mortality spiral, 225, **226**
moving rope systems (MRS), 356–357, **356**, 358, 359

mowers, lawn, 200, **200**, **205**
mulches and mulching, 56–57, 200, **200**, 209, **233**, 234, 308–309
    application, **73**, 127
    construction site, 284, 286–287, **287**
    and fertilization, 93, **93**
    overmulching, 128
    transplants, 122, 127–128
    water conservation, 73
    wood chip, 308–309, **309**
municipal arborists, 295
municipal tree maintenance, **306**, 307
mushrooms, 258–259, 261
mycorrhizae, **11**, 12, 50–51, **51**, 90, 211

N

naked buds, 4
native species, 106
natural braces, 145–146
natural pruning system, 142
natural redirects, 361
naturalized species, 106
neck injuries, 334
necrosis, 192, 225, **226**
needles
    diagnosis of disorders, 199
    discolored, **208**
    identification, 32–33, **33–34**
neem, 224
nematodes, 208
nickel, 89, 90
nicotine, 224
nitrogen, 12, 72, 83, 84–85, **85**, 88, 89, 90
    deficiency, 85
    fertilization, pests promoted by, 84(t)
    fertilizer, 84, 224
    mobility in soil, 91
nitrogen fixation, 84
nodes, **7**, 8, 31
nomenclature, scientific, 28–29

noninfectious agents, 191
North American Migratory Bird Treaty Act (US), 309
notch cut, 369
notches, 328, **328**
nuisances, 301
nursery stock, 107–108, **108**, 115–118, **115**, 205, 232, 234
nutrient cycling, **45**, 51, 83
nutrient limitation, 86
nutrient uptake, and pH, 48–49, **85**
nutrition, 83
    deficiencies, 85, **85**, 86, 89, 158, **192**, 200, 234
    essential elements, 12–13, 72, 83–86, **83**
    nutrient demand profiles, 84(t)
nuts, 170, 178

O

oak wilt, 210, 232
occupancy rate, 252–253
Occupational Safety and Health (OSH) Act, U.S., 320
Occupational Safety and Health Administration (OSHA), 320
oozing, 192
open fracture, 335
open-face notch, 328–329, **328**
opposite leaf arrangements, 31–32, **32**
order (taxonomy), 27, 28(t)
ordinances, **306**, 310–311
organic amendments, 56
organic fertilizer, 88
organic horizons, 45, **45**
organic matter, 45, 49, 53, 56, **56**, 73, 85, 86, 93, 234
organic mulches, 73, 127
organs, 4
OSHA *see* Occupational Safety and Health Administration
osmosis, 14
overirrigation, 71
overpruning, 158

oxygen, 12, 13, 52, 74, **74**, 118, 200, 204
ozone, **299**

P

palisade layer, **9**
palm skirt, 331, **332**
palmately compound leaf, 30, **31**
palms, 4, **107**
    biology, 18–19
    bracing, 125, **130**
    identification, 34, **35**
    nutrient deficiencies, 85, **85**, 200
    pruning, 157–159, **157**
    safety, 331, **332**
    transplanting, 125, **125**
parasites, 235, **235**
parenchyma, 5, 6
parent material, 43–44
parking lots, 67, 72, **299**
parks, 298
part size, and consequences of failure, 255
partial-thickness burns, 335–336
pathogens, 191, 196, 209–212, 229, 235
pattern of abnormality, 193–195
pavement, suspended, 54, **55**
paving, 53
percolation, 52, 66
periderm, 4
periodical cicadas, 206
permanent wilting point, 52, **65**, 66, 203
permits, 124, 311
personal protective equipment (PPE), 321–322, **321**, 329, 333, **333**, 345, **345**
pesticide resistance, 237–238
pesticides, 221, 224, 235, 235–239, **237**, **238**, **239**
pests and pest control, 191
    exotic, 212
    growth regulators, 239
    Integrated Pest Management, 227–228, **227**

pests and pest control (*continued*)
  invasive pest management, 308
  key, 231
  lab samples, **231**
  life cycles, **229**
  management, 232–239
  monitoring, 229–230
  prevention, 231, 232
  resistance to, 103(t), 105
  resurgence, 237
  secondary pest outbreak, 237, **238**
  susceptibility to, 105–106, 232
petioles, 10, **30**
petiolule, **30**
pH, soil, 48–49, **48**, 57, 200
  analysis, 86
  nutrient uptake, 48–49, **85**
  site selection, 102(t)
  tree selection, 100, 102
phelloderm, **4**
phenology, 228, **228**
phenology calendar, 230
phenols, 224
phloem, **3**, 4, **4**, 6, **6**, **7**, 14
phosphorus, 72, 84, **85**, 88, 89
photosynthates, 12–13, 14, 224
photosynthesis, 9, 12–13, **12**, 19, 146
  nitrogen and, 84
  resource allocation, 223–224, **223**
phototropism, **15**, 16
phylum (taxonomy), 27, 28(t)
physical activity, 298, **301**
physical health, 298
physical injuries, 200–201, **200**
physiology, tree, 12–17
Phytophthora, 72, 232
phytoplasmas, 212
phytotoxicity, 72, 207, 39
pine needles, 73, 127
pines, 14, 16, 32–33, **33**–**34**, 193
pinnately compound leaf, 30, **31**
pith, **3**, **5**
planning
  construction, 278–282
  urban forestry, 303–310

plant growth hormones, 15
plant growth regulators, 15, 16, 159
Plant Health Care (PHC), 221, **221**
  definition and philosophy, 222
  healthy plants, 222–224
  key stressors and plants, 231
  management options, 233–239
  management strategies, 231–233
  monitoring, 228–231, **228**, **229**
  process, 225–231
plant information, 228, **228**
plant selection *see* tree selection
planting, 115
  care, 125–131
  holes, 118–119, **118**, 120, **120**, 125
  specifications, 132
  techniques, 118–122
  transplanting, 122–125
  and utilities, 119
pleaching, 142, **143**
points of contact, 326
poisonous plants, 331
pole pruners, 361
pole saws, 361
pollarding, 142, 143, **143**
pollutants
  air, **298**, **299**
  soil, 90
  tree damage from, 203
pore space (soil), 45, 46–48, **47**
portable watering device, 122
potassium, 12, 84, **85**, 88, 89
power lines *see* utilities
predators, **234**, 235
prescription fertilization, 86, **87**
primary growth, 4
primary suspension point, 358
procambium, **3**
professional collaboration, 304, 309–310
propagation of tropical trees, 18
property owners, 265, **265**, 266
property values, 297
propping, 180–181, **181**
protection factors, and consequences of failure, 255

protection zones, 280–282, **281**, 285
protective clothing, 321, **321**, 323, 325
pruning, 9, 103, 141, **141**, 167, 177, 200, 235
  amount of, 155–156
  of conifers, 159
  for construction injury treatment, 286
  cuts, 152–155, **153**, **154**, **155**
  of live branches, 150, 157
  of mature trees, 156
  mechanical root pruning, 283
  of palms, 157–159, **157**
  for pest management, 235
  plant growth regulators, 159
  risk mitigation, 266–267, **267**
  root, 122–123, **123**
  of shrubs, **158**, 159
  specifications, 160
  structural, 144, **145**, 159
  systems, 142–143
  timing of, 156–157, **156**, 232
  tools, 155
  training young trees, 146
  transplants, 130–131, **131**
  and tree biology/architecture, 150–152
  wound dressings, 155
pruning objectives, 143–144, **144**
  clearance, 147, **147**
  crown density, reducing, 148–149
  crown restoration, 149, **149**
  improving structure, 144–146
  managing wildlife habitat, 149–150
  risk of failure, reducing, 146–147
  size management, 148
Prusik hitch, 351
pseudobark, **18**
pubescence, 67
public outreach, urban forestry, 309–310
pull lines, 327, **329**
pulleys, 346–347, **346**, 361, 365
pyrethrin, 224

Index

Q

quaking aspen (*Populus tremuloides*), 8
quick links, 346
quick-release fertilizers, 90, 221

R

rabbits, 209
radial transport, 14–15
radial trenching, 287–288, **288**
ram's horn, 263
rays (transport cells), 6, **7**, 14
reaction wood, 261
reaction zone, 17
reactive force (chain saw), 326
reclaimed water, 72–73
recreation, 298, **301**
recycled water, 72–73
recycling of wood waste, 308–309
red maple (*Acer rubrum*), 15, 100
redirects, 360–361, **361**
reduction cuts, 147, 148, 153, **153**
regulations
    digging, 119
    pesticide, 236, 237
    safety, 319–320, **319**
    urban forestry, 310–312
    water-use, 70
reproduction, 18, 224
rescue, aerial, 331, **360**, 362–363, **363**
rescue pulleys, 346–347
residual risk, 267
resilience, 303
resistance-recording drill, 263, **263**
resource allocation (tree), 223–224, **224**
respiration, **12**, 13
response growth, 253–254, 261–263, **262**
restoration pruning, 149
retail revenues, 297
retreat path, 327, **327**
reusing of wood waste, 308–309

rhizosphere, 50, **50**
rigging line, 347
rigging point, 364, **366**, 367
rigging systems, 327, 363–364, **364**
    from below, 368, **369**
    cutting techniques, 368–369, **369**
    equipment, 365–366
    forces in, 364–365
    knots, 350–355
    low-and-slow principle, 364
    strategy, 369
    techniques, 367–368, **367**, **368**, **369**
    working-load limits, 349
right-of-way zones, 310
ring porous, 5
ring spot, 212
risk, 248, 300
    acceptable, 265
    perception, 265
    residual, 267
risk assessment, 247–248, **247**, 248–249, 286
    see also likelihood of failure; likelihood of impact
    advanced assessment (level 3), 251
    aerial inspection, 263
    basic assessment (level 2), 250–251, **250**
    decay, 258–263
    guiding premises of, 249
    internal decay, 263–264
    likelihood matrix, **247**
    limited visual assessment (level 1), 249–250, **249**
    risk analysis, 264–265
    risk-rating matrix, **248**
    root excavation, 264
    tree assessment, 255–256
risk management, 247–248, 265–266
    mitigation, 266–267
    plan, 306
    policy statement, **306**
    rating systems, 267
    residual risk, 267

risk management (*continued*)
    and tree maintenance, 306
    and urban forestry, 306, **306**
risk managers, 266, 301
risk rating, 265, 267
risk-rating matrix, **248**
rodents, 209
root ball, 199
    anchoring, 129–130, **129**
    frozen, 122
    measuring, 120, **120**
    before planting, 116, **120**
    and transplanting, 122, 123, **123**, 124, 126
    and tree selection, 107
root cap, 4, **10**, 50, **50**
root collar, 356
    and disorder diagnosis, 196
    excavation, 264
    inspection, 251
root crown, 11
root death, 66
root exudates, 50
root hairs, 11
root initiation zone, 19
root mats, 19
root rot, 71, 74, 232, 260, **260**
root systems
    and abiotic disorders, 199
    aeration, **287**
    anatomy, 10–12, **10**, **11**
    circling, 108, 116–117, **121**, 205
    construction injury, 276–278, **277**, **278**
    cutting, 285
    diseases, 210
    and disorder diagnosis, **195**, 196
    girdling, 68, **108**, 117, 204–205, **204**
    palms, 19
    and planting, 118–119, **118**
    problems, and tree failure, 258
    pruning, 122–123, **123**, 285
    shallow, 67
    and soil, 45–46, 52
    and tree selection, 105
    tropical trees, 18

Index

root systems (*continued*)
   tunnelling, 280
root tips, **10**, 11
root turnover, 45
root:shoot ratio, 225
rope angle, 360
rope slings, 365–366
rope snaps, 346
rope support systems, 171, **171**, 174, 174(t)–175(t), 177
rope walking, 359
ropes, 348–349, **349**
   inspection of, **347**, **348**
   installation, 359–360
   moving rope systems, 356–357, **356**, 358, 359
   stationary rope systems, 346, 357–358, **357**, 359, 361
rot *see* fungi
rotenone, 224
routes
   access, 283
   escape, 327, **327**
running bowline, 353, 367
runoff, 52, 69, 72, 90, 296, **297**
rust fungi, 211

S

sabal palm (*Sabal palmetto*), 125
safety, 319
   aerial lifts, 330–331, **330**, **331**
   and blocking line of sight, 300
   chain saw, 325–326, **325**, **326**
   chipper, 329–330, **330**
   climbing, 361–362
   communication, 322–323, **322**, **323**
   education, 319, **319**
   emergency response, 331–338
   fire and vehicle, 323
   first-aid, 331, 332–338
   guying, 179–180
   palm, 331, **332**
   personal protective equipment, 321–322, **321**, 329, 333, **333**, 345, **345**

safety (*continued*)
   standards, 319–321
   training, 319, **319**, 323, 325, 330
   tree felling and removal, 327–329, **327**, **328**, **329**
safety cones, 323
sales comparison approach (appraisal/valuation), 302
saline soils, 49, **72**, 73
salts, 49
   chloride, 72
   deicing, 72
   sodium, 72, **72**
   tolerance, 103(t)
sampling, 86, **87**, 88
sand, 46, **46**, 47, 52, **65**, 66, 68, 125
sanitation, landscape, 231, 235
sap, 4, 157
sapsuckers, 209
sapwood, 6, **7**
sapwood rot, 260, **260**
saturation, 65–66, **65**
saucer-shaped planting hole, 118, **118**
scabbard, 361
scaffold branches, 107, 146
Schwäbisch hitch, 351
screw links, 365
second-degree burns, 335–336
secondary growth, 4
secondary macronutrients, 89
secondary pest outbreak, 237, **238**
secondary xylem, 4
7-strand, common-grade cable, 168, **170**
shackles, 365
shakes, 17
shallow root system, 67
shear plane cracks, 257
shearing, 154, **158**, 159
sheaves, 365
sheet bend, 353
Shigo, Alex, 3, 17
ship auger, 168
shock
   electric, 324
   transplant, 126

shock-loading, 365, **365**, 368
shoots, **3**, 8, 107
shrubs, 89
   pruning of, 142, 154, **158**, 159
   rejuvenation, 159
signage, 266, 282, **282**, 323
signs vs. symptoms of diseases/disorders, 191–192, **192**, **193**
silt, 46, **46**, 47
simple leaf, 30, **30**
single rope technique (SRT) *see* stationary rope systems (SRS)
sink (phloem transport), 14
sinker roots, 11
site
   analysis, 101
   conditions, 231
   considerations, 101–103, **101**, **102**, 102(t)
   and drainage, 75–76, **75**
   examination, and disorder diagnosis, 195
   information, 228, **228**
   and likelihood of failure, 254
   match trees and, 99–106, 232
   problems, and abiotic disorders, 199–200
16-strand rope, 348
size
   diversity, 307
   management, 148, 150
   part size, and consequences of failure, 255
   of transplants, **126**
   and tree selection, 103(t)
skeletonization, leaf, **205**, 206
slings, 365–366
sling-terminated synthetic rope, 174(t)–175(t)
slip knot, 355
sloughing of palm skirts, 331, 332
slow-release fertilizer, 88–89, 90
smoking, 323
snap cut, 369–370, **369**
soaker hoses, 71
social benefits of trees, 298–299

social costs of trees, 301
sodic soils, 49
sodium, 72, **72**
soft rot fungi, 259
soft tourniquets, 333, **334**
soil, 43
   and abiotic disorders, 199–200
   aeration, 75
   biological properties, 50–51, **50**
   chemical analysis, 86
   chemical properties, 48–49, **48**, **49**, **56**, 57, 102(t)
   ecology of, 43
   food web, 50, **50**
   improvement, 55–57
   moisture, 51–53, 65, **65**, 72, 99
      see also irrigation
   pest management options, 234
   physical properties, **43–44**, 43–48, **45**, **46–47**, 102(t)
   profiles, 44–45, **44**
   sampling, **87**
   stress, 54
   urban, 53–55, **53**, **54**, **55**
   volume, 67–68, 201
soil additives, 73, 93
soil amendments
   drainage improvement, 55–56, **56**, 76
   and fertilization, 91
   water conservation, 73
soil analysis, 57, 86, **87**, 88, 93, 102, 200
soil cells, 54–55
soil compaction, 55, 199–200
   construction site, 277, **277**, **288**
   definition, 48
   reducing, 56, 73, 284
   site selection, 102, 103
   urban soils, 53, **54**, **199**
soil erosion, 56, 73, 74, 296
soil food web, 50, **50**
soil grade, 74–75, 284–285, **284**, 277–278
soil moisture reservoir, 68
soil profiles, 44, **44**
soil salinity, 73

soil tests see soil analysis
sonic tomography, 264
source (phloem transport), 14
specialized pruning system, 142–143
species (taxonomy), 28, 28(t), 307
species diversity, 232, 307
species failure profiles, 255
specific epithet, 28, 28(t)
specifications
   construction, 282
   planting, 132
   pruning, 160
   urban forestry, **311**, 312
spider mites, 208, **208**
spliced eye, 349
splint, 335, **335**
split-tails, 346
spongy layer, **9**
sprains, 335
sprinkler irrigation, **69**, 69, 72
sprouts, 8, 143, 149, 159
spruce spider mites, **208**
spruces, 32, 33, 105
squirrels, 209
stakeholders, 280, 306
staking, 128–129, **128**, **129**
standards
   see also American National Standards Institute
   safety, 319–321
   urban forestry, 312
starch, 13
starvation, 150
static loads, 365
stationary rope systems (SRS), 346, 357–358, **357**, 359, 361
steel cable systems, 167–171, **168**
stems, 6–9, **30**
   leaders, 131, 144, **145**, 146
   palm, 18–19, **18**, **19**
stings, insect, 331, 336–337
stippling, 192
stomata, **9**, 10, 13, 14, 67, **67**
storage areas for, **283**
storm damage, 202
storm-response plan, **307**

stormwater runoff, 296, **297**
stress
   acute, 197
   chronic, 197, **197**
   soil, 54
   tree, 195, **195**, 197, **197**, 199, **199**, 223–224, 225, **224**, **225**, 276
   water, 52, 67–68, 203, **225**
string trimmers, 200, 205
structural cells, 54–55, **55**
structural defects, 255–256
structural pruning, 144, **145**, 159
structural soils, 54–55
structure (soil), 46–48, 66, 73
stunting, 192
subordination, 146
subspecies (taxonomy), 29
substrate, 116
subsurface application of fertilizers, 90–91
sucking insects, 207
sugar maple (*Acer saccharum*), 28(t), 30
sulfur, 12, 72, **85**
   deficiency, 89
   mobility in soil, 91
sunscald, 201, 278
superficial burns, 335
support systems, 167, **167**
   see also cabling
   bracing, 178–179, **178**, **179**
   cabling installation techniques, 175–177
   guying, 179–180, **180**
   inspection and maintenance, 181
   propping, 180–181, **181**
   risk mitigation, 267, **267**
   rope support systems, 171, **171**, 174, 174(t)–175(t), 177
   steel cable systems, 167–171, **168**
suppression, pest, 23
surface application of fertilizers, 90–91
surface drainage, 76
susceptibility to pests, 105–106
suspended pavement, 54, **55**
sustainability, 57, **99**, **295**, 303

swages, 170–171
sweating, 116
sweetgum (*Liquidambar styraciflua*), 15, 16
swing, 360, 368, **368**
symbiosis, **11**, 12, 50–51, 93, 212
symptoms
    construction site damage, 276, **276**
    monitoring, 229, **229**
    of plant disorders, 191–192, 198–199
    vs. signs, 191–192, **192**, **193**
synthetic burlap, 117
systemic herbicides, 203, **203**
systemic pesticides, 236, **237**

T

taglines (rigging), 327, 368, **368**
tannins, 224
taper, 107, 262
taproot, 11
target zone, 252
targets, 251–252, **251**, 264
tautline hitch, 350
taxonomy, 27–28, 28(t)
temperature, 13, 56, 107, 130
    and abiotic disorders, 201, **201**
    and hardiness, 105
    soil, 73
tensile strength, 349, 366
tensiometers, 72, **72**
tension wood, 261
terminal buds, 7, **7**, 125, 196
terminations (cabling), 170–171
texture (soil), 46, **46**, **47**, 52, **52**, **53**
thimbles, 170, **170**
thinning, 148
third-degree burns, 335–336
threaded rods
    with amon-eye nuts, 168, **168**
    lag-threaded rods, 178, **178**
    machine-threaded rods, 178, **178**
three-cut method, **153**, 154
3-strand rope, 348
thresholds, 225, 228–229

throwlines, 347, 359
tie-in point, 347, **356**, 358–359, **358**, 363, **363**
tilling (soil), 55
timber hitch, 354
time frame of and likelihood of failure, 253, **254**
tip-tying, 367–368, **367**
tissues, 4
tomography, 264, **264**
tools and equipment
    see also hardware
    aerial lifts, 330–331, **330**, **331**
    air excavation, 264, 287–288, **288**
    chippers, 329–330, **329**
    climbing, 345–348
    construction, 276, 277, 278, 285
    decay detection, 263–264, **263**, **264**
    fall protection and work positioning, 345, **346**
    felling wedges, 327
    gasoline-powered, 323
    hand tools, 324
    mechanical tree spades, 124, **124**
    for monitoring, 230, 263
    personal protective equipment, 321–322, **321**, 329, 333, **333**, 345, **345**
    pruning, 155, 361
    rigging, 365–366, **366**
    risk assessment, 250–251
    saws, 321, 325–326, **325**, **326**, 333–334, **334**, 361–362
    sterilization of, 123, 155, 231
    storage areas for, **283**
topiary, 142, **143**
topping, **147**, 148, **149**, 150, 266, 311
topsoil, 45
tourniquets, 333, **334**
tracheids, 5
trademark names, 29
traffic control measures, 323
training
    emergency response, 331, 362
    safety, 319, **319**, 323, 325, 330

training (pruning), 146
translocation, 14
transpiration, **67**, 68
    leaf function, 9–10, 13–14
    water loss, **69**
transplant shock, 126
transplants
    care, 125–131
    palms, 125, **125**
    recovery and reestablishment, 126, **126**
    size, **126**
    techniques, 122–125
transverse cracks, 257
trapping, 230, **230**
TRAQ *see* Tree Risk Assessment Qualification
tree establishment *see* transplants
tree inventories, 279, 304–305, **304**, **305**
tree islands, 284, **284**
tree officers, 295
tree ordinances, 310–311
tree preservation, 266, 278–279, **279**
    assessment and evaluation of trees, 279–280, **279**
    construction specifications, 282
    tree protection plan, 280–281, **280**
tree preservation orders (TPOs), 311–312
tree protection during construction, 282
    access to site, limiting, 283
    communication, 285–286, **286**
    effects of grade changes, minimizing, 284–285, **284**
    erection of barriers, 282–283
    reducing compaction, 284
    root cutting, 285
tree protection plan, 280–281
tree protection zones (TPZ), 280–282, **281**, 285
tree removal, 267, 327–329, **327**, **328**, **329**
Tree Risk Assessment Qualification (TRAQ), 248

Index

tree selection, 99, **99**
   climate change, 107
   nursery stock, 107–108, **108**, 115–118, **115**
   and pest control, 232
   site considerations, 101–103, **101**, **102**, 102(t)
   site matching, 99–106, 232
   tree considerations, 103–106, 103(t), **104–105**
tree services, 222
tree spades, 124, **124**
tree stabilization, 128–130, **128**, **129**
tree wardens, 295
tree wells, 284–285, **284**
treehoppers, **207**
trenching, 276, 287–288, **288**
triangular cabling, 176, **176**
tropical trees, 18
tropisms, 15–16
trunk formula technique, 303
trunks, 6, 8–9
   construction injury, **275**, 276
   and disorder diagnosis, 195–196
   flare, 11, 107, 108, **108**, 118–119, **119**, 121, 196, 276
   frost cracks, 201
   injury, 68, 107, 286
tub grinder, 308–309
tunneling, 280
turfgrass, 91, 200
turgidity, 67, **68**
12-strand rope, 348
24-strand rope, 348–349
twigs, 6, **7**, 66, 195, **229**
two-lined chestnut borer, 197

U

unavailable water, 52, **52**, **65**, 66
United States Forest Service, 302
uprooting, 67
urban forestry, 295–296
   assessment, 304–305
   benefits and costs of trees, 296–301
   climate change, 302

urban forestry (*continued*)
   key issues in management, 306–309
   management plans, 305–306
   planning and management, 303–310
   professional collaboration, public outreach, and education, 309–310
   regulatory and legal issues, 310–312
urban foresters, 295, **295**, 304
urban forests, 295
   valuation and appraisal, 302–303, **302**
urban settings, 15, 275
   see also construction
   biological control of pests, 235
   critical root zones in, 281
   drainage problems in, 76
   pruning in, 147
   soil pH in, 49
   weather-related problems in, 201
   and tree selection, 103, 106
   tree services, 222
urban soils, 53–55, **53**, **54**, **55**
U-shaped codominant configuration, 144
utilities, 101, **101**, 119, 159, 251–252, 324

V

Valdôtain tresse (Vt), 351
valuation, 302–303, **302**
vandalism, 103, 129, 200
varieties (taxonomy), 29
vascular discoloration, 192
vascular plants, 27
vascular system, 3, **5**, 4–6, **9**, 14
vascular wilt disease, 68, 210, **210**
vectors, 207
vehicle safety, 323
veins, leaf, 10
venomous insect bites, 336–337
vertical mulching, 288
verticillium wilt, **210**

vessels, 5
vigor, 222–223
viruses, 212
vista pruning, 147
visual assessment (risk), 249–250, **249**, 253
vitality, 222–223, 263, 282
volatilization, 85, 90–91
voles, 209
V-shaped codominant configuration, 144

W

washers, 170, 178
wasps, **235**, 366
water, 12, 14, 65
   see also irrigation
   and abiotic disorders, 203–204
   conservation, 71–73
   and fertilizer uptake, 90
   portable watering device, 122
   quality, 72
   recycled/reclaimed, 72–73
   soil, 51–53, 66
   stress, 52, 67–68, 203, **225**
   for transplants, 126–127, **127**
   and tree selection, 103
   utilization by trees, 65–68
water budget, 70
water injection, high-pressure, 70
water management, 65
   conservation, 71–73, **71**
   flooding and drainage, 73–76, **74**, **75**, **76**
   irrigation, 68–71, **69**, **72**
   soil water, 51–52, **65**, 66
water potential, 14
water-holding capacity, 52, 66, 102
water-insoluble nitrogen (WIN), 89
watersprouts, 148, **149**, 150, 159, 195, **288**
weakly attached branches, 257
weather
   and abiotic disorders, 201–202, **201**, **202**, 203
   exposure, 278

webbing slings, 365–366
wedges, felling, 327
wetting patterns, **65**
wheel bugs, **234**
white rot fungi, 259, **259**, 260
whorled leaf arrangements, 32, **32**
wildlife, 149–150, 209, 224, 267, 296, 309, **309**
wilt diseases *see* vascular wilt disease
wilting, 67, **68**, 192, 198
wind, 68, 69, 202, 278
windbreaks, 296
wire baskets, 117, **117**
wire crimp, 171
wire rope (aircraft cable), 168, 171
wire rope clamps, 171
witch's broom, 192, **194**
wood chips, 73, 127, 308–309, **309**
wood waste, 297, 308–309
woodpeckers, 209
work plans, 322, 356
work positioning, 345, **346**
work zone, 322, 323, 327
working-load limit (WLL), 349
work-positioning lanyards, 345, **346**, 360, 362
wounds *see* injuries
woundwood, 151, 262–263
wraps, 130

X

xeriscape, 71
xerophytes, 204
xylem, **3**, 4–6, **4**, **6**, **13**, 14

Y

young trees, training, 146

Z

zinc, **85**, 89
zone of rapid taper, 11
zoning, 310